AF361915

Nanoscale Ferroic Materials—Ferroelectric, Piezoelectric, Magnetic, and Multiferroic Materials

Nanoscale Ferroic Materials—Ferroelectric, Piezoelectric, Magnetic, and Multiferroic Materials

Editors

Zhenxiang Cheng
Changhong Yang
Chunchang Wang

MDPI • Basel • Beijing • Wuhan • Barcelona • Belgrade • Manchester • Tokyo • Cluj • Tianjin

Editors

Zhenxiang Cheng
University of Wollongong
Australia

Changhong Yang
University of Jinan
China

Chunchang Wang
Anhui University
China

Editorial Office
MDPI
St. Alban-Anlage 66
4052 Basel, Switzerland

This is a reprint of articles from the Special Issue published online in the open access journal *Nanomaterials* (ISSN 2079-4991) (available at: https://www.mdpi.com/journal/nanomaterials/special_issues/nano_ferroic).

For citation purposes, cite each article independently as indicated on the article page online and as indicated below:

LastName, A.A.; LastName, B.B.; LastName, C.C. Article Title. *Journal Name* **Year**, *Volume Number*, Page Range.

ISBN 978-3-0365-5943-8 (Hbk)
ISBN 978-3-0365-5944-5 (PDF)

Contents

Preface to "Nanoscale Ferroic Materials—Ferroelectric, Piezoelectric, Magnetic, and Multiferroic Materials"

Ferroic materials, including ferroelectric, dielectric, piezoelectric, magnetic, and multiferroic materials, have broad applications in current modern society. Based on their rich physics and the related properties, ferroic materials can have applications for different purposes. For example, based on the switching of domain, ferroelectric and magnetic materials can be used for information storage. Based on the piezoelectric property, piezoelectric materials can be used as sensor/actuator/energy harvesters. Based on the coupling of magnetic order and ferroelectric order, multiferroic materials can have applications as magnetic field sensors and for information storage with the advantages of high storage density and low power consumption. Recently, nanoferroic materials have found new applications based on their ferroic property (low dimension limited size effect) and high surface area, such as photocatalysis.

In this scenario, the present book offers novel research insights recently published as a part of the *Nanomaterials* Special Issue "Nanoscale Ferroic Materials—Ferroelectric, Piezoelectric, Magnetic, and Multiferroic Materials", with the aim to collect the most recent advances on nanoscale ferroic materials as well as their possible practical uses in different fields of interest.

Zhenxiang Cheng, Changhong Yang, and Chunchang Wang
Editors

Editorial

Editorial for the Special Issue "Nanoscale Ferroic Materials—Ferroelectric, Piezoelectric, Magnetic, and Multiferroic Materials"

Changhong Yang [1], Chunchang Wang [2] and Zhenxiang Cheng [3,*

[1] Shandong Provincial Key Laboratory of Preparation and Measurement of Building Materials, University of Jinan, Jinan 250022, China
[2] Laboratory of Dielectric Functional Materials, School of Materials Science & Engineering, Anhui University, Hefei 230601, China
[3] Institute for Superconducting and Electronic Materials, Australian Institute of Innovative Materials, Innovation Campus, University of Wollongong, Squires Way, North Wollongong, NSW 2500, Australia
* Correspondence: cheng@uow.edu.au

Citation: Yang, C.; Wang, C.; Cheng, Z. Editorial for the Special Issue "Nanoscale Ferroic Materials—Ferroelectric, Piezoelectric, Magnetic, and Multiferroic Materials". *Nanomaterials* **2022**, *12*, 2951. https://doi.org/10.3390/nano12172951

Received: 15 August 2022
Accepted: 19 August 2022
Published: 26 August 2022

Publisher's Note: MDPI stays neutral with regard to jurisdictional claims in published maps and institutional affiliations.

Ferroic materials, including ferroelectric, piezoelectric, magnetic, and multiferroic materials, are receiving great scientific attentions due to their rich physical properties. They have shown their great advantages in diverse fields of application, such as information storage, sensor/actuator/transducers, energy harvesters/storage, and even environmental pollution control. At the same time, nanostructure has assumed an increasing importance as the key to the miniaturization of solid-state electronics, decrease of the power consumption, and reduction of production cost. At present, ferroic nanostructures have been widely acknowledged to advance and improve currently existing electronic devices, as well as to develop future ones.

This Special Issue, comprising ten research articles, covers the characterization of crystal and microstructure, the design and tailoring of ferro/piezo/dielectric, magnetic, and multiferroic properties, and the presentation of related applications. These papers present various kinds of nanomaterials, such as the ferroelectric/piezoelectric thin films, i.e., PMNPT, $BiFeO_3$, $BiOCl/NaNbO_3$, $CH_3NH_3PbI_3$, $Hf_{0.5}Zr_{0.5}O_2$, and $CuInP_2S_6$; dielectric storage thin film, i.e., $Ba_{0.5}Sr_{0.5}TiO_3/0.4BiFeO_3$-$0.6SrTiO_3$, $Bi(Fe_{0.93}Mn_{0.05}Ti_{0.02})O_3$-$CaBi_4Ti_4O_{15}$ solid solution; dielectric gate layer, i.e., $Sm_2O_3/Al_2O_3/InP$; and magnonic metamaterials, i.e., ferromagnet/heavy metal bilayer. These nanomaterials are expected to have applications in ferroelectric non-volatile memory, ferroelectric tunneling junction memory, energy-storage pulsed-power capacitor, metal oxide semiconductor field-effect-transistor device, humidity sensor, environmental pollutant remediation, and spin-wave devices. The purpose of this Special Issue is to communicate the recent developments in the research of nanoscale ferroic materials. In the following part, we present a brief overview of the published articles and hope to provide a useful resource for potential readers.

Aiming at obtaining ferroelectric thin films with strong polarization feature, effective methods of chemical doping and regulating the preparation process are applied. Kim et al. fabricated $Hf_{0.5}Zr_{0.5}O_2$ thin films via plasma-enhanced atomic layer deposition. A high remanent polarization with $2P_r$ of 38.2 $\mu C/cm^2$ and an excellent fatigue endurance of 2.5×10^7 cycles were obtained by regulating the deposition temperature, post-annealing temperature and RF plasma discharge time [1]. Among the studies about $Hf_xZr_{1-x}O_2$, the work in this study showed relatively good remanent polarization and fatigue endurance performances despite being under the lowest deposition temperature. Feng et al. [2] focus their work on the effects of annealing temperature and Fe-doping concentration on the crystallinity, microstructure, ferroelectric and dielectric properties of PMN-PT thin films. 2% Fe-doped PMN-PT thin film annealed at 650 °C exhibits high (111) preferred orientation, high remanent polarization (P_r = 23.1 $\mu C/cm^2$) and low coercive voltage (Ec = 100 kV/cm).

BiFeO$_3$ thin films were doped with Sm by Wang et al. [3]. They discussed the changes of electrical characteristics in terms of phase transition and oxygen vacancies.

Van der Waals (vdW) layered ferroelectric materials have become a promising research branch in condensed matter physics. As the main factor affecting the ferroelectric property is the strong depolarization in ultrathin ferroelectric film, herein, Jia et al. [4] investigated the ferroelectric and piezoelectric properties at different thicknesses. Furthermore, the constructed Pt/CuInP$_2$S$_6$/Au ferroelectric tunnel junction with a 2 nm-thick functional layer has superior continuous current modulation and self-rectification functions.

In recent years, ferroelectric energy storage thin film is emerging as a key enabler for advanced pulsed power systems. However, such applications are hindered by the low energy storage density and small efficiency, so some strategies need to be explored to further improve its energy storage performance. Wang et al. [5] design a multilayer structure of Ba$_{0.5}$Sr$_{0.5}$TiO$_3$/0.4BiFeO$_3$-0.6SrTiO$_3$ on flexible mica substrate. A good balance between energy density and efficiency can be reached by utilizing the interface engineering. In addition, the authors describe a cost-effective and facile fabrication of flexible ferroelectric thin films. The energy storage performance of bismuth layer-structured ferroelectric (BLSF) compounds of CaBi$_4$Ti$_4$O$_{15}$ modified by Bi(Fe$_{0.93}$Mn$_{0.05}$Ti$_{0.02}$)O$_3$ was studied by Liu et al. [6]. In their work, Bi(Fe$_{0.93}$Mn$_{0.05}$Ti$_{0.02}$)O$_3$ played a very important role in enhancing the polarization and optimizing the shape of ferroelectric hysteresis loop.

Magnetic metamaterials are deemed to be a promising candidate as a high-quality information carrier. Since the graded magnonic refractive index can be created by modification of the material properties, Zhuo et al. [7] theoretically propose a ferromagnet/heavy metal bilayer magnonic metamaterial, and modulate the refractive index of spin waves with the inhomogeneous Dzyaloshinskii–Moriya interaction (DMI). The authors further study spin-wave refraction and reflection at the interface between two magnetic media with different DMI and build a generalized Snell's law of spin waves.

Last but not least, three papers of this Special Issue address the applications involving piezoelectric photocatalyst, humidity sensor, and metal oxide semiconductor. Piezoelectric catalysis is an efficient and environmentally friendly dye degradation method. Li et al. [8] synthesized the BiOCl/NaNbO$_3$ piezoelectric composites, and investigated the mechanism of piezocatalysis, photocatalysis, and their synergy effect of BN-3 composite. This work provides a new strategy to improve the environmental pollutant remediation efficiency of piezoelectric composites. Perovskites with the formula of AMX$_3$ have a potential use for probes for sensing of humidity due to the heavy dependence of electrical characteristics on environmental moisture. Therefore, Zhao et al. [9] carried out relevant research on the humidity sensitivity of CH$_3$NH$_3$PbI$_3$ perovskite. The CH$_3$NH$_3$PbI$_3$-based humidity sensor shows the excellent humidity sensitive properties, and the authors ascribed it to water-molecule-induced enhancement of the conductive carrier concentration. As the COMS feature sizes continue to decrease, it is urgent to use a high-k material to replace the current SiO$_2$ gate dielectrics. Lu et al. [10] systematically investigated the effect of atomic layer deposition-derived laminated interlayer on the interface chemistry and transport characteristics of sputtering-deposited Sm$_2$O$_3$/InP gate stacks. They demonstrated that Sm$_2$O$_3$/Al$_2$O$_3$/InP stacked gate dielectric is a promising candidate for InP-based metal oxide semiconductor field-effect-transistor devices in the future.

In this Special Issue, the range of themes addressed is certainly not exhaustive. With the scope of application of nanostuctured ferroic materials broadening rapidly, further developments on the aspects of theoretical simulation, synthetic technique, advanced characterization, and multi-functional combination are expected.

Author Contributions: C.Y. wrote this editorial letter. Z.C. and C.W. provided their feedback, which was assimilated into the letter. All authors have read and agreed to the published version of the manuscript.

Funding: C.Y. acknowledges the National Natural Science Foundation of China (Grant No. 51972144), and C.W. thanks the National Natural Science Foundation of China (Grant Nos. 12174001 and 51872001).

Acknowledgments: The guest editors thank all the authors for submitting their work to the Special Issue and for its successful completion. A special thank you to all the reviewers participating in the peer-review process of the submitted manuscripts and for enhancing their quality and impact. We are also grateful to the editorial assistants who made the entire Special Issue creation a smooth and efficient process.

Conflicts of Interest: The authors declare no conflict of interest.

References

1. Kim, H.-G.; Hong, D.-H.; Yoo, J.-H.; Lee, H.-C. Effect of Process Temperature on Density and Electrical Characteristics of $Hf_{0.5}Zr_{0.5}O_2$ Thin Films Prepared by Plasma-Enhanced Atomic Layer Deposition. *Nanomaterials* **2022**, *12*, 548. [CrossRef] [PubMed]
2. Feng, C.; Liu, T.; Bu, X.; Huang, S. Enhanced Ferroelectric, Dielectric Properties of Fe-Doped PMN-PT Thin Films. *Nanomaterials* **2021**, *11*, 3043. [CrossRef] [PubMed]
3. Wang, Y.; Li, Z.; Ma, Z.; Wang, L.; Guo, X.; Liu, Y.; Yao, B.; Zhang, F.; Zhu, L. Phase Structure and Electrical Properties of Sm-Doped $BiFe_{0.98}Mn_{0.02}O_3$ Thin Films. *Nanomaterials* **2022**, *12*, 108. [CrossRef] [PubMed]
4. Jia, T.; Chen, Y.; Cai, Y.; Dai, W.; Zhang, C.; Yu, L.; Yue, W.; Kimura, H.; Yao, Y.; Yu, S.; et al. Ferroelectricity and Piezoelectricity in 2D Van der Waals $CuInP_2S_6$ Ferroelectric Tunnel Junctions. *Nanomaterials* **2022**, *12*, 2516. [CrossRef] [PubMed]
5. Wang, W.; Qian, J.; Geng, C.; Fan, M.; Yang, C.; Lu, L.; Cheng, Z. Flexible Lead-Free $Ba_{0.5}Sr_{0.5}TiO_3$/0.4$BiFeO_3$-0.6$SrTiO_3$ Dielectric Film Capacitor with High Energy Storage Performance. *Nanomaterials* **2021**, *11*, 3065. [CrossRef] [PubMed]
6. Liu, T.; Wang, W.; Qian, J.; Li, Q.; Fan, M.; Yang, C.; Huang, S.; Lu, L. Excellent Energy Storage Performance in $Bi(Fe_{0.93}Mn_{0.05}Ti_{0.02})O_3$ Modified $CaBi_4Ti_4O_{15}$ Thin Film by Adjusting Annealing Temperature. *Nanomaterials* **2022**, *12*, 730. [CrossRef] [PubMed]
7. Zhuo, F.; Li, H.; Cheng, Z.; Manchon, A. Magnonic Metamaterials for Spin-Wave Control with Inhomogeneous Dzyaloshinskii–Moriya Interactions. *Nanomaterials* **2022**, *12*, 1159. [CrossRef] [PubMed]
8. Li, L.; Cao, W.; Yao, J.; Liu, W.; Li, F.; Wang, C. Synergistic Piezo-Photocatalysis of $BiOCl$/$NaNbO_3$ Heterojunction Piezoelectric Composite for High-Efficient Organic Pollutant Degradation. *Nanomaterials* **2022**, *12*, 353. [CrossRef] [PubMed]
9. Zhao, X.; Sun, Y.; Liu, S.; Chen, G.; Chen, P.; Wang, J.; Cao, W.; Wang, C. Humidity Sensitivity Behavior of $CH_3NH_3PbI_3$ Perovskite. *Nanomaterials* **2022**, *12*, 523. [CrossRef] [PubMed]
10. Lu, J.; He, G.; Yan, J.; Dai, Z.; Zheng, G.; Jiang, S.; Qiao, L.; Gao, Q.; Fang, Z. Interface Optimization and Transport Modulation of Sm_2O_3/InP Metal Oxide Semiconductor Capacitors with Atomic Layer Deposition-Derived Laminated Interlayer. *Nanomaterials* **2021**, *11*, 3443. [CrossRef] [PubMed]

nanomaterials

MDPI

Article

Effect of Process Temperature on Density and Electrical Characteristics of Hf$_{0.5}$Zr$_{0.5}$O$_2$ Thin Films Prepared by Plasma-Enhanced Atomic Layer Deposition

Hak-Gyeong Kim, Da-Hee Hong, Jae-Hoon Yoo and Hee-Chul Lee *

Department of Advanced Materials Engineering, Korea Polytechnic University, Siheung 15073, Korea; hakkyung5280@hanmail.net (H.-G.K.); ghdekgml9768@naver.com (D.-H.H.); dbwogns96@gmail.com (J.-H.Y.)
* Correspondence: eechul@kpu.ac.kr

Abstract: Hf$_x$Zr$_{1-x}$O$_2$ (HZO) thin films have excellent potential for application in various devices, including ferroelectric transistors and semiconductor memories. However, such applications are hindered by the low remanent polarization (P$_r$) and fatigue endurance of these films. To overcome these limitations, in this study, HZO thin films were fabricated via plasma-enhanced atomic layer deposition (PEALD), and the effects of the deposition and post-annealing temperatures on the density, crystallinity, and electrical properties of the thin films were analyzed. The thin films obtained via PEALD were characterized using cross-sectional transmission electron microscopy images and energy-dispersive spectroscopy analysis. An HZO thin film deposited at 180 °C exhibited the highest o-phase proportion as well as the highest density. By contrast, mixed secondary phases were observed in a thin film deposited at 280 °C. Furthermore, a post-annealing temperature of 600 °C yielded the highest thin film density, and the highest 2P$_r$ value and fatigue endurance were obtained for the film deposited at 180 °C and post-annealed at 600 °C. In addition, we developed three different methods to further enhance the density of the films. Consequently, an enhanced maximum density and exceptional fatigue endurance of 2.5 × 10^7 cycles were obtained.

Keywords: HZO; PEALD; ferroelectric memory; deposition temperature; film density; remanent polarization; fatigue endurance

Citation: Kim, H.-G.; Hong, D.-H.; Yoo, J.-H.; Lee, H.-C. Effect of Process Temperature on Density and Electrical Characteristics of Hf$_{0.5}$Zr$_{0.5}$O$_2$ Thin Films Prepared by Plasma-Enhanced Atomic Layer Deposition. *Nanomaterials* **2022**, *12*, 548. https://doi.org/10.3390/nano12030548

Academic Editors: Zhenxiang Cheng, Changhong Yang and Chunchang Wang

Received: 31 December 2021
Accepted: 2 February 2022
Published: 5 February 2022

Publisher's Note: MDPI stays neutral with regard to jurisdictional claims in published maps and institutional affiliations.

1. Introduction

Since the report of ferroelectric behavior in HfO$_2$-based thin films, studies have been conducted on HfO$_2$ thin films doped with different elements. Particularly, Hf$_x$Zr$_{1-x}$O$_2$ (HZO) thin films, which exhibit ferroelectricity even at thicknesses of a few nanometers, have gained increasing attention [1]. Among the diverse types of available ferroelectric materials, metal oxides have attained considerable technological importance owing to their compatibility with current complementary metal–oxide–semiconductor (CMOS) technology as well as large-scale integration. Therefore, active research has been underway for the application of HZO thin films to a variety of devices such as ferroelectric transistors, synapse devices, and ferroelectric tunneling junctions [2–6].

For the practical application of HZO thin films to semiconductor memories, it is necessary to overcome the issues of remanent polarization (P$_r$) and fatigue endurance. Pb(Zr,Ti)O$_3$-based materials having the crystal structure of perovskites, which are ferroelectric materials that have been studied extensively, exhibit a low wake-up effect, and show stable characteristics with exceptional fatigue endurance over 10^{10} cycles [7–9]. In this context, studies on improving the properties of HZO have been actively underway. In particular, research on the effects of crystal structure, oxygen defects inside thin films, grain size, and interface engineering using electrodes on changes in electrical properties has been mainly reported [10–14]. HZO thin films have a variety of crystalline phases such as tetragonal (t-, P4$_2$/nmc), monoclinic (m-, P2$_1$/c), and orthorhombic (o-, Pca2$_1$) phases, of which the o-phase exhibits ferroelectric properties. However, the m-phase has been generally reported to be the stable phase of HZO [10,15], and research has been conducted to achieve a high ratio and stability of the o-phase in HZO films [16–18].

For HZO deposition, thermal atomic layer deposition (THALD) is mainly used. Moreover, there has been insufficient investigation on the properties of HZO thin films deposited by plasma-enhanced atomic layer deposition (PEALD) [19]. PEALD is capable of the high-density deposition of thin films and has the advantage of enabling low-temperature deposition [20–22]. The regions wherein the o-, t-, and m-phases of HZO are formed vary depending on grain size and temperature. Therefore, density improvement in the deposition process is expected through the stabilization of the o-phase and increasing the grain size through low-temperature deposition using PEALD [10,23]. In addition, to the best of the authors' knowledge, there is no report on the relationship between the PEALD process temperature and changes in the electrical properties with respect to HZO density.

In this study, the initial process conditions for fabricating PEALD HZO thin films were set according to the deposition temperature, and the effect of the deposition temperature on the density and crystallinity of the thin films was analyzed. In addition, the optimal conditions for fabricating HZO thin films by PEALD were derived by examining the thin film density and crystallinity according to the post-annealing temperature. Furthermore, the effects of the variation of the HZO thin film density with the process temperature on the crystallinity of the o-phase exhibiting ferroelectricity, as well as on the electrical characteristics such as polarization hysteresis loops (P-E loops) and fatigue endurance, were investigated. Finally, process improvement methods for obtaining HZO thin films with high density and excellent electrical properties at a low deposition temperature of 100 °C were determined, and the results were comparatively analyzed.

2. Materials and Methods

2.1. HZO Thin Film Deposition by PEALD

For HZO thin film deposition, a substrate comprising a 50 nm TiN bottom electrode deposited on a SiO_2(100 nm)/Si wafer was used. HZO thin film deposition was performed by PEALD (iOV-dx2, iSAC research, Hwaseong, Korea) using the experimental setup illustrated in Figure 1, and tetrakis(ethylmethylamido)-hafnium (TEMA-Hf, iChems, Hwaseong, Korea) and tetrakis(ethylmethylamido)-zirconium (TEMA-Zr, iChems, Hwaseong, Korea) were used as the precursors of HfO_2 and ZrO_2, respectively. To fabricate the HZO thin films, HfO_2 and ZrO_2 were alternately deposited in a 1:1 ratio, and this cycle was repeated until a thin film with a thickness of 10 nm was obtained. To obtain an optimal HZO thin film, deposition was performed in a temperature range of 100 to 280 °C. O_2 gas was injected as a reactant, and oxidation was induced through a 200 W plasma discharge to form oxides. The detailed PEALD process conditions are outlined in Table 1. To prepare the top electrode for the evaluation of the electrical properties, a shadow mask was used, and a TiN electrode with a diameter of 200 μm was deposited at a thickness of 50 nm by reactive sputtering. Next, as shown in Table 2, crystallization of the HZO thin films was performed by post-annealing using rapid thermal annealing (RTA). Post-annealing was performed for 30 s in a temperature range of 500 to 700 °C in a nitrogen ambient of 5 Torr.

2.2. Characterization of HZO Thin Films

The thickness and refractive index of the deposited single oxides of HfO_2 and ZrO_2 and the HZO thin films were evaluated using an ellipsometer (Elli-SE, Ellipso technology, Suwon, Korea). The shape of the thin film cross-section and the elemental composition were analyzed using transmission electron microscopy (TEM) (NEO ARM, JEOL, Tokyo, Japan) and energy-dispersive spectroscopy (EDS) (JED-2300T, JEOL, Tokyo, Japan), respectively. The crystalline structure of the HZO thin films was measured by high-resolution X-ray diffraction (HR-XRD) (Smartlab, Rigaku, Tokyo, Japan) in Bragg–Brentano geometry, and the density of the thin film was calculated through X-ray reflectometry (XRR) analysis on the same instrument. Electrical properties such as the P-E curve and fatigue endurance of the thin film were evaluated using a TF analyzer (TF-2000E, aixACCT, Aachen, Germany) connected to a microprobe station (APX-6B, WIT, Suwon, Korea). Hysteresis loop measurements were performed at a frequency of 1 kHz with a triangle pulse of ± 3 V. Fatigue endurance measurements were conducted by the continuous application of a square pulse of ± 3 V at 10 kHz along with a 1 kHz triangle pulse that was applied five times at each time point to measure remnant polarization.

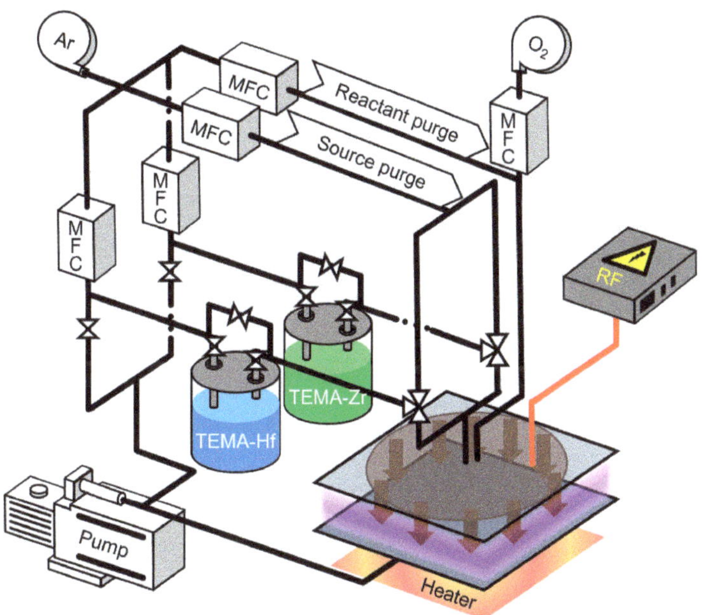

Figure 1. Schematic of PEALD setup used in this study.

Table 1. Conditions of HZO thin film deposition by PEALD and processes constituting one cycle.

Deposition Temperature Maintenance Gas Flow Pressure	100−280 °C 600 sccm 1.3 Torr

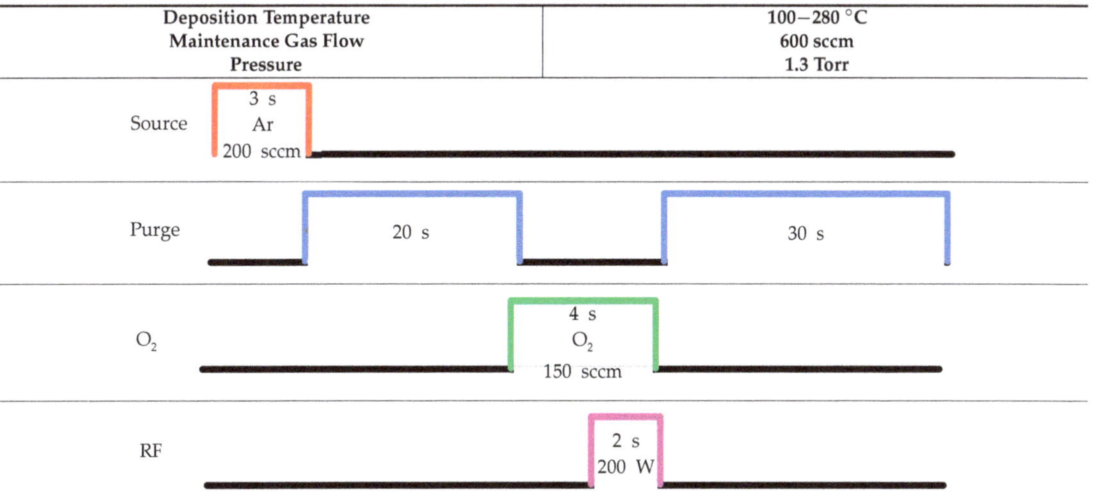

Table 2. Conditions for rapid thermal annealing of HZO thin films after TiN top electrode deposition.

Annealing Temperature Ambient Pressure	500−700 °C N_2 Atmosphere 5 Torr

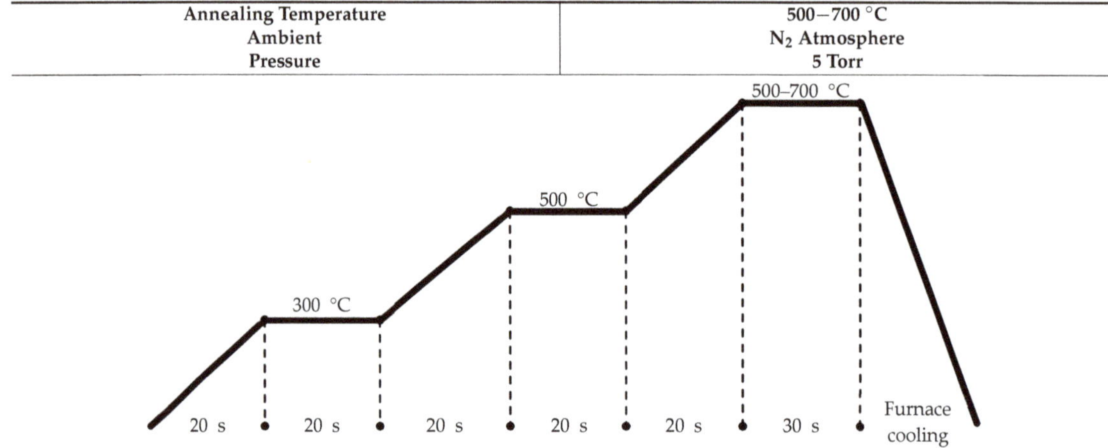

3. Results and Discussion

Prior to the start of HZO thin film deposition, the growth conditions of single thin films of HfO_2 and ZrO_2 were confirmed. Each thin film showed self-limiting behavior when the source was injected for more than 2.5 s in the previous experiment. Accordingly, as shown in Table 1, the injection time of the source was set to 3 s to allow sufficient time. Figure 2 presents the results of analyzing the change in growth per cycle (GPC) and refractive index according to the number of cycles at various substrate temperatures. The results of GPC illustrated in Figure 2a,b show that the deposition thickness value is high in the initial cycles (that is, 10 cycles or less). This phenomenon is due to an overestimated measurement error that occurred during the planarization process because of the native roughness and curvature of the substrate at the initial stage; after 10 cycles, the thickness is almost constant regardless of temperature [24,25]. Furthermore, from the refractive index results shown in Figure 2c,d, a trend of change in the refractive index according to the number of cycles can be observed. Notably, the refractive index approaches 2.0 and 2.1, the bulk refractive index values of HfO_2 and ZrO_2, respectively, with an increase in the number of cycles at the different substrate temperatures. All the thin films deposited at 100 °C had a low refractive index, and the difference was particularly pronounced in the case of the ZrO_2 thin film. It can be inferred from the Lorentz–Lorenz relation that the thin film density decreased at low deposition temperatures [26]. Based on the results in Figure 2, HfO_2 and ZrO_2 were deposited at rates of 0.123 nm and 0.112 nm per cycle, respectively, at the deposition temperature of 180 °C. The two materials were alternately deposited in each cycle, repeating the process 42 times, resulting in a $Hf_{0.5}Zr_{0.5}O_2$ thin film with a thickness of approximately 10 nm.

Figure 3a is a cross-sectional image obtained by the high-resolution TEM (HR-TEM) of a PEALD HZO thin film that was deposited at 180 °C and underwent post-annealing at 600 °C. The thickness of the thin film was approximately 10 nm, and an o-phase crystalline structure was mainly observed. The disappearance of the o-phase structure near the interface is thought to be because of interface instability due to TiN diffusion [13,27]. Figure 3b shows the EDS-based elemental composition profiles of the same thin film, and it can be observed that some of the Ti and N atoms diffused into the HZO thin film. In addition, nitrogen and carbon contamination was observed inside the HZO thin film owing to the TEMA precursors. In particular, the carbon contamination was considerable; herein, carbon is considered to be a residual impurity because the precursor is not completely decomposed during deposition [27–30]. Figure 3c shows the change in the XRD patterns of the PEALD HZO thin films according to the deposition temperature in the substrate temperature range of 100–280 °C. The peaks at 28.5 ° and 31.6 ° represent the m-phase, and the peaks at 30.5 ° and 35.4 ° represent the (111) and (200) planes of the o-phase [16]. At all deposition temperatures, the proportion of o-phase was greater than that of the m-phase, and it can be observed that the phase transformation to the o-phase was successfully achieved during the post-annealing at 600 °C. The intensity of the XRD peak corresponding to the o-phase was the highest at 180 °C, and it decreased as the deposition temperature decreased or increased further. In particular, in the case of deposition at a high temperature of 280 °C, secondary phases such as the m-phase were included.

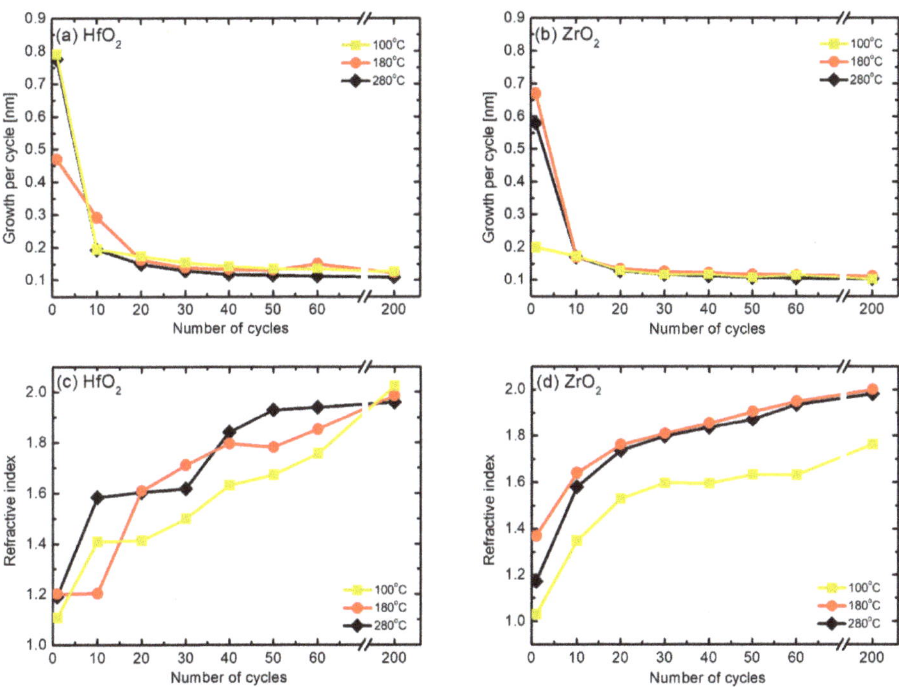

Figure 2. Changes in the (**a**,**b**) growth per cycle (GPC) and (**c**,**d**) refractive index of HfO$_2$ and ZrO$_2$ single thin films according to the number of cycles at different substrate temperatures.

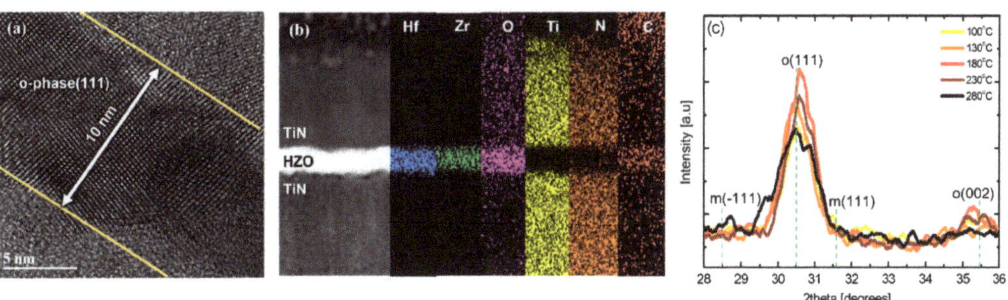

Figure 3. (**a**) Cross−sectional high−resolution TEM (HR−TEM) image and (**b**) EDS composition cross-sectional profile and (**c**) XRD pattern change with respect to substrate temperature in the range of 100–280 °C for PEALD HZO thin films deposited at 180 °C.

Figure 4a presents the XRR data of the HZO thin film deposited at 180 °C, and the inset graph corresponds to raw data showing the reflectivity according to the X-ray incident angle. The density of the thin film is calculated based on the initial angle at which the reflectivity decreases. The thickness of the thin film is simulated through the oscillation period, and the thickness and the density of the deposited thin films constituting the sample can be calculated as shown in the outer graph. Figure 4b outlines the density change according to the substrate temperature of the PEALD HZO thin film obtained by this method. The density of the thin film was the highest (8.18 g/cm^3) at the substrate temperature of 180 °C. This density exceeds the theoretical density of HZO [31]. In a multilayered structure, the density of each thin film is calculated by the reflectivity at each interface, but an error may occur if the thin film is too thin or the interface is not distinct. In this study, the deposited thin films were compared according to calculated density.

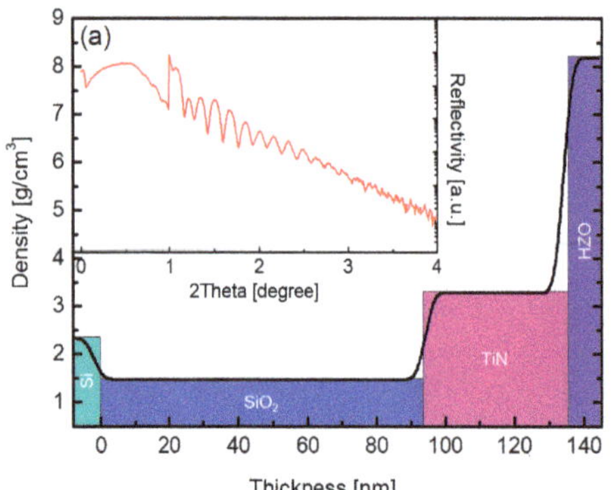
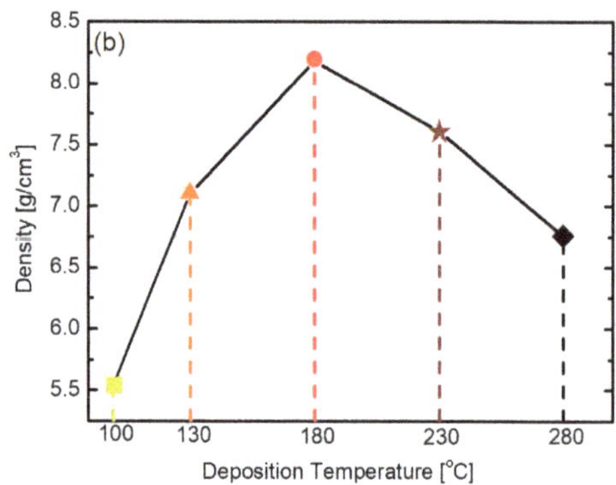

Figure 4. (**a**) XRR data of PEALD HZO thin film deposited at 180 °C and (**b**) HZO thin film density according to substrate temperature in the range of 100–280 °C.

The density gradually decreased as the deposition temperature decreased or increased further. This trend is consistent with that of the o-phase peak intensity, as observed in the XRD patterns in Figure 3c. The decrease in density at low and high temperatures can be explained by the equation of Langmuir's adsorption isotherm [32]. θ, the fraction of the surface covered by the absorbate, can be expressed as a function of time as shown in Equation (1).

$$\frac{d\theta}{dt} = \gamma_a P_i (1 - \theta) - \gamma_d \theta \tag{1}$$

Here, γ_a denotes the adsorption coefficient, γ_d is the desorption coefficient, and P_i is the pressure. Here, the adsorption coefficient and the desorption coefficient are exponentially proportional to the temperature with respect to the activation energy required for adsorption and desorption, respectively.

$$If \ \frac{d\theta}{dt} = 0 \ (equilibrium), \ \theta = \frac{P_i}{P_i + \gamma_d/\gamma_a} \tag{2}$$

Based on the assumption that the same pressure process applies in an equilibrium state where the adsorption rate becomes 0, if P_i is set to a constant value, the equilibrium value of θ is highly dependent on temperature, as shown in Equation (2). At low temperatures, the value of γ_a becomes small; consequently, sufficient chemisorption does not occur, and the space where adsorption does not occur remains empty. Furthermore, as θ decreases, the density of the thin film decreases. At high temperatures, γ_d increases, and it is thought that owing to the empty space formed by the atoms desorbed during the deposition process, both the values of θ and the density of the thin film decrease. Under the conditions of this experiment, the substrate temperature of 180 °C results in the minimum $\frac{\gamma_d}{\gamma_a}$ value and the highest fraction of the surface covered by the absorbate; therefore, this condition is thought to be the optimal deposition condition that yields the highest density. In addition, the secondary phases, including the m-phase, appearing at a substrate temperature of 280 °C may cause density reduction [33,34].

Although the m-phase is the most stable phase of HZO thin films, the formation of the m-phase is suppressed while the ratio of the ferroelectric o-phase is increased because of thermal stress caused by the difference in the thermal expansion coefficient between the HZO thin film and the TiN electrode during the post-annealing process [35,36]. Figure 5 shows the changes in the XRD patterns and density of samples of the HZO thin films deposited at 180 °C, the optimal substrate temperature, which underwent post-annealing at 500 to 700 °C. At all annealing temperatures, HZO thin films almost purely consisting of o-phases without the m-phase and secondary phases were obtained. In the case of the sample obtained via 500 °C post-annealing, the X-ray peak intensity was slightly weak, but in the samples obtained via post-annealing at 600 °C or higher, the X-ray peak intensity was strong. With regard to the density, the sample annealed at 600 °C showed the highest value; thus, the optimum annealing temperature was determined to be 600 °C. It was confirmed that the HZO thin film was densified, with crystallization, through the post-annealing process. It can be inferred from the results shown in Figure 5b that the density of the thin film may decrease as the interdiffusion between the TiN electrode and the HZO thin film increases under high-temperature post-annealing conditions.

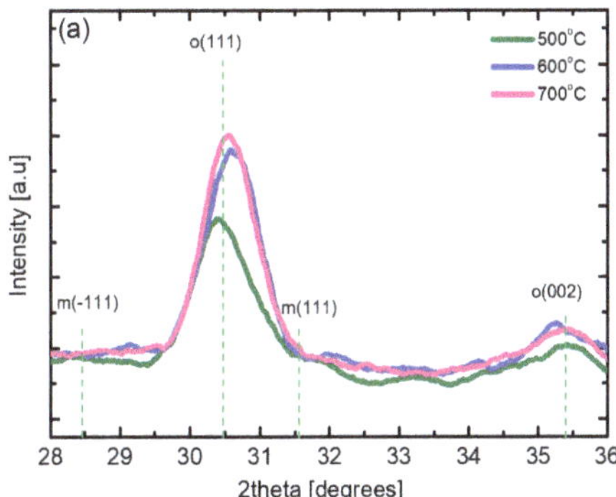

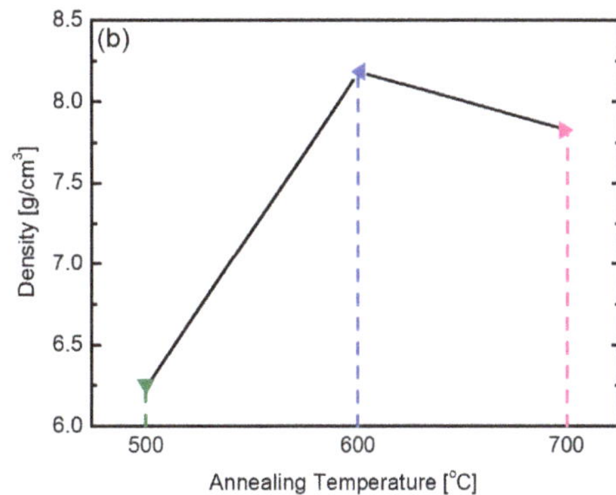

Figure 5. (**a**) XRD pattern and (**b**) density according to post-annealing temperature of PEALD HZO thin films deposited at 180 °C.

The polarization characteristics of PEALD HZO thin films deposited at various substrate temperatures were evaluated. Figure 6a shows the PE hysteresis curves measured after 10^5 cycles for each sample in which wake-up had occurred and the value of coercive field ($2E_c$) was stabilized. Figure 6b shows the dynamic polarization switching current with respect to the electric field. The value of remanent polarization $2P_r$ measured based on each P-E hysteresis curve increased significantly from 12 μC/cm^2 to 38.2 μC/cm^2 as the deposition temperature increased from 100 °C to 180 °C. Further, as the strength of the coercive field ($2E_c$) increased from 1 MV/cm to 1.97 MV/cm, the total area of the hysteresis curve increased significantly. Thereafter, as the deposition temperature increased to 280 °C, both the $2P_r$ and $2E_c$ values decreased, and this trend was consistent with the X-ray intensity of the o-phase and the density of the thin film. The maximum remanent polarization value of 38.2 μC/cm^2 of the sample obtained at the deposition temperature of 180 °C is higher than the values reported in previous papers [14,16,19]. In Figure 6b, the maximum polarization current is observed near the coercive field. In the sample obtained at a deposition temperature of 180 °C, a dynamic switching current density of up to 8.8×10^{-3} A/cm^2 was measured in an electric field of 1 MV/cm, and an almost symmetrical current pattern was also observed in negative electric fields. The trend of the polarization characteristics according to the deposition temperature is thought to be related to an increase in defects inside the HZO thin films deposited at low and high temperatures, as discussed above. These defects can limit the growth of the grain size and cause pinning of the switching of the ferroelectric domain under the external electric field [37–40].

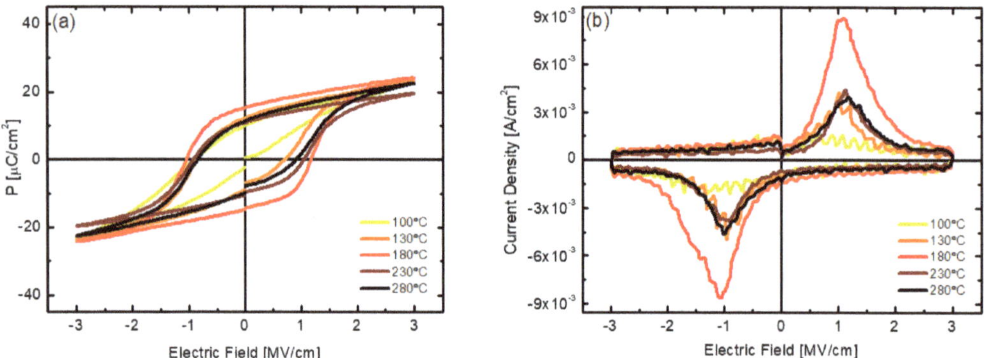

Figure 6. (**a**) P−E hysteresis curve and (**b**) polarization switching current curve with respect to electric field of HZO thin films deposited at various substrate temperatures.

Figure 7 shows the results of analyzing the electrical properties of the HZO thin films deposited at 180 °C with post-annealing at various temperatures. The best $2P_r$ value and the largest dynamic switching current density were obtained at an annealing temperature of 600 °C. The HZO thin film annealed at 500 °C showed a $2P_r$ value of 21.6 μC/cm^2 and a dynamic switching current density of up to 3.85×10^{-3} A/cm^2, showing inferior characteristics compared to those of the samples annealed at 600 °C or higher. In addition, the P-E hysteresis curve and polarization switching current curve show asymmetry according to the sign, which indicates that a built-in potential is formed inside the thin film. From this, it can be inferred that at a low post-annealing temperature of 500°C, the distribution of defects inside the film was not symmetrical or that the phase change to the o-phase was not fully completed inside the thin film [41,42].

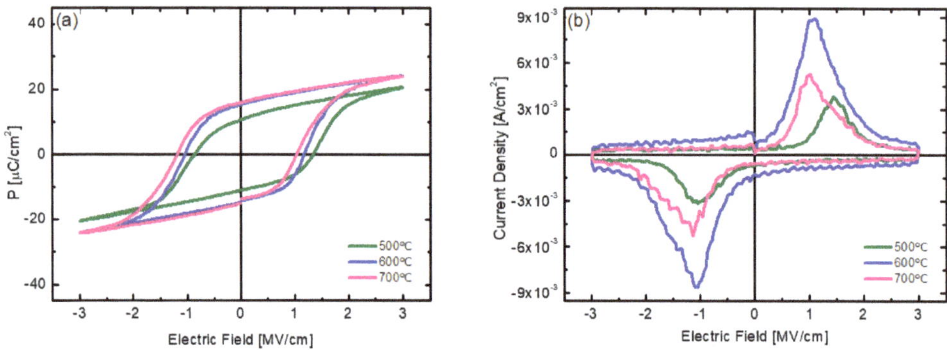

Figure 7. (**a**) P−E hysteresis curve and (**b**) polarization switching current curve with respect to electric field of HZO thin films deposited at substrate temperature of 180 °C with different post-annealing temperatures.

Figure 8 compares the results of fatigue endurance evaluation of the HZO thin films according to the above-mentioned (**a**) deposition temperature and (**b**) post-annealing temperature. In Figure 8a, the HZO thin films deposited at 180 °C and 230 °C showed the highest level of endurance of 1.6×10^7 cycles. The cases of deposition at the lowest temperature of 100 °C and the highest temperature of 280 °C showed relatively low endurances of 2.5×10^5 cycles and 1.6×10^6 cycles, respectively. In addition, in the case of these two samples, a wake-up phenomenon, in which the $2P_r$ value increased with the number of cycles, was clearly exhibited. Fatigue endurance is caused by the accumulation of impurities and oxygen vacancies inside the thin film at the electrode interface or crystal defects, and it is reported that the wake-up effect occurs in the process of redistribution of oxygen vacancies according to the application of an electric field [43,44]. It is inferred that samples with low density in the previous experiment contain many defects, resulting in low fatigue endurance or a marked wake-up effect. Figure 8b shows the results according to the annealing temperature, and it can be observed that the endurance of the thin film annealed at 600 °C is the highest. In addition, as the annealing temperature increases, the wake-up effect is improved; notably, the sample annealed at 700 °C shows the best improvement of the wake-up effect. However, the sample annealed at 700 °C showed the lowest fatigue endurance (1.6×10^6 cycles), which is attributed to diffusion at the electrode interface [45,46].

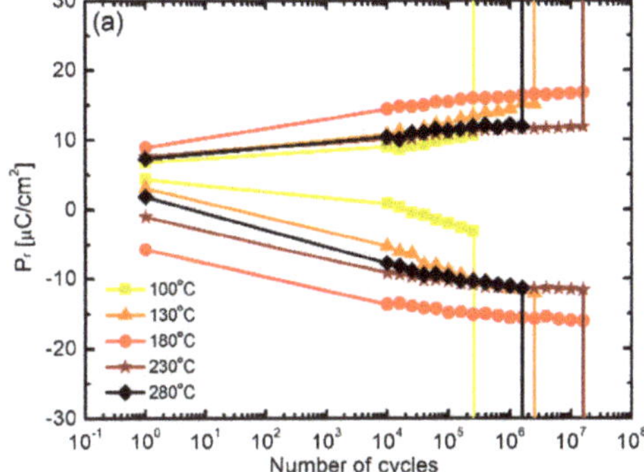
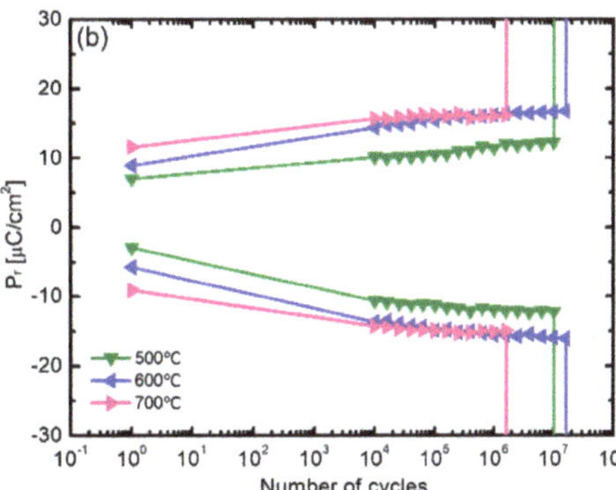

Figure 8. (**a**) Comparison of fatigue endurance of HZO films fabricated at different deposition temperatures after 600 °C annealing and (**b**) comparison of fatigue endurance of HZO films, deposited at a substrate temperature of 180 °C, with respect to annealing temperature.

The preparation methods and electrical properties of HZO films are summarized in Table 3 to compare our work with previous studies. The HZO thin film prepared at optimized PEALD conditions in this study showed relatively good remanent polarization and fatigue endurance performances despite being under the lowest deposition temperature.

Table 3. Summary and comparison of the preparation methods and electrical properties of HZO films.

Ref.	Growth Method	Electrode	Deposition Temperature	Annealing Temperature	2Pr ($\mu C/cm^2$)	Fatigue Endurance (Number of Cycles)
Our work	PEALD	TiN	180 °C	600 °C	38.2	1.6×10^7
[19]	PEALD	TiN	250 °C	450 °C	35	1.6×10^5
[10]	THALD	TiN	280 °C	600 °C	29	–
[11]	THALD	Ru	280 °C	500 °C	36	1.2×10^{11}
[12]	THALD	TiN	250 °C	400 °C	48	–
[13]	THALD	W	250 °C	720 °C	42	1×10^4

In the case of the thin film deposited at the substrate temperature of 280 °C in the previous experiment, the presence of secondary phases such as the m-phase was confirmed. Therefore, it can be inferred that for increasing the ratio of the o-phase in HZO thin films, low-temperature deposition is advantageous. However, the density was greatly reduced in the thin films deposited at low temperatures. Therefore, this study investigated process improvement methods that can increase the density of thin films deposited at low temperatures. Four types of process improvement experiments (A to D) were performed for deposition at a substrate temperature of 100 °C, followed by RTA at 600 °C, and each process is outlined as follows. Process A is the process of improving θ, the fraction of the surface covered by the absorbate, by using the discrete feeding method (DFM). In the DFM, a purge step was included in the source injection step to refine the process, thereby removing the impurities and byproducts in the precursor injection step. This increased the initial chemisorption efficiency and fraction of the surface covered by the absorbate. In this experiment, the purge step was executed twice in the middle of the process. In process B, the plasma discharge time and oxygen injection time were increased by 4 s each; thus, the discharge time in this experiment was 6 s, and the oxygen injection time was 8 s. In process C, the total flow rate in the chamber was increased from 600 sccm to 900 sccm while maintaining the pressure. Finally, in process D, all of the aforementioned process improvement methods (A to C) were applied in combination.

Figure 9 shows the changes in the XRD patterns and thin film density of the HZO thin films obtained with the various process improvement methods. In the results corresponding to processes A to D, both the o-phase peak intensity and the thin film density are significantly higher than those of the HZO thin film deposited at 100 °C without applying the process improvement. The crystallinity and density of the thin film improved upon applying process A. Specifically, θ, the fraction of the surface covered by the absorbate, increased because the application of the DFM eliminated unnecessary physical adsorption, thus stably providing the chemical adsorption sites to the precursor [25]. By applying process B, the density was significantly improved to 9.7 g/cm^3, which is thought to be due to the reduction of oxygen vacancies based on the increase in the reaction time corresponding to the precursor-reactant reaction [20,47]. The results of process C confirmed that both the crystallinity and density were improved, and it is believed that the reduced boundary layer and increased diffusion rate due to the increase in the flow rate of the precursor enhanced the fraction of the surface covered by the absorbate of the precursor [48,49]. For process D, the o-phase peak intensity and thin film density were lower than those of processes A and C, which is inferred to be due to the interaction between process parameters. The results confirmed that by applying these process improvement methods, the θ for low-temperature deposition can be improved, and with an increased RF power supply time to promote the reaction with oxygen radicals, the oxygen vacancies can be reduced, resulting in properties similar to those of thin films deposited at high temperatures.

The electrical properties of the low-temperature-deposited HZO thin films obtained with different process improvement methods were measured. Figure 10a shows the P-E hysteresis curves, and the HZO thin film obtained with process C showed the highest $2P_r$ value, 18.6 μC/cm^2. Although this value was higher than the $2P_r$ value of 12 μC/cm^2 of the thin film deposited at 100 °C without applying any process improvement, it was significantly lower than that of the thin film deposited at 180 °C. This is thought to be because although the physical properties of the thin films obtained by applying the process improvement methods were similar to those of the thin films deposited at 180 °C, the formation of a ferroelectric domain inside the thin films was hindered, and further investigation is needed to identify the cause. As shown by the fatigue endurance measurement results in Figure 10b, dielectric breakdown occurred at 10^7 cycles or more, indicating that the service life was greatly improved through the process improvement. In particular, when process B was applied, the highest endurance of 2.5×10^7 cycles was measured, which was superior to that of the thin film deposited at 180 °C. This increase in endurance is thought to be because the oxygen vacancies inside the thin film, which cause fatigue, were greatly reduced by the effects of the applied process improvement.

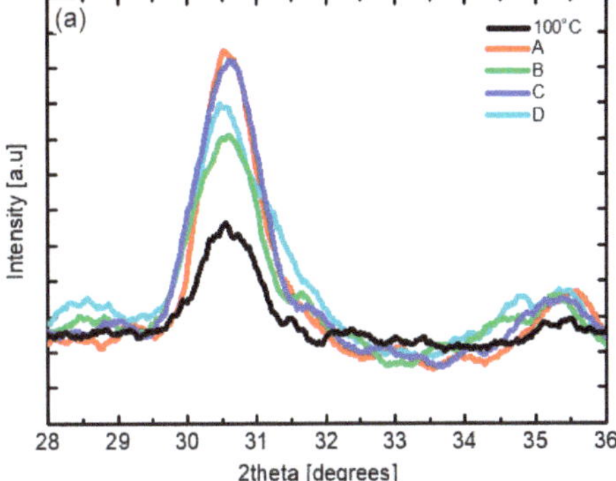 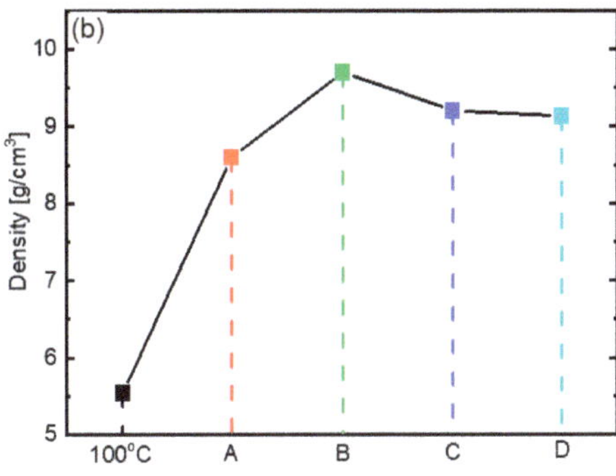

Figure 9. (**a**) XRD patterns and (**b**) thin film density according to the process improvement method of low-temperature-deposited HZO thin films; A: use of discrete feeding method, B: increase in RF plasma time, C: increase in gas flow, D: combined application of A, B, and C.

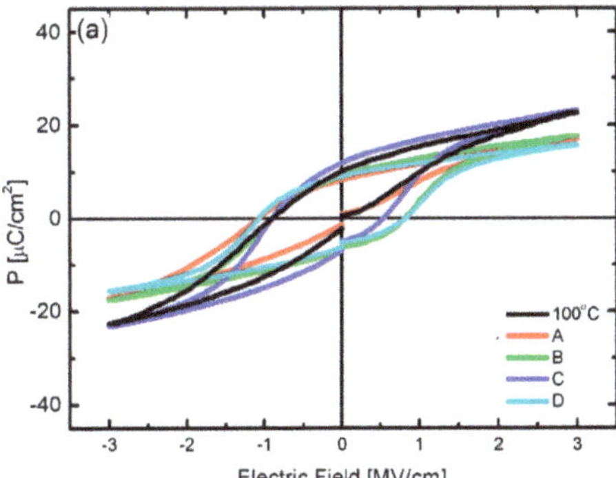 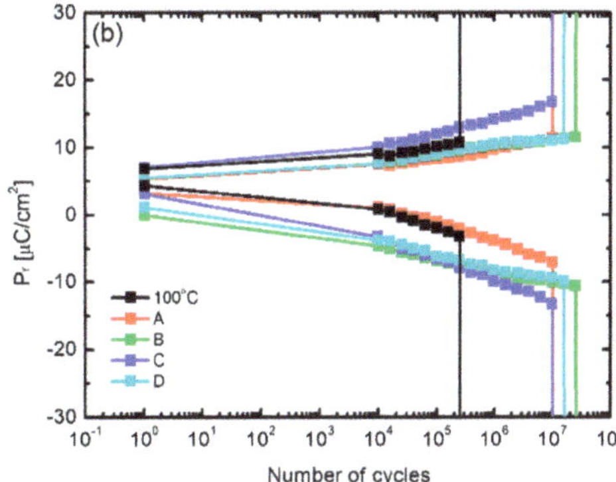

Figure 10. (**a**) P$-$E hysteresis curves and (**b**) fatigue endurance characteristics according to the process improvement method of low-temperature-deposited HZO thin films; A: use of discrete feeding method, B: increase in RF plasma time, C: increase in gas flow, D: combined application of A, B, and C.

4. Conclusions

In this study, the characteristics and electrical properties of ferroelectric HZO thin films obtained by PEALD were evaluated according to the deposition temperature and annealing temperature. Further, we developed and applied various processes to improve these characteristics and electrical properties. First, since the growth per cycle (GPC) according to the deposition temperature of HfO_2 and ZrO_2 was constant, it was possible to deposit HZO thin films with similar deposition rates at all temperatures. The thickness of the deposited HZO thin films, o-phase crystalline structure, and elemental composition profiles were examined through cross-sectional TEM images and EDS analysis. The X-ray intensity of the o-phase of the thin film deposited at the substrate temperature of 180 °C was the highest, and mixed secondary phases such as the m-phase were observed in the thin film deposited at 280 °C. Density analysis of the thin films showed that the HZO thin film deposited at 180 °C had the highest density and a decrease in density was observed in the thin films deposited at temperatures lower and higher than 180 °C. To investigate the density change according to annealing temperature, the thin film samples were annealed in the temperature range of 500–700 °C, and a post-annealing temperature of 600 °C yielded the highest thin film density. The $2P_r$ value of the thin film fabricated with the deposition temperature of 180 °C and post-annealing at 600 °C was the highest, 38.2 $\mu C/cm^2$, and the fatigue endurance was also the highest under these conditions, 1.6×10^7 cycles. Three methods were proposed to enhance the density of the low-temperature-deposited thin films. For the low-temperature-deposited thin films with an increased RF plasma discharge time, the enhanced maximum density and an excellent fatigue endurance of 2.5×10^7 cycles were obtained.

Author Contributions: Conceptualization, H.-G.K. and H.-C.L.; methodology, H.-G.K.; software, J.-H.Y.; validation, H.-G.K. and H.-C.L.; formal analysis, D.-H.H.; investigation, H.-G.K.; resources, H.-C.L.; data curation, J.-H.Y.; writing—original draft preparation, H.-G.K.; writing—review and editing, H.-C.L.; visualization, D.-H.H. and J.-H.Y.; supervision, H.-C.L.; project administration, H.-C.L.; funding acquisition, H.-C.L. All authors have read and agreed to the published version of the manuscript.

Funding: This study was supported by the Industrial Technology Innovation Program (Grant No. 20006408) funded by the Ministry of Trade, Industry, and Energy (MOTIE) and the Priority Research Centers Program (2017R1A6A1A03015562) through the National Research Foundation (NRF) funded by the Ministry of Education.

Institutional Review Board Statement: Not applicable.

Informed Consent Statement: Not applicable.

Data Availability Statement: The data presented in this study are contained within the article.

Conflicts of Interest: The authors declare no conflict of interest.

References

1. Böscke, T.S.; Müller, J.; Bräuhaus, D.; Schröder, U.; Böttger, U. Ferroelectricity in Hafnium Oxide Thin Films. *Appl. Phys. Lett.* **2011**, *99*, 102903. [CrossRef]
2. Kim, M.K.; Lee, J.S. Ferroelectric Analog Synaptic Transistors. *Nano Lett.* **2019**, *19*, 2044–2050. [CrossRef] [PubMed]
3. Ni, K.; Sharma, P.; Zhang, J.; Jerry, M.; Smith, J.A.; Tapily, K.; Clark, R.; Mahapatra, S.; Datta, S. Critical Role of Interlayer in $Hf_{0.5}Zr_{0.5}O_2$ Ferroelectric FET Nonvolatile Memory Performance. *IEEE Trans. Electron. Devices* **2018**, *65*, 2461–2469. [CrossRef]
4. Kim, M.K.; Lee, J.S. Synergistic Improvement of Long-Term Plasticity in Photonic Synapses Using Ferroelectric Polarization in Hafnia-Based Oxide-Semiconductor Transistors. *Adv. Mater.* **2020**, *32*, e1907826. [CrossRef] [PubMed]
5. Li, K.-S.; Chen, P.-G.; Lai, T.-Y.; Lin, C.-H.; Cheng, C.-C.; Chen, C.-C.; Wei, Y.-J.; Hou, Y.-F.; Liao, M.-H.; Lee, M.-H. Sub-60mV-Swing Negative-Capacitance FinFET without Hysteresis. In Proceedings of the 2015 IEEE International Electron Devices Meeting (IEDM), Washington, DC, USA, 7–9 December 2015; Volume 4, pp. 22.6.1–22.6.4.
6. Ambriz-Vargas, F.; Kolhatkar, G.; Broyer, M.; Hadj-Youssef, A.; Nouar, R.; Sarkissian, A.; Thomas, R.; Gomez-Yáñez, C.; Gauthier, M.A.; Ruediger, A. A Complementary Metal Oxide Semiconductor Process-Compatible Ferroelectric Tunnel Junction. *ACS Appl. Mater. Interfaces* **2017**, *9*, 13262–13268. [CrossRef]
7. Nakamura, T.; Nakao, Y.; Kamisawa, A.; Takasu, H. Preparation of $Pb(Zr, Ti)O_3$ Thin Films on Electrodes Including IrO_2. *Appl. Phys. Lett.* **1994**, *65*, 1522–1524. [CrossRef]
8. Park, M.H.; Lee, Y.H.; Mikolajick, T.; Schroeder, U.; Hwang, C.S. Review and Perspective on Ferroelectric HfO_2-Based Thin Films for Memory Applications. *MRS Commun.* **2018**, *8*, 795–808. [CrossRef]
9. Moazzami, R.; Hu, C.; Shepherd, W.H. Endurance Properties of Ferroelectric PZT Thin Films. In Proceedings of the International Technical Digest on Electron Devices, San Francisco, CA, USA, 9–12 December 1990; pp. 417–420.
10. Park, M.H.; Lee, Y.H.; Kim, H.J.; Kim, Y.J.; Moon, T.; Do Kim, K.D.; Hyun, S.D.; Mikolajick, T.; Schroeder, U.; Hwang, C.S. Understanding the Formation of the Metastable Ferroelectric Phase in Hafnia–Zirconia Solid Solution Thin Films. *Nanoscale* **2018**, *10*, 716–725. [CrossRef]
11. Cao, R.; Liu, Q.; Liu, M.; Song, B.; Shang, D.; Yang, Y.; Luo, Q.; Wu, S.; Li, Y.; Wang, Y.; et al. Improvement of Endurance in HZO-Based Ferroelectric Capacitor Using Ru Electrode. *IEEE Electron. Dev. Lett.* **2019**, *40*, 1744–1747. [CrossRef]
12. Kim, S.J.; Narayan, D.; Lee, J.-G.; Mohan, J.; Lee, J.S.; Lee, J.; Kim, H.S.; Byun, Y.-C.; Lucero, A.T.; Young, C.D.; et al. Large Ferroelectric Polarization of $TiN/Hf_{0.5}Zr_{0.5}O_2/TiN$ Capacitors Due to Stress-Induced Crystallization at Low Thermal Budget. *Appl. Phys. Lett.* **2017**, *111*, 242901. [CrossRef]
13. Kashir, A.; Kim, H.; Oh, S.; Hwang, H. Large Remnant Polarization in a Wake-Up Free $Hf_{0.5}Zr_{0.5}O_2$ Ferroelectric Film through Bulk and Interface Engineering. *ACS Appl. Electron. Mater.* **2021**, *3*, 629–638. [CrossRef]
14. Park, M.H.; Lee, D.H.; Yang, K.; Park, J.-Y.; Yu, G.T.; Park, H.W.; Materano, M.; Mittmann, T.; Lomenzo, P.D.; Mikolajick, T.; et al. Review of Defect Chemistry in Fluorite-Structure Ferroelectrics for Future Electronic Devices. *J. Mater. Chem. C* **2020**, *8*, 10526–10550. [CrossRef]

15. Materlik, R.; Künneth, C.; Kersch, A. The Origin of Ferroelectricity in $Hf_{1-x}Zr_xO_2$:A Computational Investigation and a Surface Energy Model. *J. Appl. Phys.* **2015**, *117*, 134109. [CrossRef]

16. Hyuk Park, M.; Joon Kim, H.; Jin Kim, Y.; Lee, W.; Moon, T.; Seong Hwang, C. Evolution of Phases and Ferroelectric Properties of Thin $Hf_{0.5}Zr_{0.5}O_2$ Films According to the Thickness and Annealing Temperature. *Appl. Phys. Lett.* **2013**, *102*, 242905. [CrossRef]

17. Saha, A.K.; Grisafe, B.; Datta, S.; Gupta, S.K. Microscopic Crystal Phase Inspired Modeling of Zr Concentration Effects in $Hf_{1-x}Zr_xO_2$ Thin Films. In Proceedings of the 2019 Symposium on VLSI Technology, Kyoto, Japan, 9–14 June 2019.

18. Cho, D.-Y.; Jung, H.-S.; Hwang, C.S. Structural Properties and Electronic Structure of HfO_2-ZrO_2 Composite Films. *Phys. Rev. B* **2010**, *82*, 094104. [CrossRef]

19. Hur, J.; Tasneem, N.; Choe, G.; Wang, P.; Wang, Z.; Khan, A.I.; Yu, S. Direct Comparison of Ferroelectric Properties in $Hf_{0.5}Zr_{0.5}O_2$ between Thermal and Plasma-Enhanced Atomic Layer Deposition. *Nanotechnology* **2020**, *31*, 505707. [CrossRef] [PubMed]

20. Kim, K.M.; Jang, J.S.; Yoon, S.G.; Yun, J.Y.; Chung, N.K. Structural, Optical and Electrical Properties of HfO_2 Thin Films Deposited at Low-Temperature Using Plasma-Enhanced Atomic Layer Deposition. *Materials* **2020**, *13*, 2008. [CrossRef]

21. Chen, Z.; Wang, H.; Wang, X.; Chen, P.; Liu, Y.; Zhao, H.; Zhao, Y.; Duan, Y. Low-Temperature Remote Plasma Enhanced Atomic Layer Deposition of ZrO_2/Zircone Nanolaminate Film for Efficient Encapsulation of Flexible Organic Light-Emitting Diodes. *Sci. Rep.* **2017**, *7*, 40061. [CrossRef]

22. Xiao, Z.; Kisslinger, K.; Chance, S.; Banks, S. Comparison of Hafnium Dioxide and Zirconium Dioxide Grown by Plasma-Enhanced Atomic Layer Deposition for the Application of Electronic Materials. *Crystals.* **2020**, *10*, 136. [CrossRef]

23. Lin, Y.-H.; Chen, W.-C.; Chen, P.-H.; Lin, C.-Y.; Chang, K.-C.; Chang, Y.-C.; Yeh, C.-H.; Lin, C.-Y.; Jin, F.-Y.; Chen, K.-H.; et al. Effect of Deposition Temperature on Electrical Properties of One-Transistor-One-Capacitor (1T1C) FeRAM Devices. *Appl. Phys. Lett.* **2020**, *117*, 023502. [CrossRef]

24. Myers, T.J.; Throckmorton, J.A.; Borrelli, R.A.; O'Sullivan, M.; Hatwar, T.; George, S.M. Smoothing Surface Roughness Using Al_2O_3 Atomic Layer Deposition. *Appl. Surf. Sci.* **2021**, *569*, 150878. [CrossRef]

25. Park, T.J.; Kim, J.H.; Jang, J.H.; Kim, U.K.; Lee, S.Y.; Lee, J.; Jung, H.S.; Hwang, C.S. Improved Growth and Electrical Properties of Atomic-Layer-Deposited Metal-Oxide Film by Discrete Feeding Method of Metal Precursor. *Chem. Mater.* **2011**, *23*, 1654–1658. [CrossRef]

26. Groner, M.D.; Fabreguette, F.H.; Elam, J.W.; George, S.M. Low-Temperature Al_2O_3 Atomic Layer Deposition. *Chem. Mater.* **2004**, *16*, 639–645. [CrossRef]

27. Qi, Y.; Xu, X.; Krylov, I.; Eizenberg, M. Ferroelectricity of as-Deposited HZO Fabricated by Plasma-Enhanced Atomic Layer Deposition at 300°C by Inserting TiO_2 Interlayers. *Appl. Phys. Lett.* **2021**, *118*, 032906. [CrossRef]

28. Park, M.H.; Kim, H.J.; Do Kim, K.; Lee, Y.H.; Hyun, S.D.; Hwang, C.S. Impact of Zr Content in Atomic Layer Deposited $Hf_{1-x}Zr_xO_2$ Thin Films. In *Ferroelectricity in Doped Hafnium Oxide: Materials, Properties and Devices*; Elsevier: Amsterdam, The Netherlands, 2019; pp. 75–101.

29. Suzuki, R.; Taoka, N.; Yokoyama, M.; Kim, S.-H.; Hoshii, T.; Maeda, T.; Yasuda, T.; Ichikawa, O.; Fukuhara, N.; Hata, M.; et al. Impact of Atomic Layer Deposition Temperature on HfO_2/InGaAs Metal-Oxide-Semiconductor Interface Properties. *J. Appl. Phys.* **2012**, *112*, 084103. [CrossRef]

30. Kim, K.D.; Lee, Y.H.; Gwon, T.; Kim, Y.J.; Kim, H.J.; Moon, T.; Hyun, S.D.; Park, H.W.; Park, M.H.; Hwang, C.S. Scale-Up and Optimization of HfO_2-ZrO_2 Solid Solution Thin Films for the Electrostatic Supercapacitors. *Nano Energy* **2017**, *39*, 390–399. [CrossRef]

31. Kisi, E.H.; Howard, C.J.; Hill, R.J. Crystal Structure of Orthorhombic Zirconia in Partially Stabilized Zirconia. *J. Am. Ceram. Soc.* **1989**, *72*, 1757–1760. [CrossRef]

32. Lim, J.-W.; Park, J.-S.; Kang, S.-W. Kinetic Modeling of Film Growth Rates of TiN Films in Atomic Layer Deposition. *J. Appl. Phys.* **2000**, *87*, 4632–4634. [CrossRef]

33. Jaffe, J.E.; Bachorz, R.A.; Gutowski, M. Low-Temperature Polymorphs of ZrO_2 and HfO_2:A Density-Functional Theory Study. *Phys. Rev. B* **2005**, *72*, 144107. [CrossRef]

34. Lowther, J.E.; Dewhurst, J.K.; Leger, J.M.; Haines, J. Relative Stability of ZrO_2 and HfO_2 Structural Phases. *Phys. Rev. B* **1999**, *60*, 14485–14488. [CrossRef]

35. Haggerty, R.P.; Sarin, P.; Apostolov, Z.D.; Driemeyer, P.E.; Kriven, W.M. Thermal Expansion of HfO_2 and ZrO_2. *J. Am. Ceram. Soc.* **2014**, *97*, 2213–2222. [CrossRef]

36. Lin, Y.-C.; McGuire, F.; Franklin, A.D. Realizing Ferroelectric $Hf_{0.5}Zr_{0.5}O_2$ with Elemental Capping Layers. *J. Vac. Sci. Technol. B Nanotechnol. Microelectron. Mater. Process. Meas. Phenom.* **2018**, *36*, 011204. [CrossRef]

37. Goh, Y.; Jeon, S. First-Order Reversal Curve Diagrams for Characterizing Ferroelectricity of $Hf_{0.5}Zr_{0.5}O_2$ Films Grown at Different Rates. *J. Vac. Sci. Technol. B Nanotechnol. Microelectron. Mater. Process. Meas. Phenom.* **2018**, *36*, 052204. [CrossRef]

38. Chen, W.-C.; Tan, Y.-F.; Lin, S.-K.; Zhang, Y.-C.; Chang, K.-C.; Lin, Y.-H.; Yeh, C.-H.; Wu, C.-W.; Yeh, Y.-H.; Wang, K.-Y.; et al. Performance Improvement by Modifying Deposition Temperature in $HfZrO_x$ Ferroelectric Memory. *IEEE Trans. Electron. Devices* **2021**, *68*, 3838–3842. [CrossRef]

39. Pešić, M.; Fengler, F.P.G.; Larcher, L.; Padovani, A.; Schenk, T.; Grimley, E.D.; Sang, X.; LeBeau, J.M.; Slesazeck, S.; Schroeder, U.; et al. Physical Mechanisms behind the Field-Cycling Behavior of HfO_2-Based Ferroelectric Capacitors. *Adv. Funct. Mater.* **2016**, *26*, 4601–4612. [CrossRef]

40. Goh, Y.; Cho, S.H.; Park, S.-H.K.; Jeon, S. Oxygen Vacancy Control as a Strategy to Achieve Highly Reliable Hafnia Ferroelectrics Using Oxide Electrode. *Nanoscale* **2020**, *12*, 9024–9031. [CrossRef] [PubMed]

41. Kim, S.J.; Mohan, J.; Lee, J.; Lee, J.S.; Lucero, A.T.; Young, C.D.; Colombo, L.; Summerfelt, S.R.; San, T.; Kim, J. Effect of Film Thickness on the Ferroelectric and Dielectric Properties of Low-Temperature (400 °C) $Hf_{0.5}Zr_{0.5}O_2$ Films. *Appl. Phys. Lett.* **2018**, *112*, 172902. [CrossRef]

42. Kim, S.J.; Mohan, J.; Summerfelt, S.R.; Kim, J. Ferroelectric $Hf_{0.5}Zr_{0.5}O_2$ Thin Films: A Review of Recent Advances. *JOM* **2019**, *71*, 246–255. [CrossRef]

43. Jiang, P.; Luo, Q.; Xu, X.; Gong, T.; Yuan, P.; Wang, Y.; Gao, Z.; Wei, W.; Tai, L.; Lv, H. Wake-up Effect in HfO_2-Based Ferroelectric Films. *Adv. Electron. Mater.* **2021**, *7*, 2000728. [CrossRef]

44. Kashir, A.; Oh, S.; Hwang, H. Defect Engineering for Control of Wake-Up Effect in HfO2-Based Ferroelectrics. *arXiv* **2020**, arXiv:2009.04714.

45. Shekhawat, A.; Walters, G.; Chung, C.-C.; Garcia, R.; Liu, Y.; Jones, J.; Nishida, T.; Moghaddam, S. Effect of Forming Gas Furnace Annealing on the Ferroelectricity and Wake-Up Effect of $Hf_{0.5}Zr_{0.5}O_2$ Thin Films. *ECS J. Solid State Sci. Technol.* **2020**, *9*, 024011. [CrossRef]

46. Kim, H.J.; Park, M.H.; Kim, Y.J.; Lee, Y.H.; Moon, T.; Do Kim, K.D.; Hyun, S.D.; Hwang, C.S. A Study on the Wake-Up Effect of Ferroelectric $Hf_{0.5}Zr_{0.5}O_2$ Films by Pulse-Switching Measurement. *Nanoscale* **2016**, *8*, 1383–1389. [CrossRef] [PubMed]

47. Knoops, H.C.M.; Faraz, T.; Arts, K.; Kessels, W.M.M. Status and Prospects of Plasma-Assisted Atomic Layer Deposition. *J. Vac. Sci. Technol. A* **2019**, *37*, 030902. [CrossRef]

48. Deng, Z.; He, W.; Duan, C.; Shan, B.; Chen, R. Atomic Layer Deposition Process Optimization by Computational Fluid Dynamics. *Vacuum* **2016**, *123*, 103–110. [CrossRef]

49. Leys, M.R.; Veenvliet, H. A Study of the Growth Mechanism of Epitaxial GaAs as Grown by the Technique of Metal Organic Vapour Phase Epitaxy. *J. Cryst. Growth.* **1981**, *55*, 145–153. [CrossRef]

nanomaterials

Article

Enhanced Ferroelectric, Dielectric Properties of Fe-Doped PMN-PT Thin Films

Chao Feng, Tong Liu, Xinyu Bu and Shifeng Huang *

Shandong Provincial Key Laboratory of Preparation and Measurement of Building Materials, University of Jinan, Jinan 250022, China; mse_fengc@ujn.edu.cn (C.F.); shandongmems@163.com (T.L.); buxinyu98@163.com (X.B.)
* Correspondence: mse_huangsf@ujn.edu.cn

Abstract: Fe-doped $0.71Pb(Mg_{1/3}Nb_{2/3})O_3$-$0.29PbTiO_3$ (PMN-PT) thin films were grown in $Pt/Ti/SiO_2/Si$ substrate by a chemical solution deposition method. Effects of the annealing temperature and doping concentration on the crystallinity, microstructure, ferroelectric and dielectric properties of thin film were investigated. High (111) preferred orientation and density columnar structure were achieved in the 2% Fe-doped PMN-PT thin film annealed at 650 °C. The preferred orientation was transferred to a random orientation as the doping concentration increased. A 2% Fe-doped PMN-PT thin film showed the effectively reduced leakage current density, which was due to the fact that the oxygen vacancies were effectively restricted and a transition of Ti^{4+} to Ti^{3+} was prevented. The optimal ferroelectric properties of 2% Fe-doped PMN-PT thin film annealed at 650 °C were identified with slim polarization-applied field loops, high saturation polarization (P_s = 78.8 μC/cm²), remanent polarization (P_r = 23.1 μC/cm²) and low coercive voltage (E_c = 100 kV/cm). Moreover, the 2% Fe-doped PMN-PT thin film annealed at 650 °C showed an excellent dielectric performance with a high dielectric constant (ε_r ~1300 at 1 kHz).

Keywords: PMN-PT thin films; preferred orientation; ferroelectric property; dielectric property

Citation: Feng, C.; Liu, T.; Bu, X.; Huang, S. Enhanced Ferroelectric, Dielectric Properties of Fe-Doped PMN-PT Thin Films. *Nanomaterials* **2021**, *11*, 3043. https://doi.org/10.3390/nano11113043

Academic Editor: Jürgen Eckert

Received: 14 October 2021
Accepted: 6 November 2021
Published: 12 November 2021

Publisher's Note: MDPI stays neutral with regard to jurisdictional claims in published maps and institutional affiliations.

1. Introduction

Lead-based materials have been researched for a series of device applications, such as actuators, non-volatile memories, transducers and sensors [1–4]. The $Pb(Mg_{1/3}Nb_{2/3})O_3$-$PbTiO_3$ (PMN-PT), as relaxor ferroelectrics, with outstanding dielectric, ferroelectric and piezoelectric properties, has been widely investigated in terms of bulk ceramics and single crystals [5–8]. However, the single crystals and ceramics have been unable to meet the requirements of integrated and miniaturized devices with the development of micro-electromechanical systems (MEMS) in recent years. Advances have been made in synthesis and modification of the PMN-PT thin films because of unique advantages, such as low synthesis temperature, small size, easy integration [9,10]. However, it is still a big challenging to prepare PMN-PT thin films with single phase, crack-free and dense microstructure. In addition, the performances of thin films are far weaker than the bulk materials because of the thickness effect of film and the clamping effect of the substrates.

In order to improve the quality of crystallization and enhance electrical properties of PMN-PT thin film, researchers have done a great deal of works on topics including site engineering, regulating of annealing process and so on. For example, rare earth element doping can effectively enhance electrical properties of PMN-PT thin films [11]; Gabor et al. have reported the process window can be expanded in order to obtain pure phase PMN-PT thin film with the help of $LaNiO_3$ layer [12]; Keech et al. prepared (001)-orientation PMN-PT films with a PbO buffer to account for Pb loss [13]; Shen et al. found the additive methanamide can enhance electrical properties of PMN-PT thin films with reduced residual stress of the film and dense microstructure [14]. Deposition methods of various kinds have been reported to synthesize PMN-PT thin films, including pulsed laser deposition, magnetron sputtering, metalorganic chemical vapor deposition and chemical solution deposition (CSD) [15–18]. CSD is believed to be a further industrial method with easy composition control of precursor solutions, excellent reproducible, low synthesis temperature and low cost. For the CSD route, it is critical to control crystallization through regulating annealing process parameters because annealing treatment is necessary to crystallize the amorphous films. Due to the electrical properties of PMN-PT thin films being strongly dependent on their grain orientation, the realization of preferential orientation is also expected to improve the electrical properties of ferroelectric thin films. Therefore, the synthesis of an ideal PMN-PT thin film should include two aspects: appropriate annealing parameters and the preferential orientation resultant thin film.

In this research, $0.71Pb(Mg_{1/3}Nb_{2/3})O_3$-$0.29PbTiO_3$ thin films, with different doping concentrations of the Fe element used as acceptor doping, were prepared on $Pt(111)/Ti/SiO_2/Si$ substrate by the CSD method. The effects of annealing temperature and doping concentration on the phase, microstructure, dielectric and ferroelectric properties of PMN-PT thin films were systematically researched. The highly (111) preferred orientation and dense microstructure were achieved in the 2% Fe-doped PMN-PT thin film annealed at 650 °C, which lead to the enhanced ferroelectric with high saturation polarization (P_s = 78.8 $\mu C/cm^2$) and remanent polarization (P_r = 23.1 $\mu C/cm^2$) as well as dielectric with a high dielectric constant (ε_r~1300 at 1 kHz).

2. Materials and Methods

The precursor solutions of $0.71Pb(Mg_{1/3}Nb_{2/3})O_3$-$(0.29$-x%$)PbTiO_3$-x%$PbFeO_3$ (PMN-PT-xFe, x = 0, 2, 4, 8) thin films were prepared by the CSD method. Trihydrated lead acetate, magnesium ethoxide, niobium ethoxide, titanium iso-propoxide and iron nitrate were used as the raw materials. The solvent was 2-methoxyethanol and acetic acid; about 5 mol% excess of lead was added to compensate the volatilization during heat treatment. The solution was stirred using a magnetic stirrer for 12 h, and further aged for 24 h to promote the uniformity of solution. The concentration of precursor solutions was adjusted to be 0.2 M; the precursor was spin-coated on $Pt(111)/Ti/SiO_2/Si$ substrate at 3000 rpm for 30 s. The obtained film was heated at 350 °C for 2 min for solvents volatilizing and amorphous films formation, and subsequently annealed at the specified annealing temperature (600~700 °C) for 5 min in air atmosphere for film densification and crystallization in a programmed rapid thermal annealing furnace. The desirable thickness was achieved by a 12 time layer-by-layer spin-coating and annealing processes with one layer thickness of about 25 nm.

Crystalline phases were identified by X-ray diffraction (XRD) measurement using Bruker D8 Advanced XRD (Berlin, Germany). The microstructures of thin films were recorded using a field emission scanning electron microscopy (SEM) and atomic force microscope (AFM, MFP-3D Origin+, Concord. MA, USA). The ferroelectric tester (TF Analyzer 3000, Aachen, Germany) was used for polarization-electric field (P–E) hysteresis loops and leakage current density of thin films. The relative permittivity (ε_r) and dielectric loss ($\tan\delta$) were measured using an impedance analyzer (Agilent 4294A, Santa Clara, CA, USA).

3. Results and Discussion

Figure 1a shows the XRD patterns of PMN-PT-2Fe thin films annealed at different temperatures. It can be seen that the PMN-PT perovskite phase coexists with a pyrochlore phase (peak at 2θ = 29.3°) in thin films treated over the entire temperature range. The larger fraction of pyrochlore phase is attributed to the lead remaining due to the inadequate annealing. The decrease of concentration for pyrochlore phase from 4.2% for 600 °C to 1.4% for 700 °C indicates the phase transition from pyrochlore to perovskite phase. Moreover, the pyrochlore phase almost disappeared which is due to fuller annealing when the annealing temperature reached 700 °C. In addition, it is worth noting that the preferred orientation of thin films also depends on the annealing temperature; the degree of preferred orientation is defined by the preferential orientation parameter α_{hkl} calculated with the following equation:

$$\alpha_{hkl} = \frac{I_{hkl}}{\sum I_{hkl}} \tag{1}$$

where, I_{hkl} is the peak intensity of (hkl) orientation peak of thin film. The α_{111} of PMN-PT-2Fe thin films with the annealing temperature of 600, 650 and 700 °C were 76%, 90% and 74%, respectively. The highly (111) preferred orientation will lead to the excellent electrical performances of PMN-PT thin films [19]. The crystallization of PMN-PT-xFe thin films as a function of doping concentration of Fe element is shown in Figure 1b. The obviously enhanced intensity of peak indicates that the introduction of Fe ion promotes the crystallization of thin film. The preferred orientation transition from (111) to (110) was observed when the doping concentration of Fe ion increased from 2% to 8%. The degree of (110) preferential orientation of PMN-PT-8Fe was up to 49%. The preferred orientation of ferroelectric thin films is subject to three factors: (i) template layer [20], (ii) formation of intermetallic phase during film crystallization [21], (iii) crystallization kinetic [22]. The (111) preferred orientation of the PMN-PT and PMN-PT-2Fe thin films is attributed to the template effect of the Pt(111) substrate [23]. However, continuing to increase the doping concentration will impact the preheat decomposition of the sol, which will change the crystal orientation [24]. When the doping concentration is up to 4%, the preferred orientation transfer from (111) to (110), indicating that effect on the preheat decomposition surpasses the template effect.

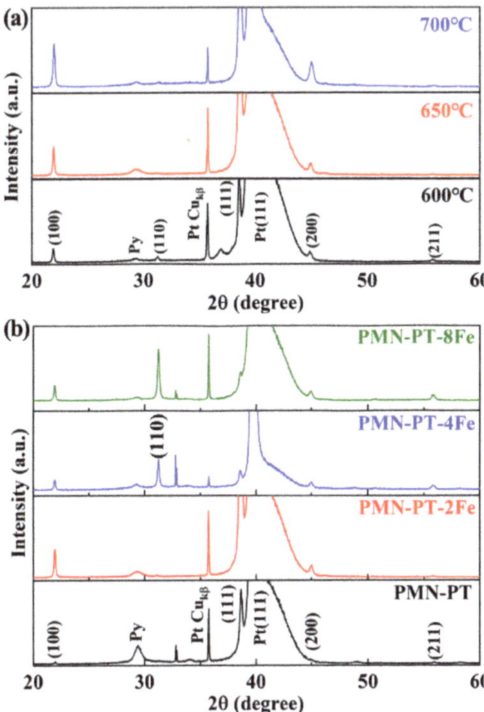

Figure 1. (a) XRD patterns of PMN-PT-2Fe thin films annealed at different temperatures, (b) XRD patterns of PMN-PT-xFe thin films annealed at 650 °C with different doping concentrations of Fe element. Note that the peaks at 21.95°, 44.98°, 38.65°, 55.81° correspond the (100), (200), (111) and (211) diffraction peaks of PMN-PT thin films, respectively. The 29.39° peak corresponds to the diffraction peak of pyrochlore structure.

Figure 2 shows the microstructures of PMN-PT-2Fe thin films annealed at different temperatures; the surface of the film, annealed at 600 °C, showed an abundance of very small features, which is due to the incomplete growth of the grain. It is clear to observe that the grains grew as the annealing temperature increased, and that the film annealed at 700 °C had the largest grains, at 100–250 nm. The uniform, dense grains, as well as the crack-free thin film were achieved at an annealing temperature of 650 °C, which is beneficial for improving electrical properties. However, the obvious cracks appeared in the film annealed at 600 and 700 °C, which leads to a deterioration in performance. As shown in cross-sectional SEM images, all films are composed of relatively dense microstructures. The PMN-PT-2Fe thin film annealed at 600 °C displays a layered structure. The typical columnar structure can be observed in the film annealed at 650 °C, which corresponds to the preferred orientation growth, as shown in XRD patterns in Figure 1a. However, the continuous increase of annealing temperature does not lead to further optimization of the column structure, accompanied by the appearance of voids. The thickness of all thin films is about 290 nm. The AFM images of PMN-PT-2Fe thin films annealed at different temperatures in the scanning area of 10 μm × 10 μm were shown in Figure 2g–i, representing the root square roughness is 4.914 nm, 3.061 nm, 8.089 nm for 600, 650 and 700 °C, respectively.

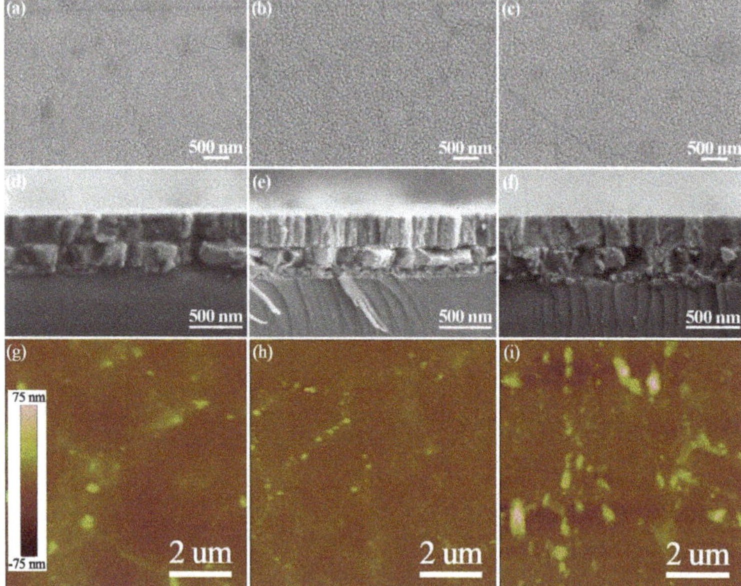

Figure 2. Microstructures of PMN-PT-2Fe thin films annealed at different temperatures. (**a,d,g**) 600 °C; (**b,e,h**) 650 °C; (**c,f,i**) 700 °C. Note the cracks are observed in (**a,c**) and the uniform, dense and crack-free structure is obtained in (**b**), the typical columnar structure is shown in (**e**).

Figure 3 shows the microstructures of PMN-PT-xFe thin films with different Fe doping concentrations. It can be seen that the fine grain was achieved in the pure PMN-PT thin film. The larger grains developed as the doping concentration of Fe ion increased. However, obvious voids can be observed between layers in thin films with 4% and 8% Fe doping, although the columnar structure still exists, which is consistent with the weakening of preferred orientation, as shown in Figure 1b. Figure 3i–l shows the AFM images of PMN-PT-xFe thin films with different Fe doping concentrations, representing the root square roughness as 3.776 nm, 3.061 nm, 3.999 nm, 4.077 nm for 0, 2, 4 and 8% doping concentrations, respectively.

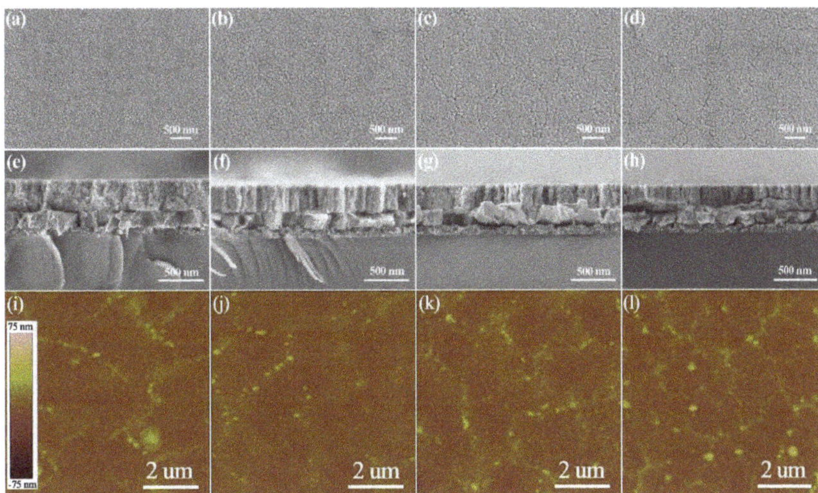

Figure 3. Microstructures of PMN-PT-xFe thin films with different doping concentration of Fe element. (**a**,**e**,**i**) x = 0; (**b**,**f**,**j**) x = 2; (**c**,**g**,**k**) x = 4; (**d**,**h**,**l**) x = 8. Note that the fine grains are obtained in (**a**), the cracks are observed in (**c**,**d**) and the uniform, dense and crack-free structure is obtained in (**b**), the layered structure is shown in (**e**), the typical columnar structure is shown in (**f**,**g**,**h**).

Figure 4a shows the leakage current density–voltage (*J–V*) curves of PMN-PT-2Fe thin films as a function of annealing temperature from 600 to 700 °C. It can be seen that the curves of all thin films exhibit asymmetry between positive and negative applied voltage, which might be attributed to the different interface states between top Au/PMN-PT-2Fe thin film and bottom Pt/PMN-PT-2Fe thin film. The leakage current density exhibits steep increasing tendency in the low voltage, which suggests that the dominant conduction mechanism is Ohmic. In the high voltage scale, the leakage current densities increase slightly, which is attributed to the contribution of space charge limited current conduction mechanism. A sudden increase in leakage current for PMN-PT-2Fe thin film annealed at 700 °C was attributed to the abnormal big grain. The shortened grain boundaries accompanied by the big grain provided shorter leakage current path [25].

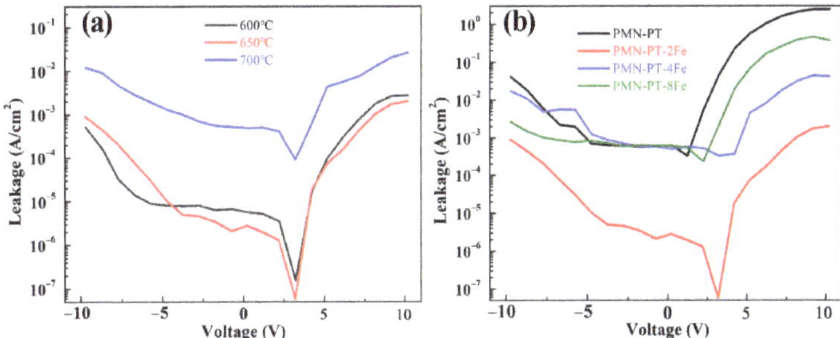

Figure 4. (**a**) Leakage current density of PMN-PT-2Fe thin films annealed at varied temperatures, (**b**) Leakage current density of PMN-PT-xFe thin films with different doping concentrations of Fe element.

The effect of doping concentration on the leakage current density of PMN-PT-xFe thin films was measured, as shown in Figure 4b. It can be seen that the pure PMN-PT thin film exhibits the highest leakage current density. The effective decrease leakage current can be obtained in 2% Fe-doped thin film. The leakage current of PMN-PT-based thin films is mainly originated from the following two aspects: (1) oxygen vacancies ($V_{O^{2-}}^{\bullet\bullet}$) due to the volatilization of PbO; (2) the transition of Ti^{4+} to Ti^{3+}. The mobile $V_{O^{2-}}^{\bullet\bullet}$ can be effectively restricted through the formation of $2(Fe_{Ti^{4+}}^{3+})'\text{-}V_{O^{2-}}^{\bullet\bullet}$ [26]. In addition, the introduction of low-valance Fe, as acceptor dopants, will prevent the transition of Ti^{4+} to Ti^{3+} [27]. Among all the films, the PMN-PT-2Fe thin film exhibits the lowest leakage current density of $\sim 10^{-3}\,A\,cm^{-2}$ in high voltage, which is more than three orders of magnitude lower than that of the pure PMN-PT thin film. However, continuing to raise the Fe element does not lead to a further decrease of the leakage current. This could be due to the fact that $(Fe_{Ti^{4+}}^{3+})'$ has exceeded the required amount to restrict $V_{O^{2-}}^{\bullet\bullet}$. Excess $(Fe_{Ti^{4+}}^{3+})'$ gathered at the grain boundary will act as leakage for the current path, which then deteriorates the leakage characteristic.

Figure 5 displays the *P–E* hysteresis loops of PMN-PT-2Fe thin films annealed at various temperatures, from 600 to 700 °C. The *P–E* hysteresis loops of PMN-PT-2Fe thin films were not well developed at a relatively low temperature (600 °C) because a large amount of non-ferroelectric pyrochlore phase existed in the thin film. With the increasing annealing temperature, the ferroelectric hysteresis loops become slim without leakage characteristics; indeed, the best ferroelectric properties were observed in the PMN-PT-2Fe thin film annealed at 650 °C. High annealing temperature improves the crystallization and reduces pyrochlore phase content. However, the hysteresis loops turn bad again at an annealing temperature of 700 °C, which is due to the more defects induced by the over volatilization with high annealing temperature. Figure 5d shows the *P–E* hysteresis loops of PMN-PT-2Fe thin films at those breakdown strengths. The sharp drop in breakdown strength for PMN-PT-2Fe thin film annealed at 700 °C is due to the significant increase of leakage current. In addition, compared with one annealed at 650 °C, the PMN-PT-2Fe thin film annealed at 700 °C displays a larger coercive electrical field, which is due to the amount of $2(Fe_{Ti^{4+}}^{3+})'\text{-}V_{O^{2-}}^{\bullet\bullet}$. The internal fields were created with the dipoles of $2(Fe_{Ti^{4+}}^{3+})'\text{-}V_{O^{2-}}^{\bullet\bullet}$, which stabilizes the domain configuration.

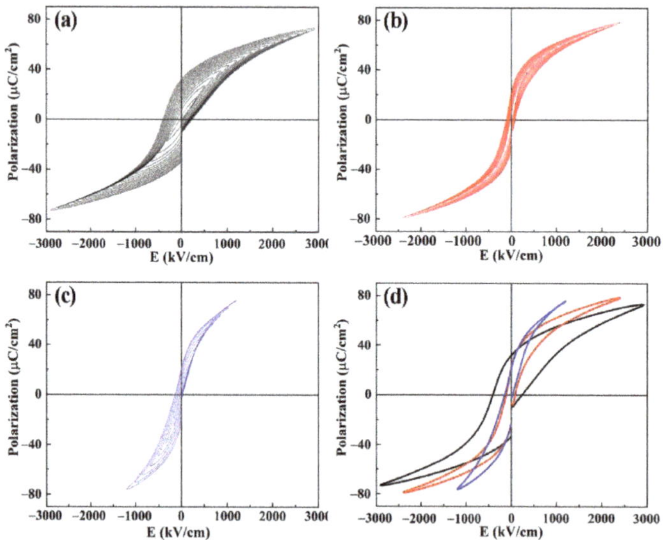

Figure 5. *P–E* hysteresis loops of PMN-PT-2Fe thin films annealed at various temperatures (**a**) 600 °C, (**b**) 650 °C, (**c**) 700 °C, (**d**) *P–E* hysteresis loops at breakdown strength.

Figure 6 shows the *P–E* hysteresis loops of PMN-PT-xFe thin films with different Fe doping concentrations. It can be seen that all the doped thin films exhibit improved *P–E* loops compared with pure PMN-PT thin film with a round shaped *P–E* loop [28]. The

PMN-PT-2Fe thin film exhibited the so-called slim loops as a feature typical for the relaxor ferroelectrics, which is due to obviously reduced leakage current density. However, this feature disappeared as the doping concentration continuously increased, owing to the excess $(Fe^{3+}_{Ti^{4+}})'$ gathered at the grain boundary, which lead to the deterioration of leakage characteristics. The optimal ferroelectric properties were exhibited in the PMN-PT-2Fe thin film with high saturation polarization (P_s = 78.8 μC/cm^2), remanent polarization (P_r = 23.1 μC/cm^2) and low coercive voltage (E_c = 100 kV/cm). At the same time, the asymmetry of hysteresis loops of the sample increases as the doping concentration increases. The increasing asymmetry is attributed to two factors: (i) internal fields created by the dipoles of $2(Fe^{3+}_{Ti^{4+}})'$-$V^{\bullet\bullet}_{O^{2-}}$, (ii) accumulation effect of excess Fe ion.

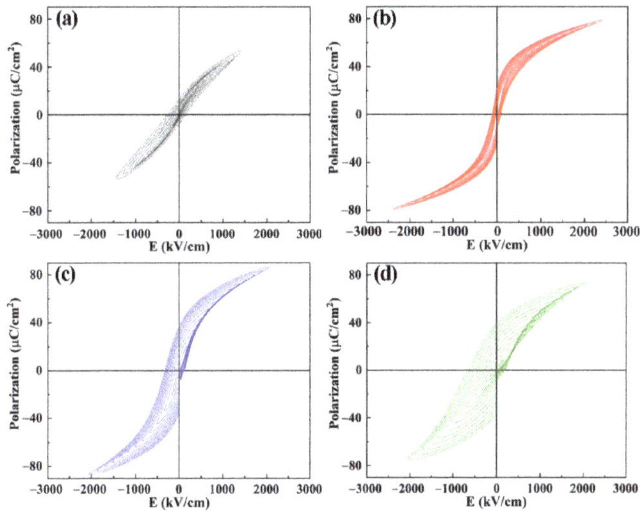

Figure 6. *PE* hysteresis loops of PMN-PT-xFe thin films with different doping concentration of Fe element (**a**) x = 0, (**b**) x = 2, (**c**) x = 4, (**d**) x = 8.

The ferroelectric nature of PMN-PT thin films can be identified from irreversible nature of dielectric performance versus electric field behavior. Voltage dependence of dielectric constant (ε_r-V) and loss tangent (tanδ-V) of PMN-PT thin films was measured at 100 Hz, as shown in Figure 7. The ferroelectric properties of PMN-PT thin films were confirmed by the butterfly-shaped curve. In order to study the nonlinear dielectric property, the corresponding dielectric tunable performance was calculated. The dielectric tunability (T) is defined as: T = [($\varepsilon_0 - \varepsilon_r$)/$\varepsilon_0$] × 100%, where ε_0 and ε_r represent the dielectric constant values at zero and the maximum applied field, respectively. According to the ε_r-V curves of the PMN-PT-2Fe thin films annealed at different temperatures as shown in Figure 7a, the calculated T values at 10 V are 52%, 73% and 70% for annealing temperature of 600, 650 and 700 °C, respectively. The increase of T with the increase of annealing temperature is attributed to the improved crystallization and grain size of the films, which is corresponding to the XRD and SEM images as shown in Figures 1 and 2 [29,30]. The dielectric loss (tanδ) shows the same tendency as the annealing temperature increases. Figure 7b shows the ε_r-V curves of PMN-PT-xFe thin films with different Fe doping concentrations. The calculated T values at 10 V are 34%, 73%, 64% and 46% for PMN-PT-xFe thin films with Fe doping of 0, 2, 4 and 8%, respectively. It can be seen that the T first increases and then decreases with the increase in doping concentration. The decrease in T is attributed to the deterioration of leakage characteristics. In addition, the transition of preferred orientation of PMN-PT thin film from (111) to (110) may be another factor that causes T to change, which corresponds to the previous literatures [31,32]. Moreover, tanδ exhibits a tendency to first decrease before increasing with the increase in doping concentration.

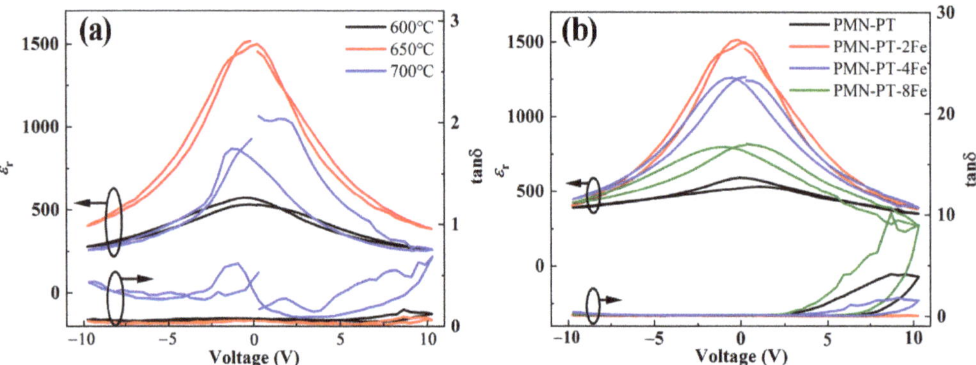

Figure 7. (**a**) Voltage dependence of ε_r and tanδ for PMN-PT-2Fe thin films annealed at different temperatures, (**b**) Voltage dependence of ε_r and tanδ for PMN-PT-xFe thin films with different doping concentrations of Fe element.

Frequency-dependent ε_r and tanδ of PMN-PT-2Fe thin films annealed at different temperatures and PMN-PT-xFe thin films with different doping concentrations of Fe element were measured from 1 kHz to 1 MHz at room temperature, as shown in Figure 8. It can be observed that the annealing temperature and doping concentration have a strong effect on its dielectric property. For all samples, ε_r decreased gradually with the increase in frequency, which is due to the fact that the polarization process of some frameworks (such as space charges) cannot be achieved [33,34]. The significantly low ε_r of PMN-PT-2Fe thin film annealed at 600 °C is attributed to presence of pyrochlore phase with low ε_r and smaller grain size of thin film, which are also evident form XRD patterns and surface SEM images [35]. The ε_r of PMN-PT-2Fe thin film increased with the increase of annealing temperature, which is due to the decrease of pyrochlore phase and the growth of crystal grain. However, continuing to raise the Fe element does not lead to a further increase of ε_r. This could be due to the fact that $\left(Fe^{3+}_{Ti^{4+}}\right)'$ has exceeded the required amount to restrict $V^{\bullet\bullet}_{O^{2-}}$. Excess $\left(Fe^{3+}_{Ti^{4+}}\right)'$ gathered at the grain boundary will act as a leakage current path, then waken ε_r. The PMN-PT-2Fe thin film annealed at 650 °C shown the largest ε_r of ~1300 which is comparable to those reported in the literature where PMN-PT thin films were grown on different substrates [36–39]. The tanδ for all samples increased with increasing frequency, which is due to the extrinsic loss and dipolar lagging phenomena [40].

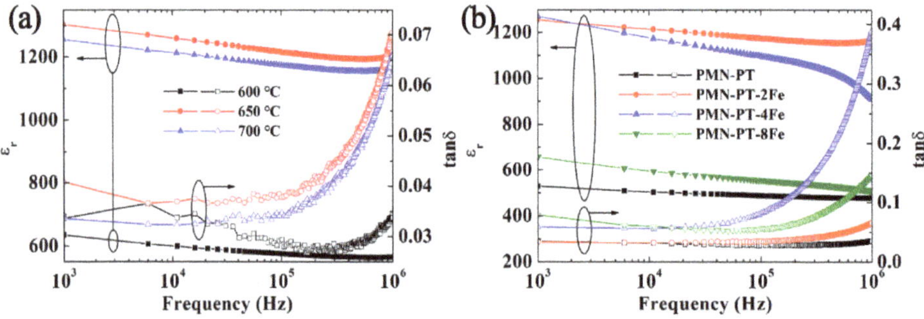

Figure 8. (**a**) The frequency dependence of ε_r and tanδ of PMN-PT-2Fe thin films annealed at different temperatures, (**b**) The frequency dependence of the ε_r and tanδ of PMN-PT-xFe thin films with different doping concentrations of Fe element.

4. Conclusions

PMN-PT thin films with different Fe doping concentrations have been synthesized on $Pt/Ti/SiO_2/Si$ substrate by the CSD technique. The crystallinity, orientation, microstructure and defect dipoles induced by ion substitution were attributed to the origin of the enhanced electrical performances; 2% Fe-doped PMN-PT thin film annealed at 650 °C showed the high (111) preferred orientation, and the preferred orientation transferred to random orientation as the doping concentration increased. The dense columnar structure was obtained in 2% Fe-doped PMN-PT thin film annealed at 650 °C. The excessive annealing temperature and excessive doping concentration will lead to appearance of cracks. In addition, compared with the pure PMN-PT thin film, the effectively enhanced leakage characteristic was obtained in 2% Fe-doped PMN-PT thin film, which is because of the reduction of movable $V_{O^{2-}}^{\bullet\bullet}$ concentration and restraint from Ti^{4+} to Ti^{3+}. The enhanced ferroelectric (P_s = 78.8 $\mu C/cm^2$, P_r = 23.1 $\mu C/cm^2$, E_c = 100 kV/cm) and dielectric properties (ε_r ~1300 at 1kHz) have been obtained in 2% Fe-doped PMN-PT thin film annealed at 650 °C. These results provide the important guiding significance for controlling the grain orientation in preparation of ferroelectric thin films and enhancing the electrical performances of ferroelectric thin films.

Author Contributions: Conceptualization, C.F. and S.H.; methodology, C.F.; validation, C.F.; formal analysis, C.F.; investigation, C.F. and T.L.; data curation, C.F. and X.B.; writing—original draft preparation, C.F.; writing—review and editing, C.F.; funding acquisition, C.F. and S.H. All authors have read and agreed to the published version of the manuscript.

Funding: This research was funded by Shandong Postdoctoral Innovative Talents Support Plan (Grant No. SDBX2020010), National Natural Science Foundation of China (Grant No. U1806221), the Project of "20 Items of University" of Jinan (Grant No. 2019GXRC017), and Synergetic Innovation Center Research Project (Grant No. WJGTT-XT2).

References

1. Wei, Y.P.; Gao, C.X.; Chen, Z.D.; Xi, S.B.; Shao, W.X.; Zhang, P.; Chen, G.L.; Li, J.G. Four-state memory based on a giant and non-volatile converse magnetoelectric effect in FeAl/PIN-PMN-PT structure. *Sci. Rep.* **2016**, *6*, 30002. [CrossRef]
2. Zhang, Z.; Xu, J.L.; Xiao, J.J.; Liu, S.X.; Wang, X.A.; Liang, Z.; Luo, H.S. Simulation and analysis of the PMN-PT based phased array transducer with the high sound velocity matching layer. *Sens. Actuat. A Phys.* **2020**, *313*, 112195. [CrossRef]
3. Singh, C.; Thakur, V.N.; Kumar, A. Investigation on barometric and hydrostatic pressure sensing properties of $Pb[(Mg_{1/3}Nb_{2/3})_{0.7}Ti_{0.3}]O_3$ electro-ceramics. *Ceram. Int.* **2021**, *47*, 6982–6987. [CrossRef]
4. Gao, X.Y.; Liu, J.F.; Xin, B.J.; Jin, H.N.; Luo, L.C.; Guo, J.Y.; Dong, S.X.; Xua, Z.; Li, F. A bending-bending mode piezoelectric actuator based on PIN-PMN-PT crystal stacks. *Sens. Actuat. A Phys.* **2021**, *331*, 113052. [CrossRef]
5. Li, C.C.; Xu, B.; Lin, D.B.; Zhang, S.J.; Bellaiche, L.; Shrout, T.R.; Li, F. Atomic-scale origin of ultrahigh piezoelectricity in samarium-doped PMN-PT ceramics. *Phys. Rev. B* **2020**, *101*, 140102. [CrossRef]
6. Ushakov, A.D.; Hu, Q.; Liu, X.; Xu, Z.; Wei, X.; Shur, Y.V. Domain structure evolution during alternating current poling and its influence on the piezoelectric properties in [001]-cut rhombohedral PIN-PMN-PT single crystals. *Appl. Phys. Lett.* **2021**, *118*, 232901. [CrossRef]
7. Deng, C.G.; Ye, L.X.; He, C.J.; Xu, G.S.; Zhai, Q.X.; Luo, H.S.; Liu, Y.W.; Bell, A.J. Reporting excellent transverse piezoelectric and electro-optic effects in transparent rhombohedral PMN-PT single crystal by engineered domains. *Adv. Mater.* **2021**. [CrossRef] [PubMed]
8. Brova, M.J.; Watson, B.H., III; Walton, R.L.; Kupp, E.R.; Fanton, M.A.; Meyer, R.J., Jr.; Messing, G.L. Templated grain growth of high coercive field CuO-doped textured PYN-PMN-PT ceramics. *J. Am. Ceram. Soc.* **2020**, *103*, 6149–6156. [CrossRef]
9. Zhang, Y.C.; Yang, Z.Z.; Ye, W.N.; Lu, C.J.; Xia, L.H. Effect of excess Pb on microstructures and electrical properties of $0.67Pb(Mg_{1/3}Nb_{2/3})O_3$–$0.33PbTiO_3$ ceramics. *J. Mater. Sci. Mater. Electron.* **2011**, *22*, 309–314. [CrossRef]
10. Baek, S.H.; Park, J.; Kim, D.M.; Aksyuk, V.A.; Das, R.R.; Bu, S.D.; Felker, D.A.; Lettieri, J.; Vaithyanathan, V.; Bharadwaja, S.S.N.; et al. Giant piezoelectricity on Si for hyperactive MEMS. *Science* **2011**, *334*, 958–961. [CrossRef] [PubMed]
11. Zhou, S.; Lin, D.B.; Su, Y.M.; Zhang, L.; Liu, W.G. Enhanced dielectric, ferroelectric, and optical properties in rare earth elements doped PMN–PT thin films. *J. Adv. Ceram.* **2020**, *9*, 98–107. [CrossRef]
12. Gabora, U.; Vengust, D.; Samardžija, Z.; Matavž, A.; Bobnar, V.; Suvorov, D.; Spreitzer, M. Stabilization of the perovskite phase in PMN-PT epitaxial thin films via increased interface roughness. *Appl. Surf. Sci.* **2020**, *513*, 145787. [CrossRef]

13. Keech, R.; Shetty, S.; Wang, K.; Trolier-McKinstry, S. Management of Lead Content for Growth of {001}-Oriented Lead Magnesium Niobate-Lead Titanate Thin Films. *J. Am. Ceram. Soc.* **2016**, *99*, 1144–1146. [CrossRef]

14. Shen, B.W.; Wang, J.; Pan, H.; Chen, J.H.; Wu, J.L.; Chen, M.F.; Zhao, R.X.; Zhu, K.J.; Qiu, J.H. Effects of annealing process and the additive on the electrical properties of chemical solution deposition derived $0.65Pb(Mg_{1/3}Nb_{2/3})O_3$-$0.35PbTiO_3$ thin films. *J. Mater. Sci. Mater. Electron.* **2018**, *29*, 16997–17002. [CrossRef]

15. Tang, Y.; Zhou, D.; Tian, Y.; Li, X.; Wang, F.; Sun, D.; Shi, W.; Tian, L.; Sun, J.; Meng, X.; et al. Low-temperature processing of high-performance $0.74Pb(Mg_{1/3}Nb_{2/3})O_3$–$0.26PbTiO_3$ thin films on $La_{0.6}Sr_{0.4}CoO_3$-buffered Si substrates for pyroelectric arrays applications. *J. Am. Ceram. Soc.* **2012**, *95*, 1367. [CrossRef]

16. Jiang, J.; Hwang, H.-H.; Lee, W.-J.; Yoon, S.-G. Microstructural and electrical properties of $0.65Pb(Mg_{1/3}Nb_{2/3})O_3$-$0.35PbTiO_3$ (PMN-PT) epitaxial films grown on Si substrates. *Sens. Actuat. B Chem.* **2011**, *155*, 854. [CrossRef]

17. Lee, S.Y.; Custodio, M.C.C.; Lim, H.J.; Feigelson, R.S.; Maria, J.P.; Trolier-McKinstry, S. Growth and characterization of $Pb(Mg_{1/3}Nb_{2/3})O_3$ and $Pb(Mg_{1/3}Nb_{2/3})O_3$–$PbTiO_3$ thin films using solid source MOCVD techniques. *J. Cryst. Growth* **2001**, *226*, 247–253. [CrossRef]

18. Kiguchi, T.; Misaka, Y.; Nishijima, M.; Sakamoto, N.; Wakiya, N.; Suzuki, H.; Konno, T. Effect of facing annealing on crystallization and decomposition of $Pb(Mg_{1/3}Nb_{2/3})O_3$ thin films prepared by CSD technique using MOD solution. *J. Ceram. Soc. Jpn.* **2013**, *121*, 236. [CrossRef]

19. Li, F.; Zhang, S.J.; Luo, J.; Geng, X.C.; Xu, Z.; Shrout, T. [111]-oriented PIN-PMN-PT crystals with ultrahigh dielectric permittivity and high frequency constant for high-frequency transducer applications. *J. Appl. Phys.* **2016**, *120*, 074105. [CrossRef]

20. Vilquina, B.; Bouregba, R.; Poullain, G.; Hervieu, M.; Murray, H. Orientation control of rhomboedral PZT thin films on $Pt/Ti/SiO_2/Si$ substrates. *Eur. Phys. J. Appl. Phys.* **2001**, *15*, 153–165. [CrossRef]

21. Chen, S.Y.; Chen, I.W. Texture development, microstructure evolution, and crystallization of chemically derived PZT thin films. *J. Am. Ceram. Soc.* **1998**, *81*, 97–105. [CrossRef]

22. Shur, V.Y.; Blankova, E.B.; Subbotin, A.L.; Borisova, E.A.; Pelegov, D.V.; Hoffmann, S.; Bolten, D.; Gerhardt, R.; Waser, R. Influence of crystallization kinetics on texture of sol-gel PZT and BST thin films. *J. Eur. Ceram. Soc.* **1999**, *19*, 1391–1395. [CrossRef]

23. Iakovlev, S.; Rätzke, K.; Es-Souni, M. Structural investigations of rare-earth doped $PbTiO_3$ thin films. *Mater. Sci. Eng. B* **2004**, *113*, 259–262. [CrossRef]

24. Sahoo, B.; Panda, P.K. Effect of lanthanum, neodymium on piezoelectric, dielectric and ferroelectric properties of PZT. *J. Adv. Ceram.* **2013**, *2*, 37–41. [CrossRef]

25. Gao, F.; Qiu, X.Y.; Yuan, Y.; Xu, B.; Wen, Y.Y.; Yuan, F.; Lv, L.Y.; Liu, J.M. Effects of substrate temperature on $Bi_{0.8}La_{0.2}FeO_3$ thin films prepared by pulsed laser deposition. *Thin Solid Films* **2007**, *515*, 5366–5373. [CrossRef]

26. Hejazi, M.M.; Taghaddos, E.; Safari, A. Reduced leakage current and enhanced ferroelectric properties in Mn-doped $Bi_{0.5}Na_{0.5}TiO_3$-based thin films. *J. Mater. Sci.* **2013**, *48*, 3511–3516. [CrossRef]

27. Wu, Y.Y.; Wang, X.H.; Zhong, C.F.; Li, L.T. Effect of Mn doping on microstructure and electrical properties of the $(Na_{0.85}K_{0.15})_{0.5}Bi_{0.5}TiO_3$ thin films prepared by sol-gel method. *J. Am. Ceram. Soc.* **2011**, *94*, 3877–3882. [CrossRef]

28. Sung, Y.S.; Kim, J.M.; Cho, J.H.; Song, T.K.; Kim, M.H.; Park, T.G. Effects of Bi non-stoichiometry in $(Bi_{0.5+x}Na)TiO_3$ ceramics. *Appl. Phys. Lett.* **2011**, *98*, 012902. [CrossRef]

29. Ha, S.; Lee, Y.S.; Hong, Y.P.; Lee, H.Y.; Lee, Y.C.; Ko, K.H.; Kim, D.-W.; Hong, H.B.; Hong, K.S. The effect of substrate heating on the tunability of rf-sputtered Bi_2O_3-ZnO-Nb_2O_5 thin films. *Appl. Phys. A* **2005**, *80*, 585–590. [CrossRef]

30. Jiang, S.W.; Jiang, B.; Liu, X.Z.; Li, Y.R. Laser deposition and dielectric properties of cubic pyrochlore bismuth zinc niobate thin films. *J. Vac. Sci. Technol. A* **2006**, *24*, 261–263. [CrossRef]

31. Han, F.F.; Hu, Y.H.; Peng, B.L.; Liu, L.J.; Yang, R.; Ren, K.L. High dielectric tunability with high thermal stability of the (111) highly oriented $0.85Pb(Mg_{1/3}Nb_{2/3})$-$0.15PbTiO_3$ thin film prepared by a sol-gel method. *J. Eur. Ceram. Soc.* **2021**, *41*, 6482–6489. [CrossRef]

32. Moon, S.E.; Kim, E.-K.; Kwak, M.-H.; Ryu, H.-C.; Kim, Y.-T.; Kang, K.-Y.; Lee, S.-J. Orientation dependent microwave dielectric properties of ferroelectric $Ba_{1-x}Sr_xTiO_3$ thin films. *Appl. Phys. Lett.* **2003**, *83*, 2166. [CrossRef]

33. Kumar, P.; Singh, P.; Singh, S.; Juneja, J.K.; Prakash, C.; Raina, K.K. Influence of lanthanum substitution on dielectric properties of modified lead zirconate titanates. *Ceram. Int.* **2015**, *41*, 5177–5181. [CrossRef]

34. Wang, F.A.; Zhou, J.G.; Wang, X.; Chen, D.; Wang, Q.S.; Dou, J.; Li, Q.; Zou, H.L. Effect of Ce doping on crystalline orientation, microstructure, dielectric and ferroelectric properties of (100)-oriented PCZT thin films via sol-gel method. *J. Mater. Sci. Mater. Electron.* **2018**, *29*, 18668–18673. [CrossRef]

35. Chen, J.; Harmer, M.P. Microstructure and dielectric properties of lead magnesium niobate-pyrochlore diphasic mixtures. *J. Am. Ceram. Soc.* **1990**, *73*, 68–73. [CrossRef]

36. Garg, T.; Kulkarni1, A.R.; Venkataramani, N. PMN-PT thin films on $La_{0.67}Ca_{0.33}MnO_3$ seeded platinized glass substrate: Phase formation, dielectric and ferroelectric studies. *Mater. Res. Express* **2018**, *5*, 096408. [CrossRef]

37. Li, W.Z.; Xue, J.M.; Zhou, Z.H.; Wang, J.; Zhu, H.; Miao, J.M. $0.67Pb(Mg_{1/3}Nb_{2/3})O_3$-$0.33PbTiO_3$ thin films derived from RF magnetron sputtering. *Ceram. Intern.* **2004**, *30*, 1539–1542. [CrossRef]

38. Chan, K.Y.; Tsang, W.S.; Mak, C.L.; Wong, K.H. Effects of composition of $PbTiO_3$ on optical properties of $(1-x)Pb(Mg_{1/3}Nb_{2/3})O_3$-$xPbTiO_3$ thin films. *Phys. Rev. B* **2004**, *69*, 144111. [CrossRef]

39. Chen, X.Y.; Wong, K.H.; Mak, C.L.; Liu, J.M.; Yin, X.B.; Wang, M.; Liu, Z.G. Growth of orientation-controlled $Pb(Mg_{1/3}Nb_{2/3})O_3$-$PbTiO_3$ thin films on Si(100) by using oriented MgO films as seeds. *Appl. Phys. A* **2002**, *74*, 1145–1149.
40. Narang, S.B.; Kaur, D. Dielectric anomaly in La modified barium titanates. *Ferroelectr. Lett.* **2009**, *36*, 20–27. [CrossRef]

Article

Phase Structure and Electrical Properties of Sm-Doped BiFe$_{0.98}$Mn$_{0.02}$O$_3$ Thin Films

Yangyang Wang [1], Zhaoyang Li [2], Zhibiao Ma [1], Lingxu Wang [1], Xiaodong Guo [3], Yan Liu [1], Bingdong Yao [1], Fengqing Zhang [1,*] and Luyi Zhu [2,*]

[1] School of Materials Science and Engineering, Shandong Jianzhu University, Jinan 250101, China; wyyedu0207@163.com (Y.W.); mzbsdjzu@163.com (Z.M.); wanglingxu@163.com (L.W.); yshy1589@163.com (Y.L.); bdyao1026@163.com (B.Y.)

[2] State Key Laboratory of Crystal Materials, Institute of Crystal Materials, Shandong University, Jinan 250100, China; lzy08201226@163.com

[3] School of Data and Computer Science, Shandong Women's University, Jinan 250300, China; sjgxd246@163.com

* Correspondence: zhangfengqing615@163.com (F.Z.); zhuly@sdu.edu.cn (L.Z.)

Abstract: Bi$_{1-x}$Sm$_x$Fe$_{0.98}$Mn$_{0.02}$O$_3$ (x = 0, 0.02, 0.04, 0.06; named BSFMx) (BSFM) films were prepared by the sol-gel method on indium tin oxide (ITO)/glass substrate. The effects of different Sm content on the crystal structure, phase composition, oxygen vacancy content, ferroelectric property, dielectric property, leakage property, leakage mechanism, and aging property of the BSFM films were systematically analyzed. X-ray diffraction (XRD) and Raman spectral analyses revealed that the sample had both R3c and Pnma phases. Through additional XRD fitting of the films, the content of the two phases of the sample was analyzed in detail, and it was found that the Pnma phase in the BSFMx = 0 film had the lowest abundance. X-ray photoelectron spectroscopy (XPS) analysis showed that the BSFMx = 0.04 film had the lowest oxygen vacancy content, which was conducive to a decrease in leakage current density and an improvement in dielectric properties. The diffraction peak of (110) exhibited the maximum intensity when the doping amount was 4 mol%, and the minimum leakage current density and a large remanent polarization intensity were also observed at room temperature (2Pr = 91.859 μC/cm^2). By doping Sm at an appropriate amount, the leakage property of the BSFM films was reduced, the dielectric property was improved, and the aging process was delayed. The performance changes in the BSFM films were further explained from different perspectives, such as phase composition and oxygen vacancy content.

Keywords: BSFM; phase transition; aging; electrical properties

Citation: Wang, Y.; Li, Z.; Ma, Z.; Wang, L.; Guo, X.; Liu, Y.; Yao, B.; Zhang, F.; Zhu, L. Phase Structure and Electrical Properties of Sm-Doped BiFe$_{0.98}$Mn$_{0.02}$O$_3$ Thin Films. *Nanomaterials* **2022**, *12*, 108. https://doi.org/10.3390/nano12010108

Academic Editor: Seiichi Miyazaki

Received: 7 December 2021
Accepted: 24 December 2021
Published: 30 December 2021

Publisher's Note: MDPI stays neutral with regard to jurisdictional claims in published maps and institutional affiliations.

1. Introduction

Only a small fraction of all magnetically polarized and electrically polarized materials are ferromagnetic or ferroelectric, and even fewer, namely multiferroic materials, have both properties [1,2]. In addition, the coupling between different properties of multiferroic materials will produce new properties, such as magnetoelectric effects. These materials have great development potential in the miniaturization and multi-functionalization of devices, as well as in a wide range of applications in the fields of magnetoelectric memory [3,4], sensors [5], and drivers [6]. Multiferroic materials are some of the most valuable multifunctional materials, and they have good application prospects in the field of multiferroic devices.

Multiferroic materials include single-phase materials and composite materials. However, few single-phase multiferroic materials have been discovered at present, and their Curie temperatures are usually low. Owing to its high Curie temperature (Tc = 1103 K) and Neal temperature (T$_N$ = 647 K), single-phase BiFeO$_3$ (BFO) exhibits ferroelectric and G-type antiferromagnetism at room temperature [7,8]. Thus, it has attracted extensive attention from materials scholars and has become a hot topic for in-depth exploration of multiferroic materials [9–12].

In BFO, Bi ions are volatile at high temperature. To balance the charge, the valence of Fe ions may change from +3 to +2 [13]:

$$Bi_{Bi}^{x} + \frac{3}{2}O_2 \Leftrightarrow Bi_2O_3 \uparrow + V_{Bi}''' + 3h^{\bullet}, \tag{1}$$

$$2Fe^{3+} + O_O^x \Leftrightarrow 2(Fe_{Fe^{3+}}^{2+})\prime + V_O^{\bullet\bullet} + \frac{1}{2}O_2. \tag{2}$$

As a result, a large number of oxygen vacancies or other defects often exist in the prepared $BiFeO_3$ samples, which increases the leakage current density of $BiFeO_3$ materials and adversely affects its performance [14]. There are many ways to improve the properties of $BiFeO_3$ materials, including element doping, solid solution, formation of a heterostructure, and control of film orientation [15–18]. Among them, many researchers adopt the element doping method to improve the performance of $BiFeO_3$ materials [19–22]. Yun et al. prepared single-phase multiferroic $BiFeO_3$ and Ho-doped $BiFeO_3$ films [23]. The ferroelectric property was enhanced, and the leakage current decreased significantly. The ferroelectric property reached 20.69 $\mu C/cm^2$ and the leakage current density was 2.89×10^{-9} A/cm^2, and these effects were attributed to the transformation from a rhombohedral structure to a coexisting cubic and orthosymmetric structure after Ho doping. Moreover, the fatigue properties of the films doped with Ho also improved, as evidenced by a 0.4% reduction in the value of the switchable polarization. Liu et al. grew a $Bi_{1-x}Eu_xFeO_3$ (BEFOx, x = 0, 0.03, 0.05, 0.07, 0.1) thin film on $LaNiO_3$-coated Si substrate by the pulse laser deposition method. As the doping amount increased, the position of the A_1-1 mode of the films shifted to a higher wave number in the Raman spectrum [24]. With the increase in Eu, the refractive index of the film increased, and the extinction coefficient and band gap width decreased. Yang et al. prepared a $BiFe_{1-x}Zn_xO_3$ (BFZO) film (x = 0%, 1%, 2%, 3%) and found that when x = 2%, the film reached the maximum remanent polarization intensity and the minimum correction field [25]. At the same time, under a low electric field, Zn doping can significantly reduce the leakage current of BFO films. In addition, the leakage mechanism changes from Ohmic conduction under a low electric field to F-N tunneling under a high electric field. Zhang et al. prepared high-quality $BiFe_{1-2x}Zn_xTi_xO_3$ (BFZTO, x = 0, 0.01, 0.02, 0.03, 0.04, and 0.05) films [26]. The authors found that the BFZTO film with x = 0.02 had uniform fine grains and high density, which can inhibit the transformation of Fe^{3+} to Fe^{2+} and, thus, greatly reduce the oxygen vacancy concentration. This film had the lowest leakage current density and the highest remanent polarization intensity. By comparing P–E hysteresis loops in different areas of $BiFe_{0.96}Zn_{0.02}Ti_{0.02}O_3$ thin films, the films have high uniformity and stable properties. Concurrently, Zn and Ti co-doping also increased the dielectric permittivity from 24.9 to 35.3 and remnant magnetization from 0.05 to 0.80 emu/cm^3 of BFZTO films. Liu et al. prepared $Bi_{0.9}Er_{0.1}Fe_{1-x}Mn_xO_3$ (BEFM$_x$O, x = 0.00–0.03) thin films by the sol-gel method [27]. By co-doping Er and Mn, the coexistence of two phases (space groups are R3c:H and R3m:R) and the reduction of oxygen vacancy and Fe^{2+} concentration in BEFM$_x$O were realized. Among all the samples, the BEFM$_{0.02}$O film had the lowest oxygen vacancy concentration, the maximum remanent polarization value, and the maximum switching current. It also exhibited excellent ferroelectric stability, which means its low concentration of oxygen vacancies had less influence on the ferroelectric domains.

Kan et al. found that doping with Sm affected the phase structure of BFO samples [28,29]. Xue et al. prepared BFO films with different Sm content by the sol-gel method and found that the rhombohedral phase to pseudo-tetragonal phase transition occurs gradually with the increase in Sm [30]. Although there are many studies on the influence of element doping on $BiFeO_3$ properties, there are few on the influence of Sm doping on the content change in the $BiFeO_3$ thin film phase structure and thus on ferroelectric properties. In addition, the literature review revealed that for Sm doping, when the doping content is less than 10 mol%, BFO has better properties than heavily doped [31,32]. It is necessary to further adjust the doping content. In this experiment, the doping amounts of Sm were 2 mol%, 4 mol%, and 6 mol%, in order to understand the influence of Sm doping on BSFM films. The performance changes were analyzed in detail from the aspects of oxygen vacancy content, grain size, relative content of the R3c phase and the Pnma phase. Additionally, the effects of different Sm content on the ferroelectric, dielectric, leakage, and aging properties of the thin film samples were systematically studied.

2. Materials and Methods

$Bi_{1-x}Sm_xFe_{0.98}Mn_{0.02}O_3$ thin films (x = 0, 0.02, 0.04, 0.06) were prepared by the sol-gel method on ITO/glass substrate. $Fe(NO_3)_3 \cdot 9H_2O$ (purity of 98.5%), $Bi(NO_3)_3 \cdot 5H_2O$ (purity of 98.5%), $Sm(NO_3)_3 \cdot 6H_2O$ (purity of 98.5%), and $MnC_4H_6O_4 \cdot 4H_2O$ (purity of 98.5%) were taken as solutes, according to stoichiometric ratio. Bi excess of 5% compensated for bismuth volatilization during high-temperature annealing. The solutes were successively added to a solvent mixture of CH_3COOH and C_2H_6O, with a volume ratio of 3:1, and stirred at room temperature at a uniform speed until completely dissolved. Then, $C_5H_8O_2$ was added to the solution as a chelating agent and stirred at room temperature for 12 h at a constant speed to obtain a red-brown and transparent precursor solution. Finally, the stable precursor solution of 0.3 mol/L was obtained by allowing the precursor solution to rest for 24 h. Then, the BSFM precursor solution was rotated onto ITO/glass substrate, and the film was coated at 3500 r/min. The wet film was dried on an electric heating plate at 250 °C to remove excess organic solvents and water. It was then placed in an annealing furnace and annealed at 550 °C. The coating was repeated, and the film was dried and annealed 10 times to obtain the desired samples.

Before testing the electrical properties of the sample, Au was sputtered on the surface of the sample to achieve the effect of conduction. We used a small-ion sputtering instrument (JS-1600, Beijing Hetong Venture Technology Co., Ltd., Beijing, China) to complete this process. The samples were characterized by an X-ray diffractometer (D8-Advance, Drudy, Germany) recorded in the 2-theta range of 20–60° with a step of 0.02°, and by a microconfocal Raman spectrometer (HR800, LabRAM, Horiba Co., Palaiseau, France) to measure in the shift range of 50–650 cm^{-1}. The Fe and O elements in the samples were analyzed by a Wscabb X-ray photoelectron spectrometer. A dielectric tester (TH2828, Xintonghui Electronics Co., Ltd., Suzhou, China) was used to test the dielectric properties of the samples in the range of 1 kHz–1 MHz with an oscillation voltage of 1 V. The ferroelectric properties at 1 kHz and leakage properties of the samples were measured via a multiferroic tester (Radiant Co., Albuquerque, NM, USA).

3. Results and Discussion

Figure 1a shows the XRD patterns of $Bi_{1-x}Sm_xFe_{0.98}Mn_{0.02}O_3$ (x = 0, 0.02, 0.04, 0.06) films deposited on ITO/glass substrate. Figure 1b,c show the local magnified diffraction peaks of the (110) and (202) crystal planes, respectively. From the XRD pattern, the generated sample had a polycrystalline perovskite structure, and the films had good crystallinity, which matches well with the JCPDS (No. 86-1518) PDF standard card. However, the peak-splitting phenomenon of the rhombohedral structure was not observed in the BFSM thin films. At 2θ = 32°, the BSFM thin film samples do not show (110)/(104) peak splitting, but preferentially grow along (110). This may be due to the structural phase transition in the BSFM films. Figure 1b,c show that with the increase in Sm doping amount, the diffraction peak of (110) and the diffraction peak of (202) gradually shift to a large angle, which may be because the radii of doped Sm^{3+} (0.96 Å) and Mn^{2+} (0.67Å) are smaller than that of Bi^{3+} (1.03 Å) and Fe^{3+} (0.64 Å) [33,34]. Lattice distortion occurred during doping, and the crystal plane spacing decreased. In addition, with the increase in Sm content, the relative strengths I = $I_{(110)}/I_{(012)}$ of BSFMx = 0.02, BSFMx = 0.04, BSFMx = 0.06 were 2.03, 3.70, 3.02, respectively. When the content was 4 mol%, the relative intensity reached the maximum value, and the intensity of diffraction peak (202) did not change significantly with doping amount.

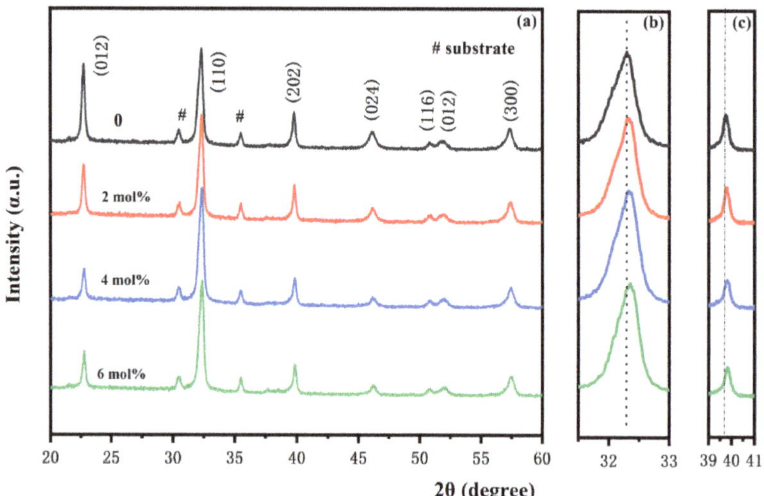

Figure 1. (**a**) XRD patterns, (**b**) magnified patterns around 30.5–33°, and (**c**) magnified patterns around 39–41° of the BSFM films with different Sm content.

The above data indicate that the crystal had structural distortion. To further study the changes in crystal structure, Rietveld refinement was conducted on all samples, and the results are shown in Figure 2a–d. All samples were used for two-phase refinement using R3c and Pnma cards from the International Crystallography database. The phase content and structural parameters of the samples are shown in Table 1. According to the experimental data, all samples have R3c and Pnma space groups. With the increase in Sm content, the content of R3c phase in the samples was 70.98%, 68.39%, 69.06%, and 67.06%. The addition of Sm reduced the content of R3c phase in the BSFM films. The lone electron pair $6S^2$ of Bi^{3+} in $BiFeO_3$ is chemically active, and it is conducive to ferroelectric distortion [20]. The substitution of rare earth element Sm for Bi may reduce the chemical activity of the lone electron pair and reduce the rhombohedral distortion of the crystal. In all Sm-doped samples, the content of the R3c phase in BSFMx = 0.04 was the highest, and the content of the non-polar orthorhombic phase Pnma was the lowest [35,36].

Table 1. Rietveld refinement parameters of the BSFM films with different Sm content.

Sample	Space Group	Fraction (%)	Lattice Parameter							R_w (%)
			a (Å)	b (Å)	c (Å)	α (°)	β (°)	γ (°)	Volume	
BSFMx = 0	R3c	70.98	5.59	5.59	13.70	90	90	120	370.41	7.90
	Pnma	29.02	5.60	16.01	11.28	90	90	90	1010.60	
BSFMx = 0.02	R3c	68.39	5.58	5.58	13.70	90	90	120	369.56	6.83
	Pnma	31.61	5.62	15.98	11.26	90	90	90	1010.75	
BSFMx = 0.04	R3c	69.06	5.58	5.58	13.66	90	90	120	368.67	7.40
	Pnma	30.94	5.58	16.06	11.30	90	90	90	1012.96	
BSFMx = 0.06	R3c	67.06	5.58	5.58	13.67	90	90	120	368.65	7.17
	Pnma	32.94	5.56	16.06	11.28	90	90	90	1007.00	

The surface differentiation features of the BSFMx (x = 0–0.06) films are shown in Figure 3a–d. The crystal grain size of the BSFMx (x = 0–0.06) films are shown in the inset. The figure shows that the grain distribution on the surface of the film without Sm doping is not uniform, and that there are many voids, which may be the reason for the volatilization of the organic solution. Furthermore, the surfaces of the Sm-doped films were uniform, compact, and well combined with the substrate, indicating that the annealing mechanism was very suitable for the growth of the BSFM films on the ITO substrate. The average grain sizes of the BSFMx (x = 0–0.06) films were 62.29, 59.53, 48.93, and 60.98 nm, indicating that Sm doping can reduce grain size. Among them, the BSFMx = 0.04 film had the smallest grain size, suggesting that appropriate Sm doping accelerated the nucleation rate and decreased the grain size [37]. The cross-sectional image of the BSFMx = 0.06 film is shown in Figure 3e. As can be seen from the figure, the film has a clear interface with the substrate, and the cross-section thickness of the film is 585 nm.

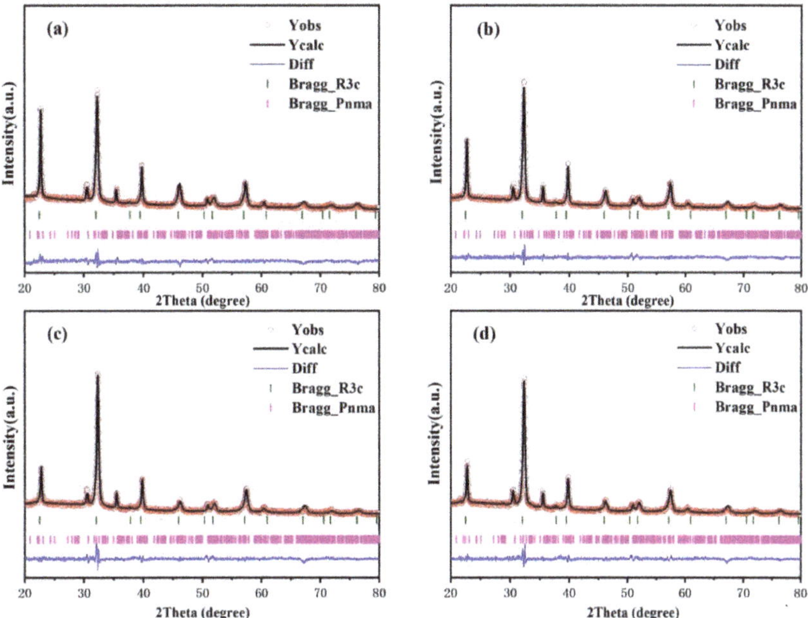

Figure 2. Rietveld refined XRD patterns of the BSFM films with different Sm content: (**a**) x = 0 mol%; (**b**) x = 2 mol%; (**c**) x = 4 mol%; (**d**) x = 6 mol%.

To further analyze the structure of Sm-doped BSFM thin films, Raman spectroscopy was used. Figure 4a shows the Raman spectra of BSFMx (x = 0−0.06) films with different Sm content in the wave number range of 50 cm^{-1}–650 cm^{-1}. The data for the crystal structure of the BSFM films (Pnma+R3c) according to XRD-refined parameters and developed by FullProf software are shown in Figure 4b,c. Figure 4d–g are the Raman spectrum fitting diagrams of each sample. Group theory analysis shows that the vibration modes of the BFO film with the R3c space group with a rhombohedral perovskite structure are $\Gamma_{Raman,R3c} = 4A_1 + 9E$ [38]. Four A_1 and nine E vibration modes analyzed by group theory were observed in the Raman spectra of the BSFM films. The Raman vibration modes extracted from the Raman fitting are listed in Table 2. From the Raman fitting diagram, the strength of mode A in the BSFM films significantly increased, which may be related to the change in Bi-O bonds caused by Sm^{3+} replacing Bi^{3+}. Owing to the Jahn–Teller distortion effect [39], the strength of modes E-8 and E-9 improved. This is because changes in Bi-O bonds lead to the distortion of the ferrite octahedron, which further changes the Fe-O bond. In addition, when the Sm element was added, the A_1-1 vibration mode shifted to a higher wave number (from 141.76 cm^{-1} to 143.31 cm^{-1}), because the frequency of the Raman vibration mode is related to the functions of ion mass and force [22]. Sm^{3+} (150.4 g) replaces Bi^{3+} (209.0 g), which leads to a blue shift in the A_1-1 mode. These results show that Sm^{3+} doping causes lattice distortion, which is consistent with XRD analysis. In addition, all samples had vibration patterns near 200, 300, 400, 490, and 620 cm^{-1}, which were consistent with the Raman frequencies of the Pnma structure [40]. In Figure 4c–f, the vibration patterns are indicated by #. The results show that in addition to the R3c phase, the Pnma phase also existed in all samples, which was consistent with the XRD refinement results.

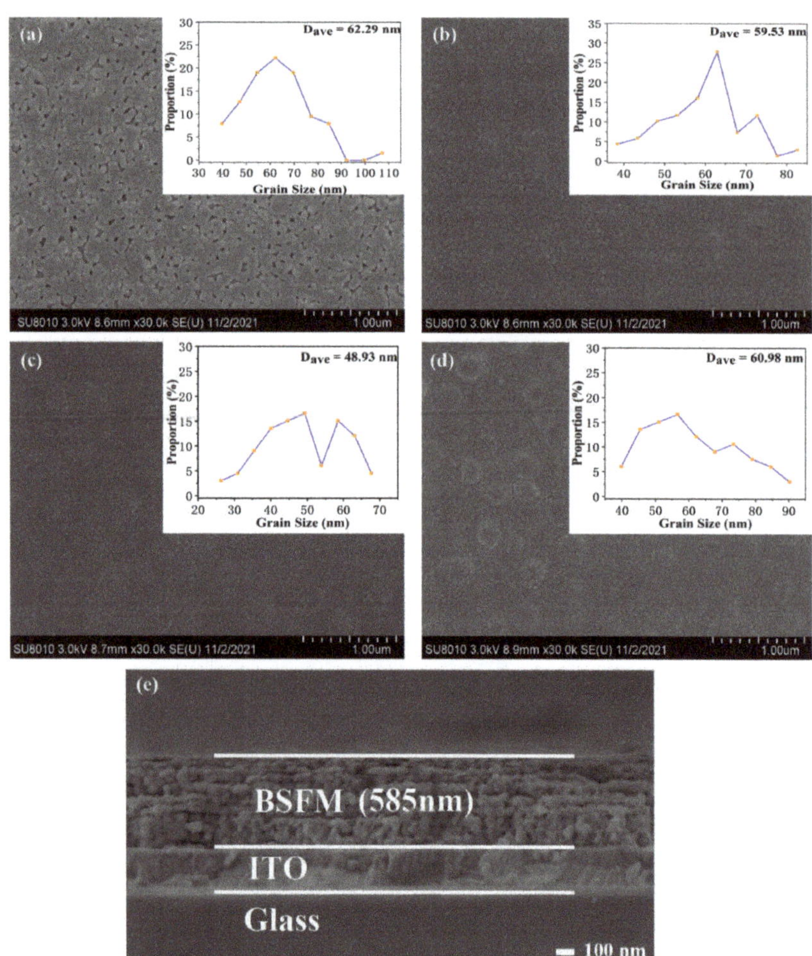

Figure 3. SEM images of the surface morphological features of the BSFM films: (**a**) x = 0 mol%; (**b**) x = 2 mol%; (**c**) x = 4 mol%; (**d**) x = 6 mol%, and (**e**) the cross-sectional image of the BSFMx = 0.06 film. The insets show the distribution of crystal grain size and the average grain diameter (D_{ave}).

According to the defect equation (Formula (2)) [13], the Fe^{2+} content can affect the oxygen vacancy content. Thus, the $Fe2p_{2/3}$ orbit of the BSFM films with different Sm doping amount was fitted, as shown in Figure 5a–d. The $Fe2p_{2/3}$ peaks were fitted into two peaks corresponding to Fe^{2+} and Fe^{3+}. From further calculations, the $Fe^{3+}:Fe^{2+}$ ratios of the BSFMx (x = 0–0.06) films were 2.33, 2.64, 2.94, and 2.77. The content of Fe^{2+} in the BSFMx = 0.04 film was the lowest, indicating that an appropriate amount of Sm doping can effectively inhibit the variation of Fe element. Generally, the content of Fe^{2+} will affect the generation of oxygen vacancies. The lower the content of Fe^{2+}, the fewer oxygen vacancies will be generated, and the fewer defects will exist in the samples.

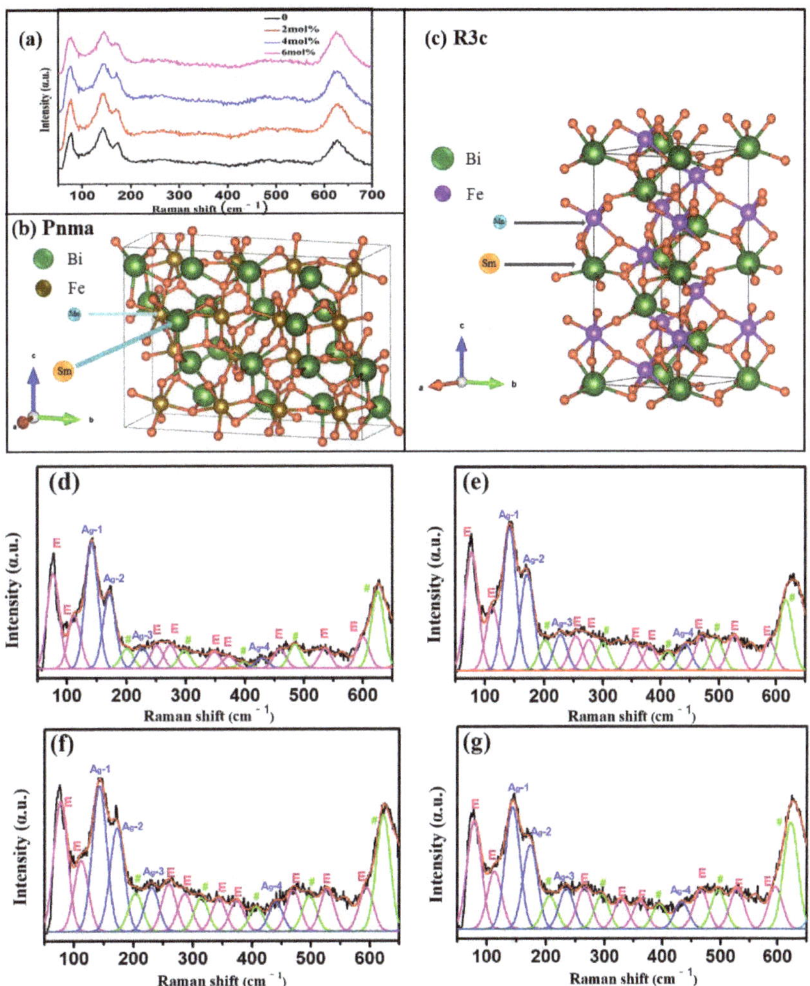

Figure 4. (**a**) Raman spectra of the BSFM films with different Sm content and crystal structure of the BSFM films: (**b**) Pnma; (**c**) R3c, Raman fitting spectra of the BSFM films with different Sm content: (**d**) x = 0 mol%; (**e**) x = 2 mol%; (**f**) x = 4 mol%; (**g**) x = 6 mol%. # refers to the Raman frequencies of the Pnma structure.

To further study the oxygen vacancy content in the BSFM films, the O1s orbital of the films was fitted, as shown in Figure 6a–d. In the figure, O1s is fitted into two peaks, which are lattice oxygen with binding energy of 529 eV (from metal) and oxygen vacancy of 531 eV (from defect) [41,42]. To calculate the oxygen vacancy ratio of the thin films, the two peaks were integrated to calculate the area, and the oxygen vacancy ratios of the four groups of samples were 0.24, 0.16, 0.14, and 0.15. According to the calculation results, the oxygen vacancy content and the $Fe2p_{3/2}$ orbital Fe^{2+} content of the BSFMx = 0.04 film were the lowest. This further confirms the view that inhibiting the transformation of Fe^{3+} to Fe^{2+} reduces the oxygen vacancy content of the sample.

Table 2. The Raman modes of the BSFM films with different Sm content.

Raman Modes (cm^{-1})	BSFMx = 0	BSFMx = 0.02	BSFMx = 0.04	BSFMx = 0.06
E	76.67	76.99	77.392	78.053
E	112.61	112.45	112.03	112.80
A$_1$-1	141.76	141.88	142.64	143.31
A$_1$-2	171.43	171.05	172.11	172.89
pnma	200.74	201.38	203.34	205.78
A$_1$-3	226.61	227.47	230.80	234.49
E	251.23	253.53	259.69	264.92
E	273.72	275.8	286.47	—
pnma	298.05	299.78	313.95	294.87
E	348.13	353.32	343.83	—
E	372.68	380.91	374.02	361.13
pnma	397.33	411.95	406.72	394.90
A$_1$-4	428.90	442.21	440.26	432.93
E	459.09	469.92	470.06	465.64
pnma	484.76	496.11	497.29	495.72
E	535.62	525.24	527.48	527.32
E	600.15	587.81	593.70	595.79
pnma	624.89	616.55	622.52	621.96

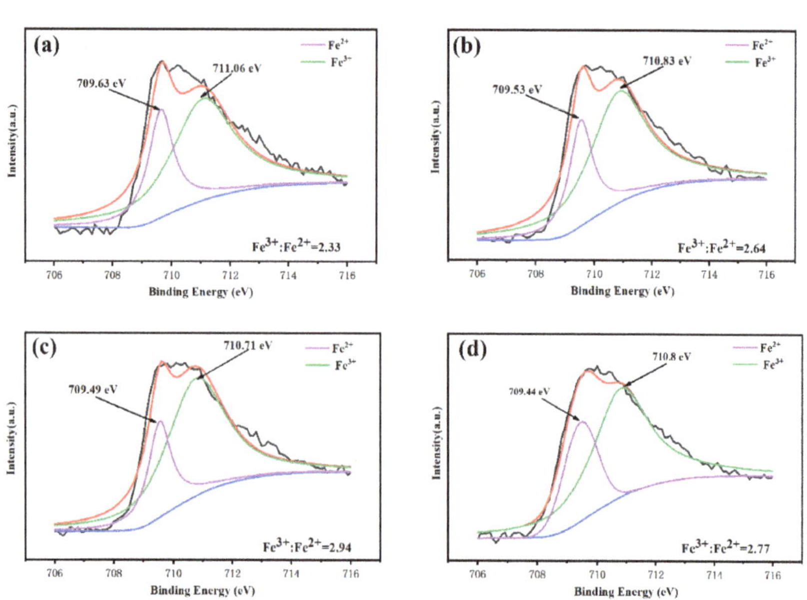

Figure 5. Fe2p$_{3/2}$ orbital XPS fitting diagram of the BSFM films with different Sm content: (**a**) x = 0 mol%; (**b**) x = 2 mol%; (**c**) x = 4 mol%; (**d**) x = 6 mol%. The black lines refer to the original curve, the red lines refer to the fitting curve, and the blue lines refer to the background line.

Figure 7 shows the leakage current density curve (J–E) of the BSFM film with different Sm content. The test electric field was 350 kV/cm. The asymmetry of leakage current of positive and negative electric fields can be attributed to the different work functions of the upper and lower electrodes. With an increasing doping amount, the leakage current density decreased first and then increased. Under the same electric field, when the doping amount was 4 mol%, the leakage current density reached the minimum value. Under a 350 kV/cm electric field, the leakage current density of the BSFMx = 0.04 film reached 8.84×10^{-5} A/cm^2. The leakage current density was about 0.5 orders of magnitude lower than that of the pure BiFe$_{0.98}$Mn$_{0.02}$O$_3$ film. Under normal circumstances, the leakage current of the sample was affected by oxygen vacancy content, which is mainly related to the high-temperature volatilization of Bi^{3+} and the variation in Fe^{3+}. It can be seen from XPS that the Fe^{2+} content and the oxygen vacancy content in the BSFMx = 0.04 film was the lowest, resulting in the minimum leakage current density. In addition, grain boundaries hinder electron migration and reduce leakage current density. According to the SEM results, the BSFMx = 0.04 film has the smallest grain size, indicating that it has more grain boundaries and greater resistance to electron migration [26,42].

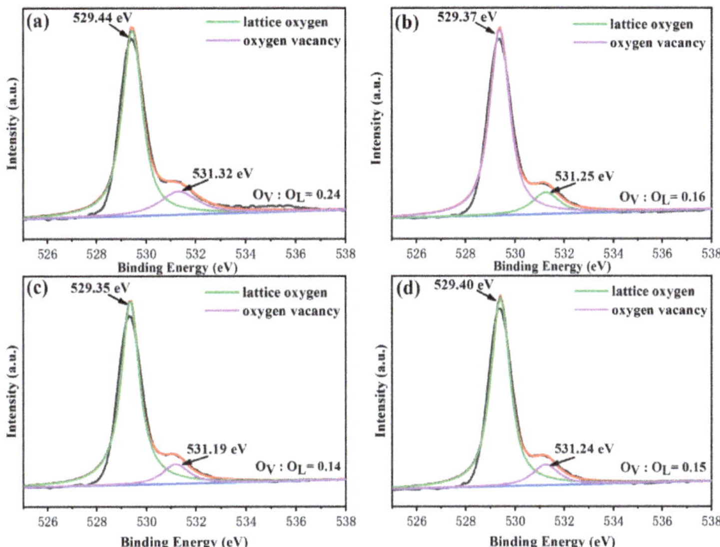

Figure 6. O1s orbital XPS fitting diagram of the BSFM films with different Sm content: (**a**) x = 0 mol%; (**b**) x = 2 mol%; (**c**) x = 4 mol%; (**d**) x = 6 mol%. The black lines refer to the original curve, the red lines refer to the fitting curve, and the blue lines refer to the background line.

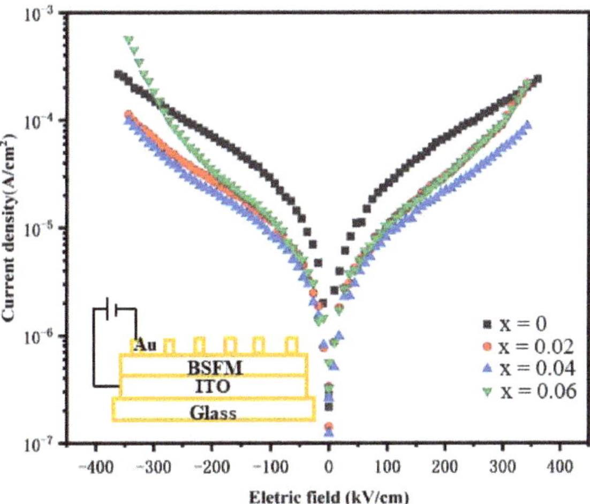

Figure 7. Leakage current density curves of the BSFM films with different Sm content.

Figure 8a shows the hysteresis loops of the BSFM films with different Sm content with an electric field of 1410 kV/cm and frequency of 1 kHz. The remanent polarization intensities of BSFMx (x = 0–0.06) under the test electric field were 111.23, 83.30, 91.86, and 73.59 $\mu C/cm^2$. The coercive field Ec was 835.22, 819.28, 830.95, and 987.55 kV/cm, respectively. When the content was less than or equal to 4 mol%, the change in the coercive field was small. The defects inhibit ferroelectric domain flipping. In all the Sm-doped samples, the remanent polarization value increased first and then decreased with increasing doping content. When the doping content was 4 mol%, the remanent polarization value was the largest, and the ferroelectric property was the best, which was consistent with the XRD analysis results. The possible reasons are as follows: (1) The leakage current density of BSFMx = 0.04 has a minimum value compared with other samples. A smaller leakage current density can improve the voltage resistance of the sample and facilitate electric domain inversion [28]. (2) According to XPS experimental data, the BSFMx = 0.04 sample had the lowest oxygen vacancy content, and the reduction of oxygen vacancy will improve its ferroelectric performance. (3) Owing to its centrosymmetric characteristics, the Pnma phase shows paramagnetic properties without ferroelectric properties, while the R3c phase mainly affects the ferroelectric property of samples [43–45].

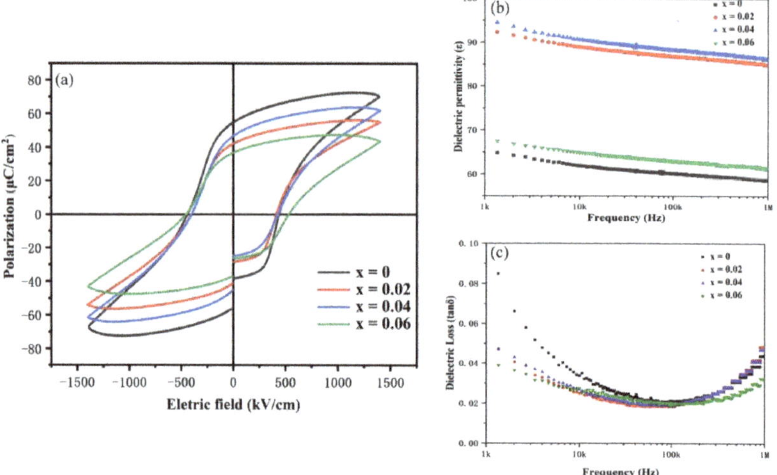

Figure 8. (**a**) P−E hysteresis loops, (**b**) dielectric permittivity, and (**c**) dielectric loss of the BSFM films with different Sm content.

Figure 8b,c show that the dielectric permittivity (ε) and dielectric loss ($\tan\delta$) of the BSFM films with different Sm contents vary with test frequency at room temperature in the range of 1 kHz–1 MHz. From the figure, the dielectric permittivity of the sample has little dependence on frequency, indicating that the sample has high intermediate frequency stability. The results show that the dielectric permittivity values of BSFMx (x = 0−0.06) at 10 kHz were 62, 89, 91, and 65. At the same frequency, the dielectric losses of BSFMx (x = 0−0.06) were 0.034, 0.025, 0.026, and 0.027. In the low frequency range (<40 kHz), Sm doping reduced the dielectric loss of the BSFM films. At the same time, Sm doping increased the dielectric permittivity of the BSFM films, and the maximum dielectric permittivity was obtained when the doping amount was 4 mol%. The existence of oxygen vacancies will distort the free volume used to replace Fe^{3+} in the Fe-O octahedron, which reduces the dielectric polarization and leads to a smaller dielectric permittivity [46]. As a result, the BSFMx = 0.04 film with the lowest oxygen vacancy content has the maximum dielectric permittivity.

BSFMx = 0.04 and BSFMx = 0.06 films were selected as representatives to explain the aging behavior of samples. Figure 9 shows the electrical hysteresis loop diagram of the BSFM films aged at room temperature for 110 d. The remanent polarization strength (2Pr) of the BSFMx = 0.04 and BSFMx = 0.06 films after aging treatment decreased by 24.9% and 41.3%, respectively, while the intensity of the coercive electric field (2Ec) decreased by 0.6% and 12.2%, respectively. This indicates that the two samples have different degrees of aging. Among them, the aging degree of the BSFMx = 0.04 film was smaller. Ren et al.

proposed the symmetric short-range order principle of point defects and inferred that the aging effect could be triggered by ion doping [47–50]. Moreover, the migration of oxygen vacancies led to the formation of complex defect dipoles in the sample. The symmetrical short-range order of oxygen vacancies created conditions suitable for reversible domain switching. In addition, the internal electric field formed during the orderly arrangement of the defective dipoles will increase the offset of the coercive field, which directly explains the asymmetry of the coercive field strength of the sample shown in Figure 9. The domains within ferroelectrics can be switched under an applied electric field, exhibiting macroscopic polarization. However, when the applied electric field is removed, the domain structure gradually shifts to a random state, resulting in the degradation of properties over time, namely, sample aging. The movement of oxygen vacancies in the sample to the domain walls creates pin centers that provide resistance to the movement of the domain walls and reduce the mobility of the domain walls. This, in turn, affects the ferroelectric properties associated with the movement of the domain walls. According to the XPS data, compared with the BSFMx = 0.06 film, the oxygen vacancy content in the BSFMx = 0.04 film was less, which means that the defect dipole content of the sample was smaller. Thus, the resistance of the domain wall to movement was smaller, leading to weak aging effects and better retention of film performance. In addition, it can be seen from SEM images that the BSFMx = 0.04 film had a smaller grain size than the BSFMx = 0.06 film. For small-grain crystals, due to the small difference between lattice symmetry and defect symmetry, the thermodynamic force driving the symmetry matching between them is weak, and this hinders the kinetic migration of oxygen vacancies and affects reversible domain switching [47,51]. Therefore, the BSFMx = 0.04 film with small grains aged slowly.

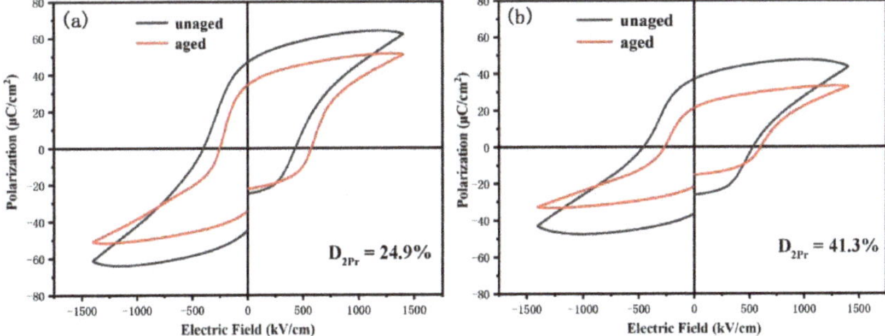

Figure 9. Comparison of hysteresis loops of the BSFM films with different Sm content measured after an interval of 110 days: (**a**) x = 4 mol%; (**b**) x = 6 mol%.

4. Conclusions

In conclusion, Sm-doped BSFM films were prepared on ITO/glass substrates by the sol-gel method. The effects of different Sm content on the leakage current density, dielectric properties, and aging properties of the BSFM films were systematically studied. Detailed explanations were made in terms of phase transition and oxygen vacancy. XRD and Raman analyses show that the samples all contained R3c and Pnma phases, and the samples with different Sm content had different phase composition. Sm doping led to lattice structure distortion and decreases in the crystal plane spacing. XPS analysis showed that the BSFMx = 0.04 thin film sample had the lowest oxygen vacancy content, indicating that an appropriate amount of Sm doping can effectively inhibit the valence of the Fe element. The decrease in oxygen vacancy increased the dielectric permittivity and the leakage current density. The minimum leakage current density of the BSFMx = 0.04 film sample was 8.84×10^{-5} A/cm^2 in a 350 kV/cm electric field. The remanent polarization intensity of the BSFMx = 0.04 film was 91.86 µC/cm^2, owing to the formation of avnonpolar orthorhombic Pnma phase during the doping process. At the same time, the aging process of BSFMx = 0.04 sample was slow, and the performance of the sample was better preserved.

Author Contributions: Conceptualization, Y.W.; Data curation, Y.W.; Formal Analysis, Y.W.; Funding acquisition, F.Z.; Investigation, Y.W.; Methodology, Y.W., Z.M. and L.W.; Project administration, F.Z. and L.Z.; Resources, X.G. and Z.L.; Software, Y.W.; Supervision, L.Z.; Validation, Y.L., B.Y.; Visualization, Y.W.; Writing-original draft, Y.W.; Writing-review & editing, Y.W., L.W., Z.M. and F.Z. All authors have read and agreed to the published version of the manuscript.

Funding: This work was supported by funding from the State Key Laboratory of Silicate Materials for Architectures (Wuhan University of Technology) (SYSJJ2020-05), and the Program of the Housing and Urban-Rural Construction Department of Shandong Province (2019-K7-10).

Data Availability Statement: Not applicable.

Conflicts of Interest: The authors declare no conflict of interest.

References

1. Martin, L.W.; Ramesh, R. Overview No. 151 Multiferroic and magnetoelectric heterostructures. *Acta Mater.* **2012**, *60*, 2449–2470. [CrossRef]
2. Eerenstein, W.; Mathur, N.D.; Scott, J.F. Multiferroic and magnetoelectric materials. *Nature* **2006**, *442*, 759–765. [CrossRef]
3. Bibes, M.; Barthelemy, A. Multiferroics: Towards a magnetoelectric memory. *Nat. Mater.* **2008**, *7*, 425–426. [CrossRef]
4. Yue, Z.W.; Tan, G.Q.; Yang, W.; Ren, H.J.; Xiao, A. Enhanced multiferroic properties in Pr-doped BiFe$_{0.97}$Mn$_{0.03}$O$_3$ films. *Ceram. Int.* **2016**, *42*, 18692–18699. [CrossRef]
5. Sreenivasulu, G.; Laletin, U.; Petrov, V.M.; Petrov, V.V.; Srinivasan, G. A permendur-piezoelectric multiferroic composite for low-noise ultrasensitive magnetic field sensors. *Appl. Phys. Lett.* **2012**, *100*, 173506. [CrossRef]
6. Hasegawa, M.; Asano, T.; Hashimoto, K.; Lee, G.C.; Park, Y.C.; Okazaki, T.; Furuya, Y. Fabrication of multiferroic composite actuator material by combining superelastic TiNi filler and a magnetostrictive Ni matrix. *Smar. Mate. Stru.* **2006**, *15*, N124–N128. [CrossRef]
7. Huang, A.; Handoko, A.D.; Goh, G.K.; Pallathadka, P.K.; Shannigrahi, S. Hydrothermal synthesis of (001) epitaxial BiFeO$_3$ films on SrTiO$_3$ substrate. *CrystEngComm* **2010**, *12*, 3806–3814. [CrossRef]
8. Lebeugle, D.; Colson, D.; Forget, A.; Viret, M.; Bonville, P.; Marucco, J.F.; Fusil, S. Room-temperature coexistence of large electric polarization and magnetic order in BiFeO$_3$ single crystals. *Phys. Rev. B* **2007**, *76*, 024116. [CrossRef]
9. Prashanthi, K.; Shaibani, P.M.; Sohrabi, A.; Natarajan, T.S.; Thundat, T. Nanoscale magnetoelectric coupling in multiferroic BiFeO$_3$ nanowires. *Phys. Status Solidi RRL* **2012**, *6*, 244–246. [CrossRef]
10. Shami, M.Y.; Awan, M.S.; Anis-ur-Rehman, M. The effect of heat treatment on structural and multiferroic properties of phase-pure BiFeO$_3$. *J. Electron. Mater.* **2012**, *41*, 2216–2224. [CrossRef]
11. Singh, S.K. Structural and electrical properties of Sm-substituted BiFeO$_3$ thin films prepared by chemical solution deposition. *Thin Solid Film.* **2013**, *527*, 126–132. [CrossRef]
12. Zhou, J.; Trassin, M.; He, Q.; Tamura, N.; Kunz, M.; Cheng, C.; Wu, J. Directed assembly of nano-scale phase variants in highly strained BiFeO$_3$ thin films. *J. Appl. Phys.* **2012**, *112*, 064102. [CrossRef]
13. Reetu, A.; Sanghi, S. Rietveld analysis, dielectric and magnetic properties of Sr and Ti codoped BiFeO$_3$ multiferroic. *J. Appl. Phys.* **2011**, *110*, 073909. [CrossRef]
14. Qi, X.; Dho, J.; Tomov, R.; Blamire, M.G.; Blamire, J.L. Greatly reduced leakage current and conduction mechanism in aliovalent-ion-doped BiFeO$_3$ Macmanus-driscoll. *Appl. Phys. Lett.* **2005**, *86*, 062903. [CrossRef]
15. Gumiel, C.; Jardiel, T.; Calatayud, D.G.; Vranken, T.; Van Bael, M.K.; Hardy, A.; Peiteado, M. Nanostructure stabilization by low-temperature dopant pinning in multiferroic BiFeO$_3$-based thin films produced by aqueous chemical solution deposition. *J. Mater. Chem. C* **2020**, *8*, 4234–4245. [CrossRef]
16. Madolappa, S.; Kundu, S.; Bhimireddi, R.; Varma, K.B. Improved electrical characteristics of Pr-doped BiFeO$_3$ ceramics prepared by sol-gel route. *Mater. Res. Express* **2016**, *3*, 065009. [CrossRef]
17. Shimada, T.; Arisue, K.; Kitamura, T. Strain-induced phase transitions in multiferroic BiFeO$_3$ (110) epitaxial film. *Phys. Lett. A* **2012**, *376*, 3368–3371. [CrossRef]
18. Yang, J.C.; He, Q.; Suresha, S.J.; Kuo, C.Y.; Peng, C.Y.; Haislmaier, R.C.; Chu, Y.H. Orthorhombic BiFeO$_3$. *Phys. Rev. Lett.* **2012**, *109*, 247606. [CrossRef]
19. Chai, Z.; Tan, G.; Yue, Z.; Yang, W.; Guo, M.; Ren, H.; Lv, L. Ferroelectric properties of BiFeO$_3$ thin films by Sr/Gd/Mn/Co multi-doping. *J. Alloys Compd.* **2018**, *746*, 677–687. [CrossRef]
20. Wen, X.L.; Chen, Z.; Liu, E.H.; Lin, X.; Chen, C. Effect of Ba and Mn doping on microstructure and multiferroic properties of BiFeO$_3$ ceramics. *J. Alloys Compd.* **2016**, *678*, 511–517. [CrossRef]

21. Karpinsky, D.V.; Pakalniškis, A.; Niaura, G.; Zhaludkevich, D.V.; Zhaludkevich, A.L.; Latushka, S.I.; Kareiva, A. Evolution of the crystal structure and magnetic properties of Sm-doped $BiFeO_3$ ceramics across the phase boundary region. *Ceram. Int.* **2021**, *47*, 5399–5406. [CrossRef]

22. Sati, P.C.; Kumar, M.; Chhoker, S. Phase evolution, magnetic, optical and dielectric properties of Zr-substituted $Bi_{0.9}Gd_{0.1}FeO_3$ multiferroics. *J. Am. Ceram. Soc.* **2015**, *98*, 1884–1890. [CrossRef]

23. Yun, Q.; Bai, Y.L.; Chen, J.; Gao, W.; Bai, A.; Zhao, S. Improved ferroelectric and fatigue properties in Ho doped $BiFeO_3$ thin films. *Mater. Lett.* **2014**, *129*, 166–169. [CrossRef]

24. Liu, J.; Deng, H.M.; Zhai, X.; Lin, T.; Meng, X.; Zhang, Y.; Chu, J. Influence of Zn doping on structural, optical and magnetic properties of $BiFeO_3$ films fabricated by the sol-gel technique. *Mater. Lett.* **2014**, *133*, 49–52. [CrossRef]

25. Yang, S.J.; Zhang, F.Q.; Xie, X.; Sun, H.; Zhang, L.; Fan, S. Enhanced leakage and ferroelectric properties of Zn-doped $BiFeO_3$ thin films grown by sol-gel method. *J. Alloys Compd.* **2018**, *734*, 243–249. [CrossRef]

26. Zhang, C.C.; Dai, J.Q.; Liang, X.L. Enhanced ferroelectric properties of (Zn, Ti) equivalent co-doped $BiFeO_3$ films prepared via the sol-gel method. *Ceram. Int.* **2021**, *47*, 16776–16785. [CrossRef]

27. Liu, Y.; Tan, G.Q.; Ren, X.X.; Li, J.; Xue, M.; Ren, H.; Liu, W. Electric field dependence of ferroelectric stability in $BiFeO_3$ thin films co-doped with Er and Mn. *Ceram. Int.* **2020**, *46*, 18690–18697. [CrossRef]

28. Kan, D.; Pálová, L.; Anbusathaiah, V.; Cheng, C.J.; Fujino, S.; Nagarajan, V.; Takeuchi, I. Universal behavior and electric-field-induced structural transition in rare-earth-substituted $BiFeO_3$. *Adv. Funct. Mater.* **2010**, *20*, 1108–1115. [CrossRef]

29. Emery, S.B.; Cheng, C.J.; Kan, D.; Rueckert, F.J.; Alpay, S.P.; Nagarajan, V.; Wells, B.O. Phase coexistence near a morphotropic phase boundary in Sm-doped $BiFeO_3$ films. *Appl. Phys. Lett.* **2010**, *97*, 152902. [CrossRef]

30. Xue, X.; Tan, G.Q.; Ren, H.J. Structural, electric and multiferroic properties of Sm-doped $BiFeO_3$ thin films prepared by the sol-gel process. *Ceram Int.* **2013**, *39*, 6223–6228. [CrossRef]

31. Tao, H.; Lv, J.; Zhang, R.; Xiang, R.; Wu, J. Lead-free rare earth-modified $BiFeO_3$ ceramics: Phase structure and electrical properties. *Mater. Des.* **2017**, *120*, 83–89. [CrossRef]

32. Zhang, F.; Zeng, X.; Bi, D.; Guo, K.; Yao, Y.; Lu, S. Dielectric, ferroelectric, and magnetic properties of Sm-doped $BiFeO_3$ ceramics prepared by a modified solid-state-reaction method. *Materials* **2018**, *11*, 2208. [CrossRef] [PubMed]

33. Liang, X.L.; Dai, J.Q. Prominent ferroelectric properties in Mn-doped $BiFeO_3$ spin-coated thin films. *J. Alloys Compd.* **2021**, *886*, 161168. [CrossRef]

34. Zhou, W.; Deng, H.; Cao, H.; He, J.; Liu, J.; Yang, P.; Chu, J. Effects of Sm and Mn co-doping on structural, optical and magnetic properties of $BiFeO_3$ films prepared by a sol-gel technique. *Mater. Lett.* **2015**, *144*, 93–96. [CrossRef]

35. Goian, V.; Kamba, S.; Greicius, S.; Nuzhnyy, D.; Karimi, S.; Reaney, I. Terahertz and infrared studies of antiferroelectric phase transition in multiferroic $Bi_{0.85}Nd_{0.15}FeO_3$. *J. Appl. Phys.* **2011**, *110*, 074112. [CrossRef]

36. Gu, Y.; Zhou, Y.; Zhang, W. Optical and magnetic properties of Sm-doped $BiFeO_3$ nanoparticles around the morphotropic phase boundary region. *AIP Adv.* **2021**, *11*, 045223. [CrossRef]

37. Li, W.; Hao, J.; Du, J.; Fu, P.; Sun, W.; Chen, C.; Chu, R. Electrical properties and luminescence properties of $0.96(K_{0.48}Na_{0.52})(Nb_{0.95}Sb_{0.05})–0.04Bi_{0.5}(Na_{0.82}K_{0.18})_{0.5}ZrO_3$-xSm lead-free ceramics. *J. Adv. Ceram.* **2020**, *9*, 72–82. [CrossRef]

38. Singh, M.K.; Jang, H.M.; Ryu, S.; Jo, M.H. Polarized Raman scattering of multiferroic $BiFeO_3$ epitaxial films with rhombohedral R3c symmetry. *Appl. Phys. Lett.* **2006**, *88*, 042907. [CrossRef]

39. Wang, Y.; Nan, C.W. Site modification in $BiFeO_3$ thin films studied by Raman spectroscopy and piezoelectric force microscopy. *J. Appl. Phys.* **2008**, *103*, 114104. [CrossRef]

40. Iliev, M.N.; Abrashev, M.V.; Lee, H.G.; Popov, V.N.; Sun, Y.Y.; Thomsen, C.; Chu, C.W. Raman spectroscopy of orthorhombic perovskitelike $YMnO_3$ and $LaMnO_3$. *Phys. Rev. B* **1998**, *57*, 2872. [CrossRef]

41. Singh, D.; Tabari, T.; Ebadi, M.; Trochowski, M.; Yagci, M.B.; Macyk, W. Efficient synthesis of $BiFeO_3$ by the microwave-assisted sol-gel method: "A" site influence on the photoelectron chemical activity of perovskites. *Appl. Surf. Sci.* **2019**, *471*, 1017–1027. [CrossRef]

42. Wang, J.; Luo, L.; Han, C.; Yun, R.; Tang, X.; Zhu, Y.; Feng, Z. The microstructure, electric, optical and photovoltaic properties of $BiFeO_3$ thin films prepared by low temperature sol-gel method. *Materials* **2019**, *12*, 1444. [CrossRef]

43. Ma, Z.B.; Liu, H.Y.; Wang, L.X.; Zhang, F.Q.; Zhang, F.; Zhu, L.; Fan, S. Phase transition and multiferroic properties of Zr-doped $BiFeO_3$ thin films. *J. Mater. Chem. C* **2020**, *48*, 17307–17317. [CrossRef]

44. Bai, H.; Li, J.; Hong, Y.; Zhou, Z. Enhanced ferroelectricity and magnetism of quenched $(1−x) BiFeO_3$-$xBaTiO_3$ ceramics. *J. Adv. Ceram.* **2020**, *9*, 511–516. [CrossRef]

45. Hanani, Z.; Merselmiz, S.; Danine, A.; Stein, N.; Mezzane, D.; Amjoud, M.B. Enhanced dielectric and electrocaloric properties in lead-free rod-like BCZT ceramics. *J. Adv. Ceram.* **2020**, *9*, 210–219. [CrossRef]

46. Lou, Y.H.; Song, G.L.; Chang, F.G. Investigation on dependence of $BiFeO_3$ dielectric property on oxygen content. *Chin. Phys. B* **2010**, *19*, 077702.

47. Ren, X.; Otsuka, K. Universal symmetry property of point defects in crystals. *Phys. Rev. Lett.* **2000**, *85*, 1016. [CrossRef]

48. Ren, X. Large electric-field-induced strain in ferroelectric crystals by point-defect-mediated reversible domain switching. *Nat. Mater.* **2004**, *3*, 91–94. [CrossRef] [PubMed]

49. Zhang, L.; Ren, X. Aging behavior in single-domain Mn-doped $BaTiO_3$ crystals: Implication for a unified microscopic explanation of ferroelectric aging. *Phys. Rev. B* **2006**, *73*, 094121. [CrossRef]

50. Zhang, L.X.; Ren, X. In situ observation of reversible domain switching in aged Mn-doped BaTiO$_3$ single crystals. *Phys. Rev. B* **2005**, *71*, 174108. [CrossRef]
51. Guo, Y.Y.; Yan, Z.B.; Zhang, N.; Cheng, W.W.; Liu, J.M. Ferroelectric aging behaviors of BaTi$_{0.995}$Mn$_{0.005}$O$_3$ ceramics: Grain size effects. *Appl. Phys. A* **2012**, *107*, 243–248. [CrossRef]

 *nanomaterials*

MDPI

Article

Ferroelectricity and Piezoelectricity in 2D Van der Waals CuInP$_2$S$_6$ Ferroelectric Tunnel Junctions

Tingting Jia [1,2,*], Yanrong Chen [1,3], Yali Cai [1,2], Wenbin Dai [1,3], Chong Zhang [1,2], Liang Yu [1,4], Wenfeng Yue [1,4], Hideo Kimura [5], Yingbang Yao [3], Shuhui Yu [1], Quansheng Guo [2,*] and Zhenxiang Cheng [6,*]

[1] Institute of Advanced Materials Science and Engineering, Shenzhen Institutes of Advanced Technology, Chinese Academy of Sciences, Shenzhen 518055, China; ry.chen@siat.ac.cn (Y.C.); yl.cai@siat.ac.cn (Y.C.); wb.dai@siat.ac.cn (W.D.); chong.zhang@siat.ac.cn (C.Z.); liang.yu@siat.ac.cn (L.Y.); wf.yue@siat.ac.cn (W.Y.); sh.yu@siat.ac.cn (S.Y.)
[2] School of Materials Science and Engineering, Hubei University, Wuhan 430062, China
[3] School of Materials and Energy, Guangdong University of Technology, Guangzhou 510006, China; ybyao@gdut.edu.cn
[4] Nano Science and Technology Institute, University of Science and Technology of China, Suzhou 215125, China
[5] School of Environmental and Material Engineering, Yantai University, Yantai 264005, China; deokimura@yahoo.co.jp
[6] Institute for Superconducting & Electronic Materials, University of Wollongong, Innovation Campus, North Wollongong, NSW 2500, Australia
[*] Correspondence: tt.jia@siat.ac.cn (T.J.); qguo@hubu.edu.cn (Q.G.); cheng@uow.edu.au (Z.C.)

Abstract: CuInP2S6 (CIPS) is a novel two-dimensional (2D) van der Waals (vdW) ferroelectric layered material with a Curie temperature of T_C~315 K, making it promising for great potential applications in electronic and photoelectric devices. Herein, the ferroelectric and electric properties of CIPS at different thicknesses are carefully evaluated by scanning probe microscopy techniques. Some defects in some local regions due to Cu deficiency lead to a CuInP2S6–In$_{4/3}$P$_2$S$_6$ (CIPS–IPS) paraelectric phase coexisting with the CIPS ferroelectric phase. An electrochemical strain microscopy (ESM) study reveals that the relaxation times corresponding to the Cu ions and the IPS ionospheres are not the same, with a significant difference in their response to DC voltage, related to the rectification effect of the ferroelectric tunnel junction (FTJ). The electric properties of the FTJ indicate Cu$^+$ ion migration and propose that the current flow and device performance are dynamically controlled by an interfacial Schottky barrier. The addition of the ferroelectricity of CIPS opens up applications in memories and sensors, actuators, and even spin-orbit devices based on 2D vdW heterostructures.

Keywords: two-dimensional materials; ferroelectric properties; scanning probe microscope; negative piezoelectricity; phase segregation

Citation: Jia, T.; Chen, Y.; Cai, Y.; Dai, W.; Zhang, C.; Yu, L.; Yue, W.; Kimura, H.; Yao, Y.; Yu, S.; et al. Ferroelectricity and Piezoelectricity in 2D Van der Waals CuInP$_2$S$_6$ Ferroelectric Tunnel Junctions. *Nanomaterials* **2022**, *12*, 2516. https://doi.org/10.3390/nano12152516

Academic Editor: Goran Drazic

Received: 4 July 2022
Accepted: 15 July 2022
Published: 22 July 2022

Publisher's Note: MDPI stays neutral with regard to jurisdictional claims in published maps and institutional affiliations.

1. Introduction

Two-dimensional (2D) ferroelectric materials are gaining extensive attention. It can effectively improve device performance when applied to devices such as memory, capacitors, actuators, and sensors [1–4]. Nowadays, people regard such kinds of 2D materials as var der Waals (vdW) layered ferroelectric materials, which benefit from covalently bonded polar or non-polar monolayers by Van der Waals forces and exhibit ferroelectric properties. However, reports of ferroelectric 2D materials at room temperature are rare. Because of the depolarization field with decreasing thickness, there is an enormous challenge in maintaining ferroelectricity in ultrathin ferroelectric films. Ferroelectricity is remained elusive in the 2D material library [5,6] Van der Waals (vdW) layered ferroelectric materials has become a promising research branch in condensed matter physics [7,8], among which copper indium thiophosphate, CuInP$_2$S$_6$ (CIPS), is one of the most representative materials because of its room temperature ferroelectricity [9]. CIPS is promised to play an important role in non-volatile memory. A recent experiment reported that the vdW ferroelectric tunnel junction

(FTJ) device based on CIPS achieved a high tunneling electroresistance (TER) ratio [10,11]. Quantum transport device simulations of an FTJ based on CIPS and graphene demonstrate that scaling of the ferroelectric layer thickness exponentially not only significantly builds up the ferroelectric tunneling ON current but also reduces the read latency, in addition to enabling the FTJs with CIPS bilayers or trilayers to read speed in nanoseconds [12]. There are also outstanding endurance and retention characteristics for FTJ devices [3].

It was reported that giant intrinsic negative longitudinal piezoelectricity was observed in 2D layered CIPS. Lu You et al. [12] tested the converse piezoelectric effects of poly(vinylidene difluoride (PVDF), CIPS, and lead zirconate titanate (PZT) using a piezoelectric microscope and concluded that the electromechanical properties of CIPS are the same as those of PVDF with a negative longitudinal piezoelectric effect but opposite to those of PZT. This abnormal electromechanical phenomenon is caused by the significant deformation sensitivity of the weak interlayer interaction and is mediated by the high displacive instability of Cu ions. Several groups have discussed the origins of negative piezoelectricity. Yubo Qi and Andrew M. Pappe [13] attributed the negative piezoelectricity to the "lag of Wannier center" effect by proposing a negative clamped-ion term in the low-dimensional layered materials. John A. Brehm et al. [14] combined first-principles calculations with local electromechanical material characterization. They predicted and verified the existence of a uniaxial quadruple potential well for Cu displacements achieved by the van der Waals gap in CIPS. This led to the explanation that the negative longitudinal piezoelectric coefficient stems from the low polarization and very high sensitivity to a strain of the Cu atoms within the layer. Due to the negative piezoelectricity, CIPS has more complex piezoelectric properties, which gives it potential application prospects for 2D vdW materials with the same complex piezoelectric behavior in calculation and energy conversion. Therefore, it is crucial to study the piezoelectric behavior of CIPS.

Recently, to achieve resistance changes by ferroelectric switching, a ferroelectric field-effect transistor (FeFET) has been proposed, which uses ferroelectric materials instead of the oxide layer in a FET [15–17]. The causesFeFET exhibits two resistive states due to the hysteresis of ferroelectric switching. Furthermore, ferroelectric tunneling junctions and ferroelectric diodes were also investigated. It is reported that the polarization-modulation of Schottky-like barriers realizes the resistive change [18,19]. Yet, the vast majority of these devices use conventional ferroelectric materials such as $PbTiO_3$ and $BaTiO_3$ [20]. Affected by the three-dimensional nature of the ferroelectric oxide lattices, it is necessary for epitaxially grown high-quality films to select the substrates with a small lattice mismatch [21]. This seriously limits the possible application of materials in ferroelectric heterostructure devices. Therefore, it may be fundamentally and practically beneficial to study weakly bonded non-oxide ferroelectric compounds. Beyond that, the pioneering work on graphene has attracted an intense search for other 2D materials [22,23].

In this work, the crystal structure and ferroelectricity in crystalline CIPS nanoflakes are investigated at room temperature. Thin layer crystals in their pristine state show ferroelectric domains that are visualized directly. Piezoresponse force microscopy (PFM) measurements show that the polarization is stable and switchable in different layers of CIPS, and a negative piezoelectric effect is observed. Furthermore, we observed the ion migration phenomenon in CIPS via electrochemical strain microscopy (ESM). Finally, to obtain insight into the nature of the TER phenomena in the CIPS—based FTJs, we investigated their current—voltage (*I–V*) characteristics with a conductive probe microscope.

2. Experimental Section

2.1. Sample Preparation and Structural Characterization

All CIPS crystals were purchased from 6Carbon Technology (Shenzhen, China). The purchased synthetic bulk crystals were mechanically exfoliated onto platinized silicon substrates to obtain the flakes. The flake thickness was measured with a three-dimensional (3D) laser confocal microscope (VK-X1100, KEYENCE CORPORATION., Itasca, IL, USA) and an atomic force microscope (AFM, Cypher S, OXFORD INSTRUMENTS, Abingdon,

Oxfordshire, UK). The microscope objective lenses used in our experiments are ×5, ×10, ×20, ×50, ×100, and the microscopic ocular is ×5. The numerical aperture and working distance of the microscope objective lens with ×100 magnification are 0.3 mm and 4.7 mm, respectively. The confocal Raman system (LabRAM HR, HORIBA Instuments(SHANGHAI) co. LTD, Shanghai, China) with 532 nm laser excitation was used to collect Raman spectra. The morphologies and structures of the as—prepared samples were collected by transmission electron microscopy (TEM, JEM-3200FS, JEOL (BEIJING) Co., Ltd. GUANGZHOU BRANCH, Beijing, China).

2.2. Piezoelectric Force Microscopy (PFM) Measurements

The PFM measurements were performed on μm—sized CIPS flakes on the crystal surface. The silver paste was used to attach the sample to the sample stage, which served as an electric back contact. PFM image scanning measurements were performed using a piezoelectric force microscope (PFM) with Au- and diamond-coated Si cantilever tips (FM-LC, 100 kHz and 8 N/m), respectively. The cantilever tip acted as a local variable electrode. The electric voltage was applied to the sample surface via the conductive PFM tip. The sample was cleaved using Scotch tape directly before the measurements. The topography of CIPS flakes was probed in AC mode, while the piezoelectric and ferroelectric responses were measured using dual AC resonance tracking PFM (DART-PFM) mode. The local piezoresponse hysteresis loops were measured 10 times at each position at multiple different arbitrary points.

2.3. Electrochemical Strain Microscopy (ESM) Measurements

Relaxation and variable ESM measurements were performed using an atomic force microscope. Conductive Au- and diamond-coated Si cantilever tips (FM-LC, 100 kHz, and 8 N/m) were used in all measurements.

2.4. Conductive Atomic Force Microscopy (C-AFM) Measurements

Local current-voltage and current phase diagram measurements were performed by C-AFM using with Au- and diamond-coated Si cantilever tips (FM-LC, 100 kHz, and 8 N/m). *I–V* curves were collected perpendicular to the CIPS samples in C-AFM mode.

3. Results and Discussion

3.1. Structural Characterization

The room-temperature crystal structure of CIPS was first discovered in 1995 using single-crystal X-ray diffraction [24]. Bulk CIPS is composed of vertically stacked, weakly interacting layers bound together by vdW interactions. The metal cations and P–P pairs fill the octahedral voids in the sulfur framework in a CIPS crystal (Figure 1a). A complete unit cell is reported to consist of two adjacent monolayers due to the site exchange between Cu and the P–P pairs from one layer to another [25]. Figure 1b shows two typical Raman spectra measured in different regions of a CIPS flake on a Pt/Si substrate. In general, we observed five Raman active modes, including δ(S–P–P) modes at 162.9 cm^{-1}, δ(S–P–S) modes around 226.1 cm^{-1}, and 263.3 cm^{-1}, an active mode from the cations at 303.2 cm^{-1}, andυ(P–P) and υ(P–S) modes at 374.3 cm^{-1} and 436.4 cm^{-1}, respectively. The tested Raman results are consistent with the former reports on CIPS crystals [26,27]. P–S bond stretching vibrations (υ) can be resolved to $A_{1g} + A_{2u} + E_u + E_g$. The S–P–S modes change the S–P–S angles, resulting in another set of $A_{1g} + A_{2u} + E_u + E_g$. The P–P bond is associated with bending the PS$_3$ groups, which accounts for $E_u + E_g$. Twisting $(P_2S_6)^{4-}$ oscillation is related to the A_{1u} mode [27]. The Raman spectra in orange showed a strong peak at ~303 cm^{-1}, which corresponds to the cations [26]. The blue curve showed a weak and broad peak at ~303 cm^{-1}, which is mainly related to the reduction of Cu content. It is believed to originate from the paraelectric phase of CIPS–IPS, where IPS stands for $In_{4/3}P_2S_6$, appearing in the flake due to Cu$^+$ deficiency [26,27]. The transmission electron microscope (TEM) images show two structures on the same CIPS flakes (Figure 1c), which correspond to the monoclinic and triclinic phases, consistent with the Raman spectrum.

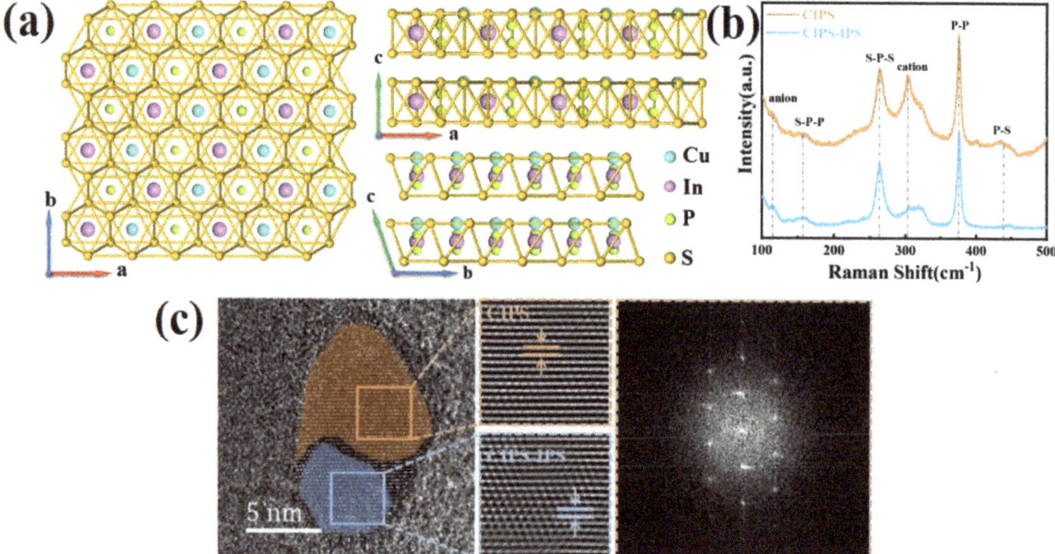

Figure 1. Material characterization of CIPS flake. (**a**) Top and side views of the CIPS crystal structure. In the atomic model, the yellow networks are S triangular networks, and the green, purple, and blue balls are P, In, and Cu atoms, respectively. (**b**) Raman spectra of CIPS flakes, including ferroelectric and paraelectric phases on the Pt substrate with 532 nm laser excitation. (**c**) The TEM characterizations of CIPS crystal include fast Fourier transform (FFT, **right**) and filtered inverse FFT (**middle**) patterns of the selected areas, respectively.

3.2. Ferroelectric Domain and Domain Switching

To approve the ferroelectricity of CIPS flakes with different thicknesses, the domain distribution of the CIPS flakes was investigated in 3 nm, 9 nm, 12 nm, 21 nm, 35 nm, and 78 nm CIPS flakes by piezoresponse force microscopy (PFM). The domain evolution of the CIPS flakes was observed with height, amplitude, and phase images, as shown in Figure 2a–f. The domain size varied with the increasing thickness of the CIPS. As shown in Figure 2a–f, the size of the ferroelectric domains increases with the thickness of the CIPS flakes. Domains vary in size but are on the order of ~100 nm and ~1 μm in diameter, respectively, different from reports in the literature [7]. In our experience, it is easy to observe clear domain structures in CIPS flakes with thicknesses in the thickness range from 20 to 40 nm, as shown in Figure 2d,e. We observed the reversed piezoelectric effect in the typical amplitude-voltage "butterfly" loops by scanning the piezoresponse hysteresis loops in dual AC resonance tracking PFM (DART-PFM) mode. There are regions where the ferroelectric inversion loop disappears in the 20 nm CIPS flakes, which may be related to the paraelectric phase caused by Cu^{2+} segregation.

According to the above results, we chose the CIPS sample with a thickness of 20 nm to investigate the ferroelectric switching behavior. Well-defined butterfly loops of the saturated PFM amplitude and the distinct 180° switching of the phase signals were observed, as shown in Figure 3a,b, indicating switchable ferroelectric polarization in the CIPS flakes with different thicknesses (28 nm and 20 nm). We divided the hysteresis loops into four sections, as shown in Figure 3a–c, and observed that the direction of the polarization switching of CIPS was consistent with that of PVDF [28], showing negative piezoelectricity, which arose from the low polarization of Cu atoms in the CIPS layer [14].

We still observed abnormal butterflies and switching of the phase loops in the 25 nm and 20 nm CIPS, however, as shown in Figure 3c,d. There are the following several possible reasons for the abnormal switching loops: One, when sweeping the electric fields in the reverse sequence, the piezoelectric response curves show asymmetric shapes. This phenomenon was attributed to the existence of remnant-injected electrons. Due to the different diffusion distances in both the positive and negative electric range, the injected electrons cannot be eliminated absolutely when applying a lower electric field [29]. Two, the abnormal hysteresis loops may also relate to negative electrostriction and electrostatic signal contributions accompanied by charge injection during scanning [30]. Three, abnormal domain switching is generated due to in-plane ionic migration in CIPS [31]. Since the Cu is deficient, when applying an electric field, a few regions of the CIPS undergo a chemical phase separation into a paraelectric $In_{4/3}P_2S_6$ phase and a ferroelectric $CuInP_2S_6$ phase [32]. Abnormal hysteresis loops may be from the local paraelectric phase of the $CuInP_2S_6$–$In_{4/3}P_2S_6$ region due to Cu deficiency [27].

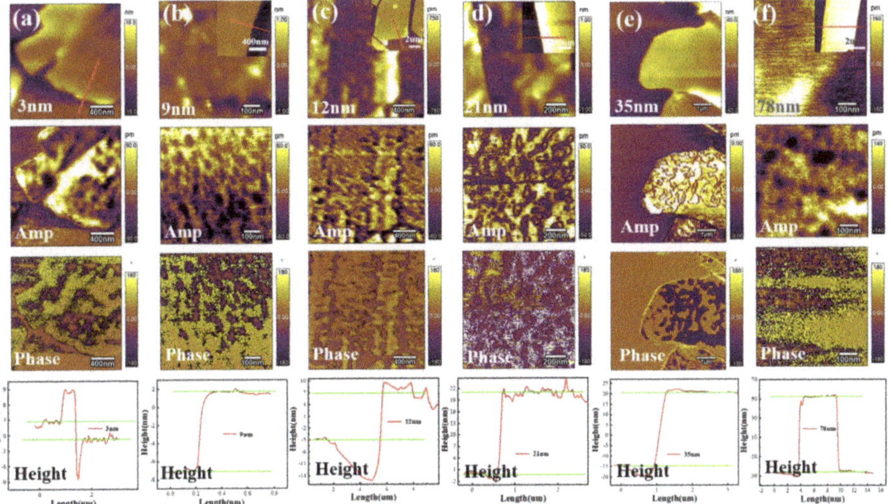

Figure 2. PFM images including height, amplitude, and phase image (upper three), and height curves (bottom row) of CIPS flakes with different thicknesses. (**a**) 3 nm, (**b**) 9 nm, (**c**) 12 nm, (**d**) 21 nm, (**e**) 35 nm, and (**f**) 78 nm.

To investigate the in-plane ionic migration, we carried out a relaxation test to analyze the local ionic dynamics of CIPS via electrochemical strain measurements (ESM), as shown in Figure 4 [33,34]. As schematically shown in Figure 4a, a direct-current (DC) voltage is applied on top of an AC voltage to induce a longer-range redistribution of ions in CIPS flakes during the probe. After removing the DC voltage, the ions relax back to their original equilibrium state. The local dynamics can be deduced from the time constant associated with the relaxation of ESM amplitude, as shown in Figure 4b. The DC voltage is applied on top of the AC following the profile shown in Figure 4a to induce a longer-range redistribution of Cu^+ ions; meanwhile, polarizing the sample over a larger scale. The faster relaxation shown in Figure 4d corresponds to the Vegard strain directly related to the ionic concentration of Cu^+. While slower relaxation in Figure 4d is because of the induced electrochemical dipoles and results from the readjustment of the negative $InP_2S_6{}^-$ in response to the redistributed Cu^+. The process takes longer over a much shorter distance. This phenomenon verifies our conjecture that the IPS paraelectric phase forms in the Cu-absent region due to the Cu ion migration, with no ferroelectric polarization switching. That is, the relaxation of Cu ions is associated with the change in ferroelectric polarization after the addition of voltage, and the longer time is related to the saturation phenomenon of polarization specific to CIPS reported in the literature.

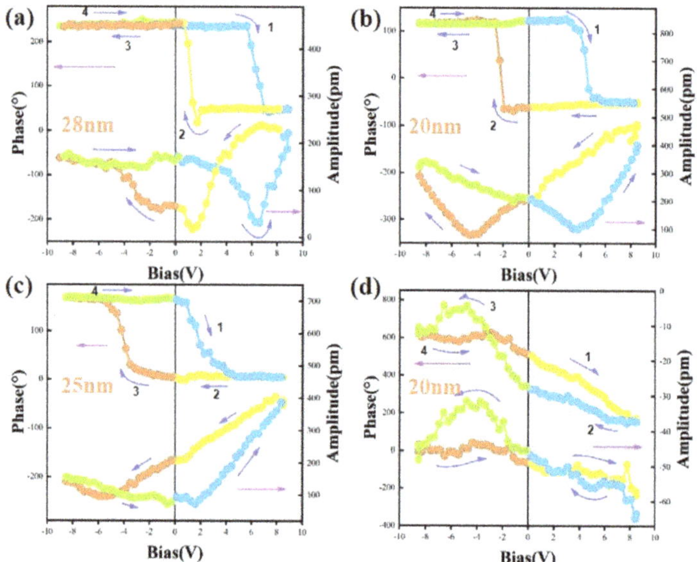

Figure 3. Ferroelectric polarization switching by PFM for CIPS flakes with different thicknesses. The PFM amplitude (green) and phase (blue) hysteresis loops during the switching process for CIPS flakes with thickness of (**a**) 28 nm, (**b**) 20 nm, (**c**) 25 nm, and (**d**) another area of the sample with the thickness of 20 nm.

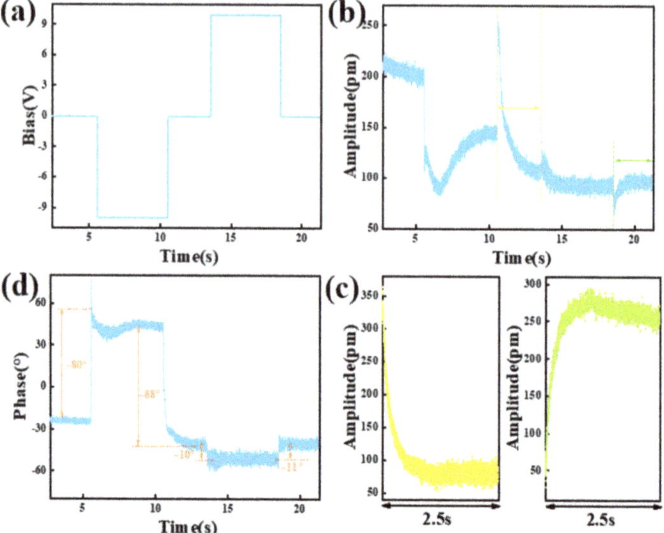

Figure 4. Relaxation dynamics in local electrochemical strain measurements of CIPS. (**a**) Illustrated DC profile applying in the relaxation measurements. (**b**,**d**) ESM amplitude-time and phase-time curves were obtained corresponding to the DC profile. (**c**) Zoomed—in relaxation curves of (**b**) were recorded after removing negative and positive DC voltage, respectively.

3.3. Electric Properties of Pt/CuInP$_2$S$_6$/Au FTJ

CIPS by itself is the only two-dimensional ferroelectric material with a ferroelectric transition temperature (T_c) just over room temperature. Based on the above studies, we have observed the paraelectric phase in CIPS flakes, but it is rarely reported whether the paraelectric phase affects the device's performance. In this work, we used conductive AFM to study the electric properties of a CIPS FTJ as shown in Figure 5. The electrical characterization of an Au/CIPS/Pt vdW FTJ is shown in Figure 5a for a device with a 2-nm-thick CIPS layer. Figure 5b,c shows the topography and current images of the CIPS flakes. The leakage current scanning was performed within a 3×3 μm^2 area at a read voltage of 10 mV. The observed local conductive path regions indicated good electrical conductivity in the CIPS flakes.

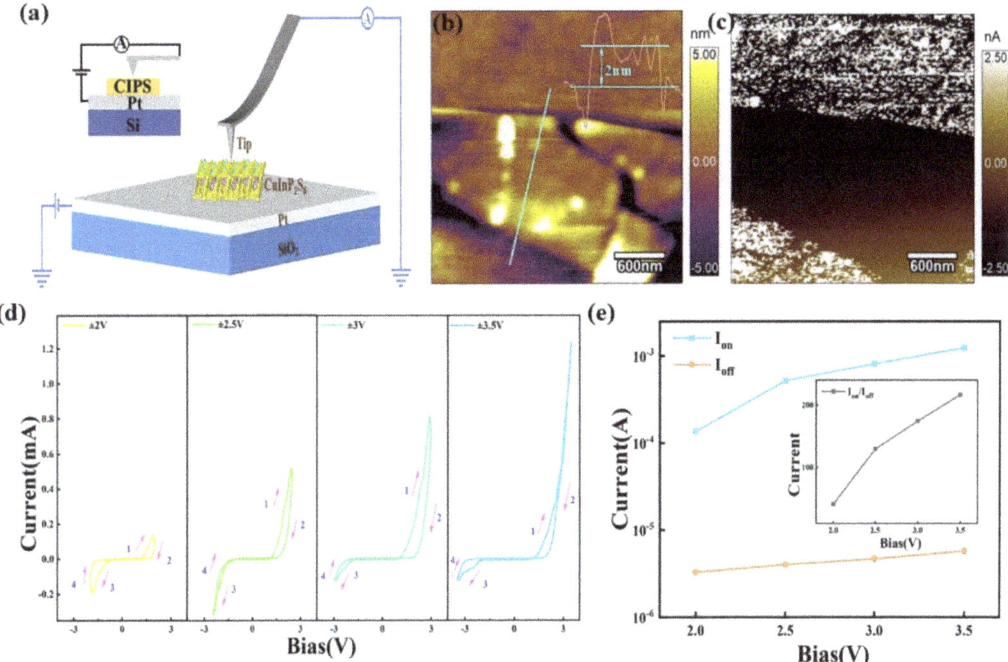

Figure 5. Electrical characterization of a Si/Pt/CIPS/Au diode with1.7 nm CIPS. (**a**) Schematic representation of the experimental setup for C-AFM measurements. (**b**) Topography image of the CIPS nano flake with a thickness of 2 nm on the Si/SiO$_2$/Ti/Pt substrate. The inset shows the height map of the CIPS flake. (**c**) The corresponding current phase diagram of CIPS flakes. (**d**) *I–V* curves measured with increasing sweep voltages, where V_{Max} is from 2 to 3.5 V. (**e**) I_{on} and I_{off} are the on and off current of the FTJ with resistance switching behavior, which are also corresponding to the low and high resistance states, read from (**d**) under different scanning voltages, with the inset the calculated switching ratio based on (**d**).

Figure 5d presents the *I–V* curves of a CIPS FTJ, measured with varying sweep ranges (V_{max} from 2 V to 3.5 V). We can observe resistive switching in both positive and negative voltage ranges, demonstrating that FTJ has superior continuous current modulation and self-rectification functions. The *I–V* curves show a nonsymmetrical contour, and worse symmetry appears with increased voltage. This corresponds to the Cu^{2+} migration process. Before the Cu^{2+} migration, a Schottky barrier must be overcome. Due to the difference in the work functions between Au and Pt, the current of the *I–V* curve under a positive voltage and a negative voltage is asymmetric. The current limiting behavior in the negative range is similar to the rectifying effect of diodes. As the applied voltage increases, the *I–V* curves show more obvious hysteresis in the positive range, which is always regarded as a resistance switching behavior. As shown in Figure 5e, the I_{on} and I_{off} correspond to the on-current and off-current when applying different voltages during the resistance switching. As shown in the inset of Figure 5e, the I_{on}/I_{off} of the FTJ with an ultra-thin CIPS film is over 200, which is comparable with the previous results [10], indicating that CIPS has good development prospects in the research on and application of nonvolatile memory devices.

4. Conclusions

In summary, we clearly observed the domain structure of CIPS with a thickness of 20–40 nm in PFM measurements and found that the hysteresis loop of CIPS showed the same negative longitudinal piezoelectric effect as PVDF, indicating the complexity of the piezoelectric response of CIPS. We used electrochemical strain microscopy (ESM) to further study Cu^{2+} segregation and discovered different relaxation times between the Cu^{2+} and an $InP_2S_6^{2-}$ ion groups, with both ions showing significant differences in their DC voltage response. These observations imply that in the regions where the ferroelectric inversion loop cannot be observed, Cu^{2+} segregation occurs, and $InP_2S_6^{2-}$ paraelectric phase is formed. In addition, the Cu^{2+} relaxation is related to the change in the ferroelectric polarization after the application of voltage, and the longer relaxation time is relevant to the distinctive ferroelectric polarization saturation in $CIPS^2$. Finally, we constructed a Pt/CIPS/Au FTJ. We observed resistance switches in the positive and negative voltage ranges, demonstrating that FTJ has superior continuous current modulation and self-rectification functions. Combined with its piezoelectric characteristics, layered ferroelectric CIPS is suitable not only for memory, but also for sensors, actuators, and even spin-orbit devices based on vdW heterostructures.

Author Contributions: Conceptualization, T.J., Q.G. and Z.C.; methodology, T.J., Y.C. (Yanrong Chen); validation, Y.C. (Yali Cai), W.D., C.Z., L.Y. and W.Y.; writing—original draft preparation, T.J. and Y.C. (Yanrong Chen); writing—review and editing, T.J., Y.Y., Q.G. and Z.C.; visualization, H.K.; formal analysis, Z.C.; investigation, H.K.; resources, T.J. and S.Y.; supervision, S.Y., T.J. and Y.C. (Yanrong Chen) contribute equally. All authors have read and agreed to the published version of the manuscript.

Funding: This research was funded by the National Natural Science Foundation of China (51702351, 51777209), the Basic and Applied Basic Research Foundation of Guangdong Province (2020B1515120019), Guangdong Provincial Key Laboratory (2014B030301014), and the Shenzhen Science and Technology Innovation Committee (JCYJ20170413152832151, KQTD20170810160424889). And The APC was funded by JCYJ20170413152832151.

Institutional Review Board Statement: Not applicable.

Informed Consent Statement: Not applicable.

Data Availability Statement: Not applicable.

Conflicts of Interest: The authors declare no conflict of interest.

References

1. Zhou, S.; You, L.; Zhou, H.; Pu, Y.; Gui, Z.; Wang, J. Van der Waals layered ferroelectric $CuInP_2S_6$: Physical properties and device applications. *Front. Phys.* **2021**, *16*, 13301. [CrossRef]
2. Zhao, Z.; Rakheja, S.; Zhu, W. Nonvolatile Reconfigurable 2D Schottky Barrier Transistors. *Nano Lett.* **2021**, *21*, 9318–9324. [CrossRef] [PubMed]
3. Yang, N.; Chen, H.-Y.; Wu, J.; Wu, T.; Cao, J.; Ling, X.; Wang, H.; Guo, J. Multiscale Simulation of Ferroelectric Tunnel Junction Memory Enabled by van der Waals Heterojunction: Comparison to Experiment and Performance Projection. In Proceedings of the 2020 IEEE International Electron Devices Meeting (IEDM), New York, NY, USA, 12–18 December 2020.
4. Huang, H.-F.; Dong, Y.-J.; Yao, Y.; Zhang, J.-Y.; Hao, X.; Gu, H.; Wu, Y.-Z. Nonvolatile tuning of the Rashba effect in the $CuInP_2S_6$/MoSSe/$CuInP_2S_6$ heterostructure. *J. Appl. Phys.* **2020**, *128*, 224105. [CrossRef]

5. Kim, D.J.; Jo, J.Y.; Kim, Y.S.; Chang, Y.J.; Lee, J.S.; Yoon, J.G.; Song, T.K.; Noh, T.W. Polarization Relaxation Induced by a Depolarization Field in Ultrathin Ferroelectric BaTiO$_3$ Capacitors. *Phys. Rev. Lett.* **2005**, *95*, 237602. [CrossRef]

6. Si, M.; Saha, A.K.; Gao, S.; Qiu, G.; Qin, J.; Duan, Y.; Jian, J.; Niu, C.; Wang, H.; Wu, W.; et al. A ferroelectric semiconductor field-effect transistor. *Nat. Electron.* **2019**, *2*, 580–586. [CrossRef]

7. Yuan, S.; Luo, X.; Chan, H.L.; Xiao, C.; Dai, Y.; Xie, M.; Hao, J. Room-temperature ferroelectricity in MoTe$_2$ down to the atomic monolayer limit. *Nat. Commun.* **2019**, *10*, 1775. [CrossRef]

8. Guan, Z.; Hu, H.; Shen, X.; Xiang, P.; Zhong, N.; Chu, J.; Duan, C. Recent Progress in Two-Dimensional Ferroelectric Materials. *Adv. Electron. Mater.* **2019**, *6*, 1900818. [CrossRef]

9. Deng, J.; Liu, Y.; Li, M.; Xu, S.; Lun, Y.; Lv, P.; Xia, T.; Gao, P.; Wang, X.; Hong, J. Thickness-Dependent In-Plane Polarization and Structural Phase Transition in van der Waals Ferroelectric CuInP$_2$S$_6$. *Small* **2020**, *16*, e1904529. [CrossRef]

10. Liu, F.; You, L.; Seyler, K.L.; Li, X.; Yu, P.; Lin, J.; Wang, X.; Zhou, J.; Wang, H.; He, H.; et al. Room-temperature ferroelectricity in CuInP$_2$S$_6$ ultrathin flakes. *Nat. Commun.* **2016**, *7*, 12357. [CrossRef]

11. Si, M.; Liao, P.Y.; Qiu, G.; Duan, Y.; Ye, P.D. Ferroelectric Field-Effect Transistors Based on MoS$_2$ and CuInP$_2$S$_6$ Two-Dimensional van der Waals Heterostructure. *ACS Nano* **2018**, *12*, 6700–6705. [CrossRef]

12. You, L.; Zhang, Y.; Zhou, S.; Chaturvedi, A.; Morris, S.A.; Liu, F.; Chang, L.; Ichinose, D.; Funakubo, H.; Hu, W.; et al. Origin of giant negative piezoelectricity in a layered van der Waals ferroelectric. *Sci. Adv.* **2019**, *5*, eaav3780. [CrossRef]

13. Qi, Y.; Rappe, A.M. Widespread Negative Longitudinal Piezoelectric Responses in Ferroelectric Crystals with Layered Structures. *Phys. Rev. Lett.* **2021**, *126*, 217601. [CrossRef]

14. Brehm, J.A.; Neumayer, S.M.; Tao, L.; O'Hara, A.; Chyasnavichus, M.; Susner, M.A.; McGuire, M.A.; Kalinin, S.V.; Jesse, S.; Ganesh, P.; et al. Tunable quadruple-well ferroelectric van der Waals crystals. *Nat. Mater.* **2020**, *19*, 43–48. [CrossRef]

15. Jiang, X.; Hu, X.; Bian, J.; Zhang, K.; Chen, L.; Zhu, H.; Sun, Q.; Zhang, D.W. Ferroelectric Field-Effect Transistors Based on WSe$_2$/CuInP$_2$S$_6$ Heterostructures for Memory Applications. *ACS Appl. Electron. Mater.* **2021**, *3*, 4711–4717. [CrossRef]

16. Wang, X.; Yu, P.; Lei, Z.; Zhu, C.; Cao, X.; Liu, F.; You, L.; Zeng, Q.; Deng, Y.; Zhu, C.; et al. Van der Waals negative capacitance transistors. *Nat. Commun.* **2019**, *10*, 3037. [CrossRef]

17. Tian, B.; Liu, L.; Yan, M.; Wang, J.; Zhao, Q.; Zhong, N.; Xiang, P.; Sun, L.; Peng, H.; Shen, H.; et al. A Robust Artificial Synapse Based on Organic Ferroelectric Polymer. *Adv. Electron. Mater.* **2019**, *5*, 1800600. [CrossRef]

18. Guo, R.; Zhou, Y.; Wu, L.; Wang, Z.; Lim, Z.; Yan, X.; Lin, W.; Wang, H.; Yoong, H.Y.; Chen, S.; et al. Control of Synaptic Plasticity Learning of Ferroelectric Tunnel Memristor by Nanoscale Interface Engineering. *ACS Appl. Mater. Interfaces* **2018**, *10*, 12862–12869. [CrossRef]

19. Van Breemen, A.J.J.M.; van der Steen, J.-L.; van Heck, G.; Wang, R.; Khikhlovskyi, V.; Kemerink, M.; Gelinck, G.H. Crossbar arrays of nonvolatile, rewritable polymer ferroelectric diode memories on plastic substrates. *Appl. Phys. Express* **2014**, *7*, 031602. [CrossRef]

20. Garcia, V.; Bibes, M. Ferroelectric tunnel junctions for information storage and processing. *Nat. Commun.* **2014**, *5*, 4289. [CrossRef]

21. Schlom, D.G.; Chen, L.-Q.; Eom, C.-B.; Rabe, K.M.; Streiffer, S.K.; Triscone, J.-M. Strain Tuning of Ferroelectric Thin Films. *Annu. Rev. Mater. Res.* **2007**, *37*, 589–626. [CrossRef]

22. Xi, X.; Zhao, L.; Wang, Z.; Berger, H.; Forro, L.; Shan, J.; Mak, K.F. Strongly enhanced charge-density-wave order in monolayer NbSe2. *Nat. Nanotechnol.* **2015**, *10*, 765–769. [CrossRef]

23. Li, L.; Yu, Y.; Ye, G.J.; Ge, Q.; Ou, X.; Wu, H.; Feng, D.; Chen, X.H.; Zhang, Y. Black phosphorus field-effect transistors. *Nat. Nanotechnol.* **2014**, *9*, 372–377. [CrossRef]

24. Maisonneuve, V.; Evain, M.; Payen, C.; Cajipe, V.B.; Molinie, P. Room-temperature crystal structure of the layered phase Cu^{I}In^{III}P$_2$S$_6$. *J. Alloy. Compd.* **1995**, *218*, 157–164. [CrossRef]

25. Li, B.; Li, S.; Wang, H.; Chen, L.; Liu, L.; Feng, X.; Li, Y.; Chen, J.; Gong, X.; Ang, K.-W. An Electronic Synapse Based on 2D Ferroelectric CuInP$_2$S$_6$. *Adv. Electron. Mater.* **2020**, *6*, 2000760. [CrossRef]

26. Ma, R.-R.; Xu, D.-D.; Guan, Z.; Deng, X.; Yue, F.; Huang, R.; Chen, Y.; Zhong, N.; Xiang, P.-H.; Duan, C.-G. High-speed ultraviolet photodetectors based on 2D layered CuInP$_2$S$_6$ nanoflakes. *Appl. Phys. Lett.* **2020**, *117*, 131102. [CrossRef]

27. Vysochanskii, Y.M.; Stephanovich, V.A.; Molnar, A.A.; Cajipe, V.B.; Bourdon, X. Raman spectroscopy study of the ferrielectric-paraelectric transition in layered CuInP$_2$S$_6$. *Phys. Rev. B* **1998**, *58*, 9119–9124. [CrossRef]

28. Katsouras, I.; Asadi, K.; Li, M.; van Driel, T.B.; Kjaer, K.S.; Zhao, D.; Lenz, T.; Gu, Y.; Blom, P.W.; Damjanovic, D.; et al. The negative piezoelectric effect of the ferroelectric polymer poly(vinylidene fluoride). *Nat. Mater.* **2016**, *15*, 78–84. [CrossRef]

29. Li, Q.; Liu, Y.; Schiemer, J.; Smith, P.; Li, Z.; Withers, R.L.; Xu, Z. Piezoresponse force microscopy studies on the domain structures and local switching behavior of Pb(In$_{1/2}$Nb$_{1/2}$)O$_3$-Pb(Mg$_{1/3}$Nb$_{2/3}$)O$_3$-PbTiO$_3$ single crystals. *Appl. Phys. Lett.* **2011**, *98*, 092908. [CrossRef]

30. Neumayer, S.M.; Eliseev, E.A.; Susner, M.A.; Tselev, A.; Rodriguez, B.J.; Brehm, J.A.; Pantelides, S.T.; Panchapakesan, G.; Jesse, S.; Kalinin, S.V.; et al. Giant negative electrostriction and dielectric tunability in a van der Waals layered ferroelectric. *Phys. Rev. Mater.* **2019**, *3*, 024401. [CrossRef]

31. Xu, D.-D.; Ma, R.-R.; Zhao, Y.-F.; Guan, Z.; Zhong, Q.-L.; Huang, R.; Xiang, P.-H.; Zhong, N.; Duan, C.-G. Unconventional out-of-plane domain inversion via in-plane ionic migration in a van der Waals ferroelectric. *J. Mater. Chem. C* **2020**, *8*, 6966–6971. [CrossRef]

32. Zhang, D.; Luo, Z.-D.; Yao, Y.; Schoenherr, P.; Sha, C.; Pan, Y.; Sharma, P.; Alexe, M.; Seidel, J. Anisotropic ion migration and electronic conduction in van der Waals ferroelectric $CuInP_2S_6$. *Nano Lett.* **2021**, *21*, 995–1002. [CrossRef] [PubMed]
33. Yu, J.; Huang, B.; Li, A.; Duan, S.; Jin, H.; Ma, M.; Ou, Y.; Xie, S.; Liu, Y.; Li, J. Resolving local dynamics of dual ions at the nanoscale in electrochemically active materials. *Nano Energy* **2019**, *66*, 104160. [CrossRef]
34. Chen, Q.N.; Liu, Y.; Liu, Y.; Xie, S.; Cao, G.; Li, J. Delineating local electromigration for nanoscale probing of lithium ion intercalation and extraction by electrochemical strain microscopy. *J. Appl. Phys. Lett.* **2012**, *101*, 063901.

Article

Flexible Lead-Free Ba$_{0.5}$Sr$_{0.5}$TiO$_3$/0.4BiFeO$_3$-0.6SrTiO$_3$ Dielectric Film Capacitor with High Energy Storage Performance

Wenwen Wang [1,†], **Jin Qian** [1,†], **Chaohui Geng** [1], **Mengjia Fan** [1], **Changhong Yang** [1,*], **Lingchao Lu** [1] **and Zhenxiang Cheng** [2]

[1]   Shandong Provincial Key Laboratory of Preparation and Measurement of Building Materials, University of Jinan, Jinan 250022, China; wangwenwen_0717@163.com (W.W.); j28_qian@163.com (J.Q.); g1902803618@163.com (C.G.); fan15176103198@163.com (M.F.); mse_lulc@ujn.edu.cn (L.L.)

[2]   Institute for Superconducting and Electronic Materials, Australian Institute for Innovative Materials, University of Wollongong, Innovation Campus, North Wollongong, NSW 2500, Australia; cheng@uow.edu.au

*   Correspondence: mse_yangch@ujn.edu.cn

†   These authors contributed equally to this work.

Abstract: Ferroelectric thin film capacitors have triggered great interest in pulsed power systems because of their high-power density and ultrafast charge–discharge speed, but less attention has been paid to the realization of flexible capacitors for wearable electronics and power systems. In this work, a flexible Ba$_{0.5}$Sr$_{0.5}$TiO$_3$/0.4BiFeO$_3$-0.6SrTiO$_3$ thin film capacitor is synthesized on mica substrate. It possesses an energy storage density of $W_{rec} \sim 62\,\mathrm{J\,cm^{-3}}$, combined with an efficiency of $\eta \sim 74\%$ due to the moderate breakdown strength (3000 kV cm^{-1}) and the strong relaxor behavior. The energy storage performances for the film capacitor are also very stable over a broad temperature range (-50–200 °C) and frequency range (500 Hz–20 kHz). Moreover, the W_{rec} and η are stabilized after 10^8 fatigue cycles. Additionally, the superior energy storage capability can be well maintained under a small bending radius ($r = 2$ mm), or after 10^4 mechanical bending cycles. These results reveal that the Ba$_{0.5}$Sr$_{0.5}$TiO$_3$/0.4BiFeO$_3$-0.6SrTiO$_3$ film capacitors in this work have great potential for use in flexible microenergy storage systems.

Keywords: flexible; film capacitor; Ba$_{0.5}$Sr$_{0.5}$TiO$_3$/0.4BiFeO$_3$-0.6SrTiO$_3$; energy storage properties

Citation: Wang, W.; Qian, J.; Geng, C.; Fan, M.; Yang, C.; Lu, L.; Cheng, Z. Flexible Lead-Free Ba$_{0.5}$Sr$_{0.5}$TiO$_3$/0.4BiFeO$_3$-0.6SrTiO$_3$ Dielectric Film Capacitor with High Energy Storage Performance. *Nanomaterials* **2021**, *11*, 3065. https://doi.org/10.3390/nano11113065

Academic Editor: Sergio Brutti

Received: 11 October 2021
Accepted: 8 November 2021
Published: 14 November 2021

Publisher's Note: MDPI stays neutral with regard to jurisdictional claims in published maps and institutional affiliations.

1. Introduction

At present, the energy crisis and environmental pollution have aroused widespread concern. In order to solve these problems, it is necessary to develop and utilize clean and sustainable energy sources and energy storage devices [1–4]. At present, advanced energy storage techniques include batteries, superconducting magnetic energy storage systems and electrochemical/dielectric capacitors [5,6]. Among them, dielectric capacitors are attracting immense research interest in pulsed power systems due to their unique features of high-power density (up to 10^8 W kg^{-1}) and short charge–discharge time (10^{-3}–10^{-6} s) [7–12].

For dielectric capacitors, the energy storage capability (recoverable energy storage density W_{rec}, energy storage efficiency η) can be calculated by [13,14]:

$$W_{rec} = \int_{P_r}^{P_m} E\,dP \tag{1}$$

$$W = \int_{0}^{P_m} E\,dP \tag{2}$$

$$\eta = \frac{W_{rec}}{W} \times 100\% = \frac{W_{rec}}{W_{rec} + W_{loss}} \times 100\% \tag{3}$$

where W, W_{loss}, E, P_m and P_r represent the total energy storage density, the energy loss density, applied electric field, the maximum polarization and the remanent polarization during the discharge process, respectively. Therefore, W_{rec} can be improved by increasing the difference between P_m and P_r and enhancing the electric breakdown strength (E_b).

Currently, commercial biaxially oriented polypropylene (BOPP) has been widely employed in power inverter capacitor systems. The bottleneck problems faced by BOPP are its limited W_{rec} of (<2 J cm^{-3}) and poor thermal stability (<80 °C), which inevitably burden the weight of the device, increase the difficulty of structure design and narrow the working temperature window [15–17]. In recent years, inorganic dielectric capacitors using oxide thin films as functional elements have been widely studied due to their relatively high W_{rec} compared with organic dielectrics. Recent investigations into the energy storage characteristics of several representative dielectric capacitors have been summarized and listed in Table 1. Obviously, energy storage properties such as W_{rec} and η have been studied in film capacitors containing BiFeO$_3$(BFO)/BaTiO$_3$(BTO)/SrTiO$_3$(STO). For example, Huang et al. reported that introducing Sr in BTO can effectively reduce the coercive field (E_c) and P_r, leading to an enhanced W_{rec} and η [18]. Pan et al. demonstrated that a giant W_{rec} of ~70 J cm^{-3}, together with a high η, can be achieved in lead-free 0.4BFO-0.6STO films through domain engineering [6]. They also designed a 0.25BFO-0.3BTO-0.45STO ternary solid solution system, in which a high W_{rec} of up to 112 J cm^{-3} and an η of ~80% were obtained coexistence of the rhombohedral and tetragonal nanodomains in a cubic matrix [19]. Moreover, in Bi doped STO, ferroelectric relaxation behavior is observed, which plays a decisive role in the high energy storage property, especially the ultra-high η [20–23].

Table 1. Comparison of energy storage performance of different types of materials.

	Materials	Substrate	P_m-P_r (μC/cm^2)	E (kV·cm^{-1})	W_{rec} (J·cm^{-3})	η (%)	T (°C)	Fatigue (Cycles)	Bending Test	Ref.
organic	P(VDF-TrFE-CTFE)		~10	4000	9		<150			[19]
	VDF/PVDF		~8	8200	27.3	67	<85			[24,25]
	P(VDF-CTFE)		9	5750	17					[26]
	P(MDA/MDI)		~3.2	8000	12	>90%	RT-180			[27]
inorganic	BST-BF	Pt/Ti/SiO$_2$/Si	~30	4800	48.5	47.57	30–120			[28]
	SBTMO	Pt/Ti/SiO$_2$/Si	34.3	1380	24.4	87	30–110			[20]
	BFO-STO	Nb:SrTiO$_3$	~45	3850	70.3	70	−50–100	10^7		[6]
	BZT/BZT	Nb:SrTiO$_3$	~33	7500	83.9	78.4	−100–200	10^6		[29]
	PBZ	Pt/TiO$_x$/SiO$_2$/Si	65	2801	40.18	64.1	−23–250			[30]
	PLZST	(La$_{0.7}$Sr$_{0.3}$)MnO$_3$/Al$_2$O$_3$(0001)	55	4000	46.3	84	27–107	10^5		[31]
	HZO	SiO$_2$/Si	30	4500	46	53	25–175	10^9		[32]
	Si-HZO	Si		3500	50	80	25–125	10^9		[33]
	BBTO	Pt/Si	40	2000	43.3	87.1	20–140	10^8		[34]
	NBT-BT/BFO	Pt/TiO$_x$/SiO$_2$/Si	43.19	2400	31.96	61	25–120			[35]
	Mn:NBT-BT-BFO	Pt/F-mica	97.8	2285	81.9	64.4	25–200	10^9	r = 2 mm or 10^3 at r = 4 mm	[36]
	PLZT	LaNiO$_3$/F-Mica	~64	1998	40.2	58	30–180	10^7	2 × 10^3 at r = 4.5 mm	[37]
	BZT	Indium Tin Oxide (ITO)/F-mica	~25	4230	40.6	68.9	−120–150	10^6	r = 4 mm or 10^3 at r = 4 mm	[38]
	BST/0.4BFO-0.6STO	Pt/F-mica	56.79	3000	62	74	−50–200	10^8	r = 2 mm or 10^4 at r = 4 mm	This work

Poly(vinylidene fluoride-trifluoroethylene-chlorofluoroethylene) (P(VDF-TrFE-CTFE), vinylidene fluoride/Poly(vinylidene fluoride) (VDF/PVDF), Poly(vinylidene fluoride- chlorofluoroethylene) (P(VDF-CTFE), poly(diaminodiphenylmethane-diphenylmethane diisocyanate) (P(MDA/MDI), 0.1BiFeO$_3$-0.9Bi$_{0.2}$Sr$_{0.7}$TiO$_3$ (BST-BF), (Sr$_{0.85}$Bi$_{0.1}$)Ti$_{0.99}$Mn$_{0.01}$O$_3$ (SBTMO), 0.4BiFeO$_3$-0.6SrTiO$_3$ (BFO-STO), BaZr$_{0.15}$Ti$_{0.85}$O$_3$/BaZr$_{0.35}$Ti$_{0.65}$O$_3$ (BZT/BZT), Pb$_{0.8}$Ba$_{0.2}$ZrO$_3$ (PBZ), Pb$_{0.97}$La$_{0.02}$Zr$_{0.66}$Sn$_{0.23}$Ti$_{0.11}$O$_3$ (PLZST), Hf$_{0.3}$Zr$_{0.7}$O$_2$ (HZO), Si-Hf$_{0.5}$Zr$_{0.5}$O$_2$ (Si-HZO), BaBi$_4$Ti$_4$O$_{15}$ (BBTO), 0.94(Bi$_{0.5}$Na$_{0.5}$)$_{0.94}$TiO$_3$-0.06BaTiO$_3$/BiFeO$_3$ (NBT-BT/BFO), 0.97(0.93Na$_{0.5}$Bi$_{0.5}$TiO$_3$-0.07BaTiO$_3$)-0.03BiFeO$_3$ (Mn:NBT-BT-BFO), Pb$_{0.91}$La$_{0.09}$(Zr$_{0.65}$Ti$_{0.35}$)$_{0.9775}$O$_3$ (PLZT), Ba(Zr$_{0.35}$Ti$_{0.65}$)O$_3$ (BZT), Ba$_{0.5}$Sr$_{0.5}$TiO$_3$/0.4BiFeO$_3$-0.6SrTiO$_3$ (BST/0.4BFO-0.6STO).

With the rapid development of electronic devices leaning toward miniaturization and integration, flexible electronics have been an active research topic in various areas due to their distinctive advantages of being portable, lightweight, foldable, stretchable and even wearable [39–44]. Flexible and microscale dielectric capacitors as energy storage components are indispensable especially in next-generation micro-electrical power systems. Nevertheless, most inorganic dielectric films are grown on rigid substrates due to the lack of suitable flexible substrates. Common flexible polymer substrates, such as polyimide (PI) or polyethylene naphthalate (PEN), have very excellent mechanical compliance but they cannot withstand the high crystallization temperature of inorganic films due to their low melting point (PI ~ 520 °C, PEN ~ 270 °C). Fortunately, the emergence of MICAtronics provides a new idea to realize flexibility in oxide functional films with two-dimensional mica as the substrate. This is due to the fact mica possesses ultrahigh melting point (1000 °C–1100 °C) and atomically flat surface, making it more compatible with the inorganic thin film preparation process.[45–47]. However, a flexible dielectric film capacitor consisting of BFO, BTO, and STO elements has rarely been reported.

Considering that $0.4BiFeO_3$-$0.6SrTiO_3$ (0.4BFO-0.6STO) is a relaxor ferroelectric with an attractive relaxor feature and $Ba_{0.5}Sr_{0.5}TiO_3$ (BST) is paraelectric with low dielectric loss and high breakdown strength [6,48,49], a multilayer structure of BST/0.4BFO-0.6STO is envisaged in this work based on the two potential energy storage elements. A series of systematic studies about the energy storage capability are undertaken on the designed film, which is deposited on flexible mica substrate using a sol-gel method. The capacitor shows a high W_{rec} of ~62 J cm^{-3} and an η of ~74% simultaneously due to its relatively high E_b of 3000 kV cm^{-1} and strong relaxor behavior. Satisfyingly, prominent mechanical-bending resistance is also realized in the flexible BST/0.4BFO-0.6STO film, in which the W_{rec} and η have no obvious deterioration under various bending radii (r = 12–2 mm) and even after 10^4 bending cycles at r = 4 mm.

2. Materials and Methods

2.1. Film Fabrication

Firstly, the flexible mica substrate coated with bottom electrode was provided for depositing dielectric thin film. The fluorophlogopite mica [$KMg_3(AlSi_3O_{10})F_2$] was purchased from Changchun Taiyuan Fluorophlogopite Co., Ltd. (Changchun, China). The mica sheet was washed with ethanol and water to get a cleaned surface. Then, a 20 nm thick Pt layer was sputtered onto the surface under a 30 mA current in Ar atmosphere of 0.05 mbar, to be used as the bottom electrode.

The multilayer BST/0.4BFO-0.6STO thin film was fabricated on Pt/mica substrate by sol-gel. Precursor solutions of BFO, BTO and STO were prepared, respectively, with the use of bismuth nitrate pentahydrate, iron nitrate nonahydrate, strontium acetate, barium acetate and tetrabutyl titanate. Ethylene glycol and acetic acid were selected as solvents to dissolve the solid raw materials. Here, 5 mol% excess bismuth was added to compensate for element volatilization during the high temperature treatment. Subsequently, the tetrabutyl titanate and acetylacetone were added into the solution. Meanwhile, 2 mol% manganese acetate tetrahydrate was added to each solution to improve the electrical resistivity of the film. The final concentration of each precursor solution was 0.15 M. Then, we used proportionable BFO and BTO solutions, separately mixed with STO, to form solutions of 0.4BFO-0.6STO and BST. The solutions were stirred with a magnetic stirrer for 12 h and further aged for another 48 h. The BST layer was first spin-coated on the substrate, and 0.4BFO-0.6STO layers were then deposited in situ on top of the BST layer. Both layers were dried at 200 °C for 2 min, successively. Subsequently, each layer was pyrolyzed on a hot plate at 300 °C for 5 min and annealed in a mini tubular furnace at 700 °C for 10 min. Both components (0.4BFO-0.6STO and BST) were spin coated alternately 10 times, with the ultimate sample consisting of twenty dielectric layers. For electrical measurements, Au top electrodes with a diameter of about 200 μm were sputtered through a shadow mask to form the capacitor structure. Finally, a simple mechanical peeling process was conducted to realize flexibility in the film by tearing off the bottom mica layer to reduce the thickness to ~10 μm.

2.2. Characterization

The crystalline structure of BST/0.4BFO-0.6STO film was monitored in the 2θ range of 20–60° by an X-ray diffractometer (XRD, D8 ADVANCE, Karlsruhe, Germany). During the XRD test, the scanning rate was 0.12 s per step, and the number of scanning steps was a total of 2054 steps. The surface morphology was studied by a tapping mode atomic force microscope (AFM, Bruker Dimension Icon, Santa Barbara, CA, US). The cross-sectional microstructure and EDS spectrum were studied by a field-emission scanning electron microscope (FESEM, ZEISS Gemini300, Oberkochen, Germany) using 2 kV acceleration voltage and 10 kV acceleration voltage, respectively. The polarization electric field (P-E) relations were examined by a standard ferroelectric tester (aixACCT TF3000, Aachen, Germany) at room temperature, at a frequency of 10 kHz. In regard to temperature-dependence polarization properties, the loops were measured from −50 to 200 °C with a temperature interval of 25 °C at 10 kHz. The frequency dependent P-E measurements were conducted at room temperature from 500 Hz to 20 kHz. The dielectric properties were characterized by way of an impedance analyzer (HP4294A, Agilent, Palo Alto, CA, USA) at a temperature range of −50 to 250 °C from 1 kHz to 100 kHz, with an oscillation voltage of 1.0 V. The temperature-related electrical measurements were carried out with the assistance of a temperature-controlled probe station (Linkam-HFS600E-PB2, London, UK) with a heating rate of 8 °C min^{-1}. The cyclic bending tests were realized by using a homebuilt stepper motor control system. The fast energy discharge behavior was evaluated by using a home-built resistance-capacitance (RC) circuit with a load resistance of 100 kΩ.

3. Results

Figure 1a shows the XRD pattern of BST/0.4BFO-0.6STO film grown on Pt/mica substrate. Visually, the film possesses a single perovskite phase with no detectable secondary phase, suggesting that the film can be well crystallized. Figure 1b shows the surface AFM image of BST/0.4BFO-0.6STO film. The average surface roughness (R_a) and root mean square roughness (R_{rms}) of the film are determined to be 2.54 nm and 2.06 nm, respectively, which may be attributed to the atomic flatness of mica substrate and high crystallinity of the film. The obtained roughness is at the same level of the reported inorganic films [36,47]. The grain size distribution of the film is analyzed using the Nano Measurer software by randomly selecting 100 grains. In addition, the average grain size value estimated from the AFM image is 53.26 nm. Figure 1c shows the cross-sectional image of multilayer film. From it, the film's thickness can be determined to be ~350 nm. Furthermore, the thickness of the bottom Pt electrode is about 20 nm. The ultimate film composition of the BST/0.4BFO-0.6STO film is determined via EDS spectrum, as displayed in Figure 1d. The atomic percentages (atom%) of O, Ti, Sr, Ba, Fe and Bi are 59.80, 16.44, 10.91, 5.14, 4.13 and 3.68, respectively, confirming a near perfect BST/0.4BFO-0.6STO stoichiometry.

The bipolar P-E loops for the BST/0.4BFO-0.6STO film in Figure 2a are measured from a low electric field to 3000 kV cm^{-1} at room temperature, at a frequency of 10 kHz. Figure 2b presents the corresponding energy storage parameters of the W, W_{rec}, W_{loss} and η at various electric fields determined by P-E loops. The W_{rec} and η extrapolated from a bipolar P-E loop under E_b (3000 kV/cm) are 62 J cm^{-3} and ~74%, respectively, which is a relatively high level among the flexible dielectric films [37,50]. It can be seen that the P_m and P_r are 63.52 μC cm^{-2} and 6.73 μC cm^{-2}, respectively, which contributed a great $\Delta P = 56.79$ μC cm^{-2}, and the result is beneficial for energy storage performance. This small P_r can be due to the fact that BiFeO$_3$-SrTiO$_3$ is a relaxor ferroelectric and BST is paraelectric.

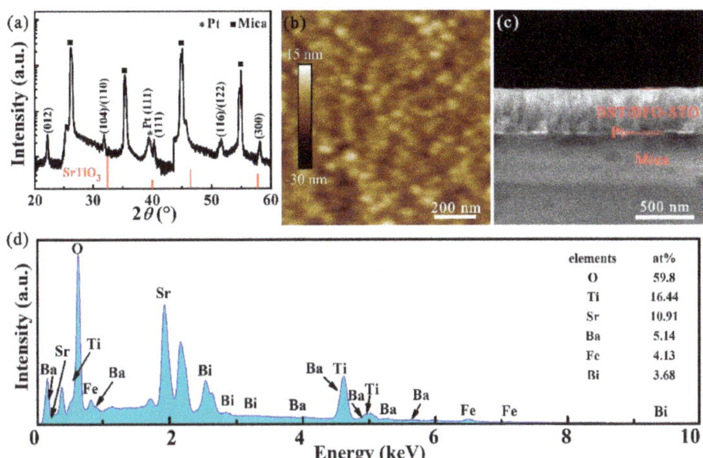

Figure 1. (**a**) X-ray diffraction pattern in the 2θ range of 20–60°, (**b**) AFM, (**c**) Cross-sectional SEM images and (**d**) EDS spectrum of BST/0.4BFO-0.6STO thin film.

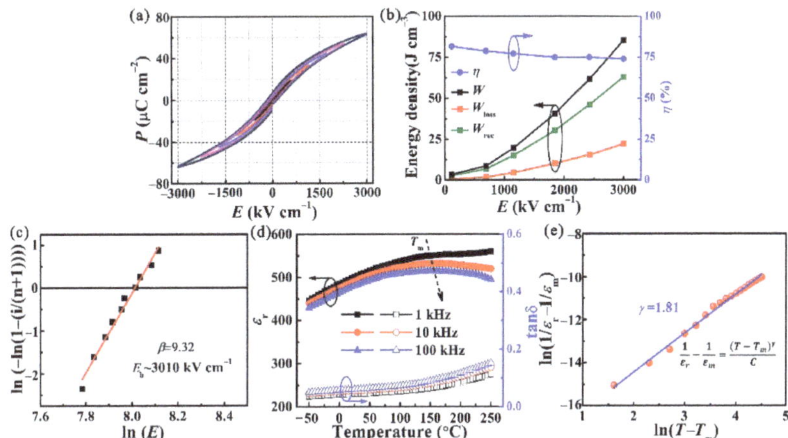

Figure 2. (**a**) The *P-E* loops for BST/0.4BFO-0.6STO under various applied electric fields. (**b**) The calculated W, W_{rec}, W_{loss} and η values as functions of the electric field. (**c**) Two-parameter Weibull analysis of dielectric breakdown strength. (**d**) Temperature-dependent ε_r and tanδ under the frequency range of 1 kHz–100 kHz and the temperature range from −50 to 250 °C. (**e**) $\ln(1/\varepsilon_r - 1/\varepsilon_m)$ as a function of $\ln(T-T_m)$.

The Weibull distribution of E_b can be obtained through the following formula:

$$X_i = Ln(E_i) \tag{4}$$

$$Y_i = Ln\left(-Ln\left(1 - \frac{i}{n+1}\right)\right) \tag{5}$$

where E_i, i and n signify the breakdown electric field, the serial number of tested specimens and the total number of tested specimens, respectively. Based on the Weibull distribution function, there exists a linear relationship between X_i and Y_i. The mean E_b for thin film can be extracted from the intersect points of the fitting lines and the horizontal axis at $Y_i = 0$. The solid fitting straight line shown in Figure 2c is the Weibull analysis result of ten data gathered from our thin film. It can be observed that the slope parameter β

is 9.32, which indicates both the good composition uniformity and high dielectric reliability of BST/0.4BFO-0.6STO [47]. The average E_b extracted by the horizontal intercept is about 3010 kV cm^{-1}. The temperature-dependent dielectric permittivity (ε_r) and loss (tan δ) of the BST/0.4BFO-0.6STO film exhibit nearly flat permittivity peaks and frequency dispersion over the range of -50 to 250 °C, as shown in Figure 2d, indicating the relaxor characteristic. Notably, a broad and smeared peak of maximum ε_r appears, especially near 150 °C. With increasing frequency, the maximum dielectric permittivity (ε_m) at T_m decreases and T_m shifts to a higher temperature, which are important signatures of relaxor behavior [48]. To evaluate the relaxor dispersion degree, a modified Curie–Weiss equation of $1/\varepsilon_r - 1/\varepsilon_m = (T - T_m)^\gamma/C$ can be used to estimate the relaxor dispersion degree, where ε_m represents the maximum dielectric constant at T_m, C is the Curie constant and γ is the relaxor diffuseness factor. Generally, $\gamma = 1$ represents a normal ferroelectric, $1 \leq \gamma \leq 2$ represents the relaxor ferroelectric behavior and $\gamma = 2$ is valid for a classical ferroelectric relaxor [49]. After calculation, the γ for the film is 1.81 in Figure 2e, further evidencing the relaxor feature.

The temperature and frequency stability, as well as the antifatigue property for the sample, are evaluated, as shown in Figure 3. Firstly, the *P-E* hysteresis loops are measured at 10 kHz under 2286 kV cm^{-1} in the temperature range of -50 to 200 °C. As illustrated in Figure 3a, the *P-E* loops almost preserve their pinched shape, and the P_m and P_r values have tiny changes. Correspondingly, the W_{rec} and η of BST/0.4BFO-0.6STO films fluctuate slightly by 11% and 5% as shown in Figure 3b, which indicates the excellent thermal stability of the energy performance of the film. In practical application, it is necessary to meet the working temperature range of capacitors; for example, when in use in the fields of hybrid electric vehicles (~140 °C), drilling operations (150–200 °C), or in outer space and high-altitude aircraft (~-50 °C) [1,51–53]. The obtained temperature range in our film can basically fulfil the requirement. Furthermore, as more attention is paid to electronics technology, the requirement of reliability under high/low frequencies is highlighted. The room temperature frequency dependent *P-E* loops are displayed in Figure 3c. When the measured frequency rises from 500 Hz to 20 kHz, the changes of the W_{rec} and η values are only 9% and 2%, respectively, as shown in Figure 3d. Furthermore, the energy storage performance of the capacitor in long-term working conditions is also a key requirement for practical application. To evaluate its long-term charging–discharging stability, the fatigue endurance of BST/0.4BFO-0.6STO film is evaluated under 10 kHz at room temperature. The *P-E* loops of samples over 10^8 charge–discharge cycles are exhibited in Figure 3e. It can be seen that there is no obvious change in the hysteresis loop. The corresponding W_{rec} and η present a negligible degradation of 6% and 2%, respectively, as shown in Figure 3f. The weak dependence of the energy storage performance on the temperature, frequency and fatigue cycles makes the BST/0.4BFO-0.6STO thin film more competent to work in different complex environments.

It is generally believed that the bending strain S can be calculated using the equation $S = (t_f + t_s)/2r$ [54,55], where t_f is the film thicknesses, t_s is the substrate thicknesses and r is the bending radius of the sample. The t_f and t_s for the BST/0.4BFO-0.6STO sample are ~350 nm and ~10 μm, respectively. Due to the limitations of stripping mica technology, the minimum bent radius of mica is 2 mm. In this curved state, the calculated S (~0.25%) is much less than the strain limit that the oxide film can withstand [56]. The mechanical stability of the BST/0.4BFO-0.6STO film is further evaluated under flex-in (compressive strain) and flex-out (tensile strain) modes at 2286 kV cm^{-1} and 10 kHz with different bending radii (from 12 mm to 2 mm), as depicted in Figure 4a,b. Then, home-made molds with different required bending radii are used to test mechanical stability. It can be seen that the *P-E* loops keep its slim feature without obvious deterioration regardless of what compressive strain or tensile strain it is under. As plotted in Figure 4c, when the bending radius decreases from 12 mm to 2 mm, the corresponding W_{rec} and η variations are both within 1%, indicating that the film possesses excellent bendability. The discharge energy density–time plots under various compressive and tensile radii are shown in Figure 4d,e.

Obviously, all curves are very similar. Figure 4f shows the bending radius dependence of the discharged energy density and the discharge speed $t_{0.9}$. The BST/0.4BFO-0.6STO film possesses a high discharged energy density (W_{dis}) of ~32 J cm^{-3}. Further, it can deliver the energy in ~40 μs without significant differences with the change of bending radius, exhibiting a fast charge–discharge rate and mechanical bending endurance.

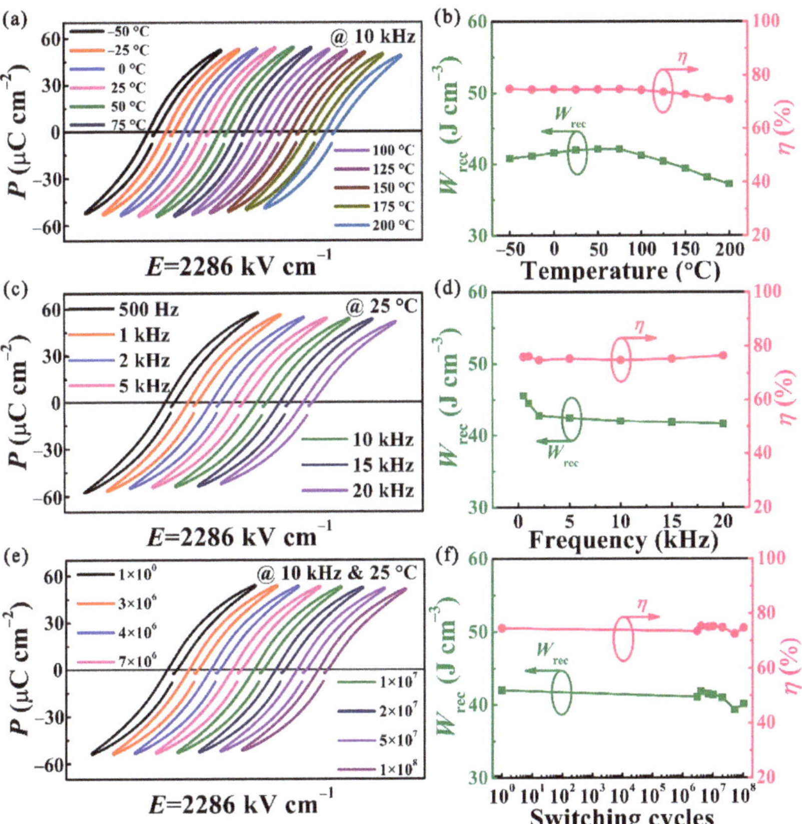

Figure 3. (a) The *P-E* curves and (b) the corresponding W_{rec} and η measured from −50 to 200 °C at 2286 kV cm^{-1}. (c) The *P-E* curves and (d) the corresponding W_{rec} and η with various frequencies measured under 2286 kV cm^{-1}. (e) The *P-E* curves and (f) the corresponding W_{rec} and η during the 10^8 fatigue cycles at 2286 kV cm^{-1}. The measurements are realized at about 76% of E_b.

Figure 5a,b presents the *P-E* loops of the BST/0.4BFO-0.6STO sample in the flat and re-flatted after experiencing repeated bending at r = 4 mm. Over the course of 10^4 cycles, nearly unchanged *P-E* hysteresis shapes are observed, guaranteeing high mechanical stability of the energy storage performances. As demonstrated in Figure 5c, the variations of the W_{rec} and η are negligible, further ascertaining its bending–endurance property. Finally, the influences of the ferroelectric fatigue endurance are investigated with r = 4 mm (Figure 5d,e). The energy storage performance is apparently undamaged even after 10^8 switching cycles at a radius as small as of 4 mm.

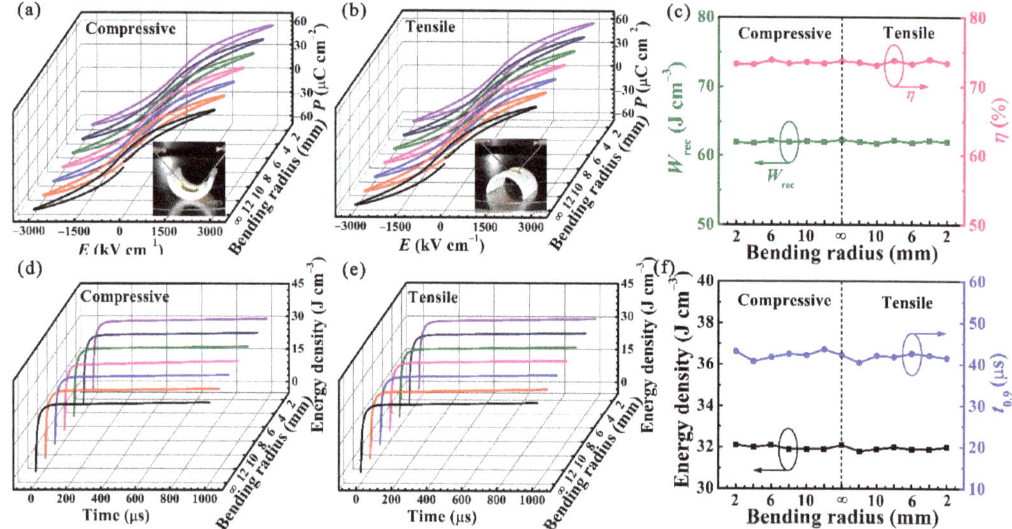

Figure 4. *P-E* loops measured at various (**a**) compressive radii and (**b**) tensile radii. The inset is the photographs of the BST/0.4BFO-0.6STO film under different bending states. (**c**) W_{rec} and η as functions of the bending radius. The energy discharge behaviors at various (**d**) compressive radii and (**e**) tensile radii. (**f**) Discharged energy density and discharge speed as functions of bending radius. (The lines in Figure 4a,b,d,e from black to purple represent the measurements of the bending radius of BST/0.4BFO-0.6STO film from ∞ to 2 mm, respectively.)

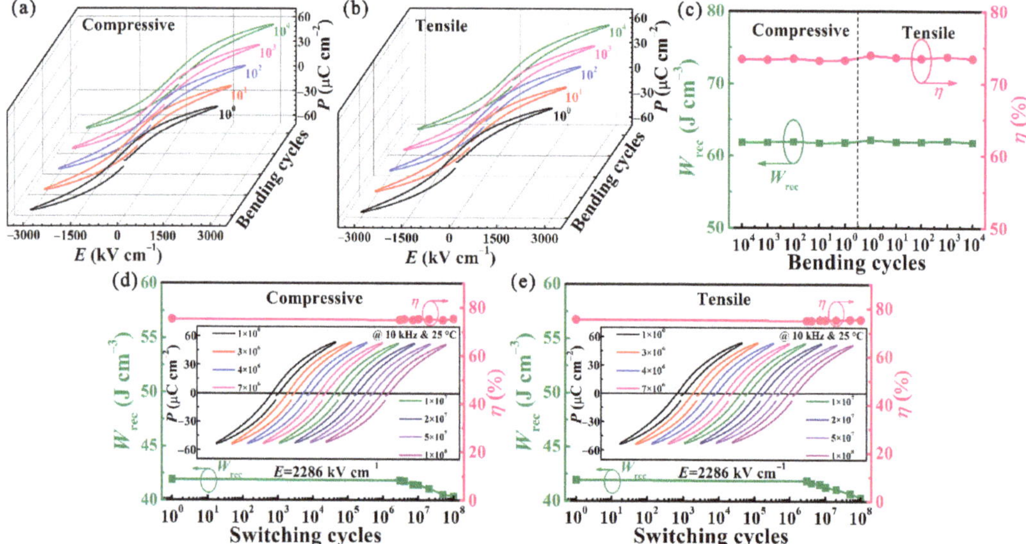

Figure 5. (**a,b**) *P-E* loops of BST/0.4BFO-0.6STO film under various bending cycles under the compressive and tensile states. (**c**) W_{rec} and η as functions of bending number. (**d,e**) W_{rec} and η as functions of switching cycle during 10^8 fatigue cycles under compressive and tensile bending states with $r = 4$ mm. The insets are the corresponding *P-E* loops. (The lines in Figure 5a,b from black to green represent the measurements of the bending cycles of BST/0.4BFO-0.6STO film from 10^0 to 10^4, respectively.)

Finally, the core parameters of E_b, W_{rec} and η for energy storage properties are compared with some previously reported representative dielectrics (Figure 6). As depicted in Figure 6a, the BST/0.4BFO-0.6STO film exhibits relatively high W_{rec} of 62 J cm^{-3} at a moderate E of 3000 kV cm^{-1}, which is much higher than HZO (46 J cm^{-3}), BST-BF (48.5 J cm^{-3}) and NBT-BT/BFO (31.96 J cm^{-3}) [28,32,35], but slightly inferior to Mn: NBT-BT-BFO (81.9 J cm^{-3}) and BFO-STO (70.3 J cm^{-3}) [6,36]. In Figure 6b, it can be seen that the obtained η of 74% in this work is lower than the reported dielectrics on rigid substrate, such as SBTMO (87%), BBTO (87.1%) and PLZST (84%) [20,31,34] but reaches a relatively high level among all the currently reported bendable inorganic dielectric film capacitors. In view of the aforesaid observations, there is still much room for improvement of the η in the flexible film capacitors, and it needs further research.

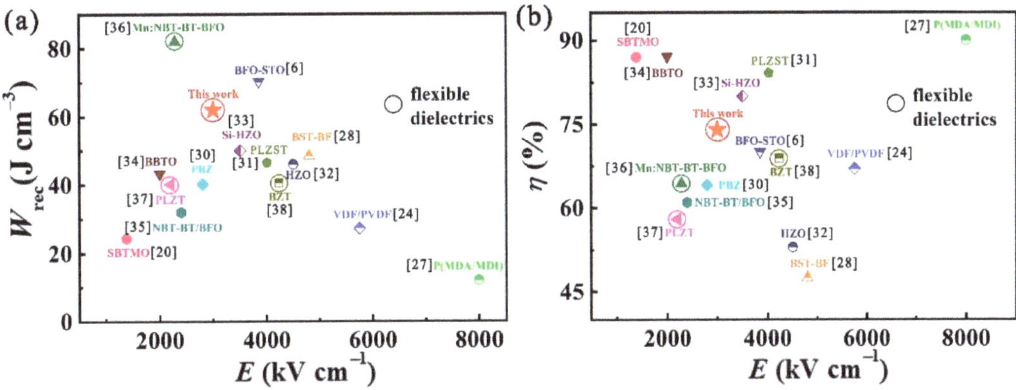

Figure 6. A comparison of the (**a**) W_{rec} and (**b**) η of the flexible BST/0.4BFO-0.6STO film capacitor reported in this study and a number of film capacitors reported previously.

4. Conclusions

In this work, a high W_{rec} of ~62 J cm^{-3} and an η of ~74% are achieved in the BST/0.4BFO-0.6STO film. The film shows superior thermal stability (from -50 to 200 °C), frequency reliability (from 500 Hz to 20 kHz) and fatigue endurance (10^8 cycles). Most importantly, prominent mechanical stability can also be obtained. The energy storage behaviors show no obvious deterioration after undergoing different bending radii (from 12 to 2 mm), and even after 10^4 bending cycles. All of these outstanding performances demonstrate that the designed flexible BST/0.4BFO-0.6STO thin film is expected to pave the way for its application in flexible energy storage electronic devices.

Author Contributions: Conceptualization, C.Y., L.L. and Z.C.; methodology, C.Y.; investigation, W.W., J.Q. and C.G.; resources, C.Y.; data curation, W.W., J.Q. and M.F.; writing—original draft preparation, W.W. and J.Q.; writing—review and editing, C.Y. and Z.C.; supervision, C.Y., L.L. and Z.C. All authors have read and agreed to the published version of the manuscript.

Funding: This work was supported by the National Natural Science Foundation of China (Grant No. 51972144), Shandong Provincial Natural Science Foundation (Grant No. ZR2020KA003), the Project of "20 Items of University" of Jinan (Grant No. 2020GXRC051) and the Australian Research Council (DP190100150).

Conflicts of Interest: The authors declare no conflict of interest.

References

1. Li, Q.; Chen, L.; Gadinski, M.R.; Zhang, S.; Zhang, G.; Li, H.U.; Iagodkine, E.; Haque, A.; Chen, L.-Q.; Jackson, T.N.; et al. Flexible high-temperature dielectric materials from polymer nanocomposites. *Nature* **2015**, *523*, 576–579. [CrossRef]
2. Yan, F.; Bai, H.R.; Zhou, X.F.; Ge, G.L.; Li, G.H.; Shen, B.; Zhai, J.W. Realizing superior energy storage properties in lead-free ceramics via a macro-structure design strategy. *J. Mater. Chem. A* **2020**, *8*, 11656. [CrossRef]
3. Zou, K.; Dan, Y.; Xu, H.; Zhang, Q.; Lu, Y.; Huang, H.; He, Y. Recent Advances in Lead-Free Dielectric Materials for Energy Storage. *Mater. Res. Bull.* **2019**, *113*, 190–201. [CrossRef]

4. Dang, Z.-M.; Yuan, J.-K.; Yao, S.-H.; Liao, R.-J. Flexible nanodielectric materials with high permittivity for power energy storage. *Adv. Mater.* **2013**, *25*, 6334–6365. [CrossRef]

5. Thackeray, M.M.; Wolverton, C.E.; Isaacs, D. Electrical energy storage for transportation- approaching the limits of, and going beyond, lithium-ion batteries. *Energy Environ. Sci.* **2012**, *57*, 7854–7863. [CrossRef]

6. Pan, H.; Ma, J.; Ma, J.; Zhang, Q.H.; Liu, X.Z.; Guan, B.; Gu, L.; Zhang, X.; Zhang, Y.-J.; Li, L.L.; et al. Giant energy density and high efficiency achieved in bismuth ferrite-based film capacitors via domain engineering. *Nat. Commun.* **2018**, *9*, 1813. [CrossRef]

7. Yan, F.; Bai, H.R.; Shi, Y.J.; Ge, G.L.; Zhou, X.F.; Lin, J.F.; Shen, B.; Zhai, J.W. Sandwich structured lead-free ceramics based on $Bi_{0.5}Na_{0.5}TiO_3$ for high energy storage. *Chem. Eng. J.* **2021**, *425*, 130669. [CrossRef]

8. Liu, M.L.; Zhu, H.F.; Zhang, Y.X.; Xue, C.H.; Ouyang, J. Energy Storage Characteristics of $BiFeO_3/BaTiO_3$ Bi-Layers Integrated on Si. *Materials* **2016**, *9*, 935. [CrossRef] [PubMed]

9. Yang, L.; Kong, X.; Li, F.; Hao, H.; Cheng, Z.; Liu, H.; Li, J.-F.; Zhang, S. Perovskite lead-free dielectrics for energy storage applications. *Prog. Mater. Sci.* **2019**, *102*, 72–108. [CrossRef]

10. Zhang, Y.; Feng, H.; Wu, X.B.; Wang, L.Z.; Zhang, A.Q.; Xia, T.C.; Dong, H.C.; Li, X.F.; Zhang, L.S. Progress of electrochemical capacitor electrode materials: A review. *Int. J. Hydrogen Energy* **2009**, *34*, 4889–4899. [CrossRef]

11. Yan, F.; Huang, K.W.; Jiang, T.; Zhou, X.F.; Shi, Y.J.; Ge, G.L.; Shen, B.; Zhai, J.W. Significantly enhanced energy storage density and efficiency of BNT-based perovskite ceramics via A-site defect engineering. *Energy Storage Mater.* **2020**, *30*, 392–400. [CrossRef]

12. Cheng, H.B.; Ouyang, J.; Zhang, Y.-X.; Ascienzo, D.; Li, Y.; Zhao, Y.-Y.; Ren, Y.H. Demonstration of ultra-high recyclable energy densities in domain-engineered ferroelectric films. *Nat. Commun.* **2017**, *8*, 1999. [CrossRef]

13. Burn, I.; Smyth, D.M. Energy Storage in Ceramic Dielectrics. *J. Mater. Sci.* **1972**, *7*, 339–343. [CrossRef]

14. Yang, C.H.; Qian, J.; Han, Y.J.; Lv, P.P.; Huang, S.F.; Cheng, X.; Cheng, Z.X. Design of an all-inorganic flexible $Na_{0.5}Bi_{0.5}TiO_3$-based film capacitor with giant and stable energy storage performance. *J. Mater. Chem. A* **2019**, *7*, 22366. [CrossRef]

15. Chen, J.; Wang, Y.F.; Yuan, Q.B.; Xu, X.W.; Niu, Y.J.; Wang, Q.; Wang, H. Multilayered ferroelectric polymer films incorporating low-dielectric constant components for concurrent enhancement of energy density and charge–discharge efficiency. *Nano Energy* **2018**, *54*, 288–296. [CrossRef]

16. Li, H.; Ai, D.; Ren, L.L.; Yao, B.; Han, Z.B.; Shen, Z.H.; Wang, J.J.; Chen, L.-Q.; Wang, Q. Scalable Polymer Nanocomposites with Record High-Temperature Capacitive Performance Enabled by Rationally Designed Nanostructured Inorganic Fillers. *Adv. Mater.* **2019**, *31*, 1900875. [CrossRef] [PubMed]

17. Sun, Z.X.; Wang, L.X.; Liu, M.; Ma, C.R.; Liang, Z.S.; Fan, Q.L.; Lu, L.; Lou, X.J.; Wang, H.; Jia, C.-L. Interface thickness optimization of lead-free oxide multilayer capacitors for high-performance energy storage. *J. Mater. Chem. A* **2018**, *6*, 1858–1864. [CrossRef]

18. Huang, Y.H.; Wang, J.J.; Yang, T.N.; Wu, Y.J.; Chen, X.M.; Chen, L.Q. A thermodynamic potential, energy storage performances, and electrocaloric effects of $Ba_{1-x}Sr_xTiO_3$ single crystals. *Appl. Phys. Lett.* **2018**, *112*, 102901. [CrossRef]

19. Pan, H.; Li, F.; Liu, Y.; Zhang, Q.; Wang, M.; Lan, S.; Zheng, Y.; Ma, J.; Gu, L.; Shen, Y.; et al. Ultrahigh-energy density lead-free dielectric films via polymorphic nanodomain design. *Science* **2019**, *365*, 578–582. [CrossRef]

20. Yang, X.R.; Li, W.L.; Qiao, Y.L.; Zhang, Y.L.; He, J.; Fei, W.D. High energy-storage density of lead-free $(Sr_{1-1.5x}Bi_x)Ti_{0.99}Mn_{0.01}O_3$ thin films induced by Bi^{3+}-V_{Sr} dipolar defects. *Phys. Chem. Chem. Phys.* **2019**, *21*, 16359–16366. [CrossRef]

21. Chen, A.; Zhi, Y. High, Purely Electrostrictive Strain in Lead-Free Dielectrics. *Adv. Mater.* **2006**, *18*, 103–106.

22. Chen, A.; Zhi, Y. Dielectric relaxor and ferroelectric relaxor: Bi-doped paraelectric $SrTiO_3$. *J. Appl. Phys.* **2002**, *91*, 1487.

23. Okhay, O.; Wu, A.Y.; Vilarinho, P.M.; Tkach, A. Dielectric relaxation of $Sr_{1-1.5x}Bi_xTiO_3$ sol-gel thin films. *J. Appl. Phys.* **2011**, *109*, 064103. [CrossRef]

24. Rahimabady, M.; Chen, S.T.; Yao, K.; Tay, F.E.H.; Lu, L. High electric breakdown strength and energy density in vinylidene fluoride oligomer/poly(vinylidene fluoride) blend thin films. *Appl. Phys. Lett.* **2011**, *99*, 142901. [CrossRef]

25. Peng, B.L.; Zhang, Q.; Li, X.; Sun, T.Y.; Fan, H.Q.; Ke, S.M.; Ye, M.; Wang, Y.; Lu, W.; Niu, H.B.; et al. Giant electric energy density in epitaxial lead-free thin films with coexistence of ferroelectrics and antiferroelectrics. *Adv. Electron. Mater.* **2015**, *1*, 1500052. [CrossRef]

26. Chu, B.J.; Zhou, X.; Ren, K.L.; Neese, B.; Lin, M.R.; Wang, Q.; Bauer, F.; Zhang, Q.M. A dielectric polymer with high electric energy density and fast discharge speed. *Science* **2006**, *313*, 334. [CrossRef]

27. Wang, Y.; Zhou, X.; Lin, M.R.; Zhang, Q.M. High-energy density in aromatic polyurea thin films. *Appl. Phys. Lett.* **2009**, *94*, 202905. [CrossRef]

28. Song, B.J.; Wu, S.H.; Li, F.; Chen, P.; Shen, B.; Zhai, J.W. Excellent energy storage density and charge-discharge performance in a novel $Bi_{0.2}Sr_{0.7}TiO_3$-$BiFeO_3$ thin film. *J. Mater. Chem. C* **2019**, *7*, 10891–10900. [CrossRef]

29. Fan, Q.L.; Liu, M.; Ma, C.R.; Wang, L.X.; Ren, S.P.; Lu, L.; Lou, X.J.; Jia, C.-L. Significantly enhanced energy storage density with superior thermal stability by optimizing $Ba(Zr_{0.15}Ti_{0.85})O_3/Ba(Zr_{0.35}Ti_{0.65})O_3$ multilayer structure. *Nano Energy* **2018**, *51*, 539–545.

30. Peng, B.L.; Zhang, Q.; Li, X.; Sun, T.Y.; Fan, H.Q.; Ke, S.M.; Ye, M.; Wang, Y.; Lu, W.; Niu, H.B.; et al. Large energy storage density and high thermal stability in a highly textured (111)-oriented $Pb_{0.8}Ba_{0.2}ZrO_3$ relaxor thin film with the coexistence of antiferroelectric and ferroelectric phases. *ACS Appl. Mater. Interfaces* **2015**, *7*, 13512. [CrossRef]

31. Lin, Z.J.; Chen, Y.; Liu, Z.; Wang, G.S.; Rémiens, D.; Dong, X.L. Large Energy Storage Density, Low Energy Loss and Highly Stable $(Pb_{0.97}La_{0.02})(Zr_{0.66}Sn_{0.23}Ti_{0.11})O_3$ Antiferroelectric Thin-Film Capacitors. *J. Eur. Ceram. Soc.* **2018**, *38*, 3177–3181. [CrossRef]

32. Park, M.H.; Kim, H.J.; Kim, Y.J.; Moon, T.; Kim, K.D.; Hwang, C.S. Thin $Hf_xZr_{1-x}O_2$ Films: A New Lead-Free System for Electrostatic Supercapacitors with Large Energy Storage Density and Robust Thermal Stability. *Adv. Energy Mater.* **2014**, *4*, 1400610. [CrossRef]

33. Lomenzo, P.D.; Chung, C.-C.; Zhou, C.Z.; Jones, J.L.; Nishida, T. Doped $Hf_{0.5}Zr_{0.5}O_2$ for High Efficiency Integrated Supercapacitors. *Appl. Phys. Lett.* **2017**, *110*, 232904. [CrossRef]

34. Song, D.P.; Yang, J.; Yang, B.B.; Wang, Y.; Chen, L.Y.; Wang, F.; Zhu, X.B. Energy Storage in $BaBi_4Ti_4O_{15}$ Thin Films with High Efficiency. *J. Appl. Phys.* **2019**, *125*, 134101. [CrossRef]

35. Chen, P.; Wu, S.H.; Li, P.; Zhai, J.W.; Shen, B. Great enhancement of energy storage density and power density in BNBT/xBFO multilayer thin film hetero-structures. *Inorg. Chem. Front.* **2018**, *5*, 2300–2305. [CrossRef]

36. Yang, C.H.; Lv, P.P.; Qian, J.; Han, Y.J.; Ouyang, J.; Lin, X.J.; Huang, S.F.; Cheng, Z.X. Fatigue-free and bending-endurable flexible Mn-doped $Na_{0.5}Bi_{0.5}TiO_3$-$BaTiO_3$-$BiFeO_3$ film capacitor with an ultrahigh energy storage performance. *Adv. Energy Mater.* **2019**, *9*, 1803949. [CrossRef]

37. Shen, B.Z.; Li, Y.; Hao, X.H. Multifunctional all-inorganic flexible capacitor for energy storage and electrocaloric refrigeration over a broad temperature range based on PLZT 9/65/35 thick films. *ACS Appl Mater. Interfaces* **2019**, *11*, 34117–34127. [CrossRef] [PubMed]

38. Liang, Z.S.; Liu, M.; Shen, L.K.; Lu, L.; Ma, C.R.; Lu, X.L.; Lou, X.J.; Jia, C.-L. All-inorganic flexible embedded thin-film capacitors for dielectric energy storage with high performance. *ACS Appl. Mater. Interfaces* **2019**, *11*, 5247. [CrossRef]

39. Saeed, M.A.; Kang, H.C.; Yoo, K.; Asiam, F.K.; Lee, J.-J.; Shim, J.W. Cosensitization of metal-based dyes for high-performance dye-sensitized photovoltaics under ambient lighting conditions. *Dye. Pigment.* **2021**, *194*, 109624. [CrossRef]

40. Saeed, M.A.; Kim, S.H.; Baek, K.; Hyun, J.K.; Lee, S.Y.; Shim, J.W. PEDOT:PSS: CuNW-based transparent composite electrodes for high-performance and flexible organic photovoltaics under indoor lighting. *Appl. Surf. Sci.* **2021**, *567*, 150852. [CrossRef]

41. Wen, Z.; Wu, D. Ferroelectric Tunnel Junctions: Modulations on the Potential Barrier. *Adv. Mater.* **2020**, *32*, 1904123. [CrossRef]

42. Zhang, Y.T.; Cao, Y.Q.; Hu, H.H.; Wang, X.; Li, P.Z.; Yang, Y.; Zheng, J.; Zhang, C.; Song, Z.Q.; Li, A.D.; et al. Flexible metal-insulator transitions based on van der waals oxide heterostructures. *ACS Appl. Mater. Interfaces* **2019**, *11*, 88284–88290. [CrossRef]

43. Bitla, Y.; Chu, Y.H. MICAtronics: A new platform for flexible X-tronics. *FlatChem* **2017**, *3*, 26–42. [CrossRef]

44. Qian, J.; Han, Y.J.; Yang, C.H.; Lv, P.P.; Zhang, X.F.; Feng, C.; Lin, X.J.; Huang, S.F.; Cheng, X.; Cheng, Z.X. Energy storage performance of flexible NKBT/NKBT-ST multilayer film capacitor by interface engineering. *Nano Energy* **2020**, *74*, 104862. [CrossRef]

45. Chu, Y.H. Van der waals oxide heteroepitaxy. *npj Quantum Mater.* **2017**, *2*, 67. [CrossRef]

46. Lv, P.P.; Yang, C.H.; Qian, J.; Wu, H.T.; Huang, S.F.; Cheng, X.; Cheng, Z.X. Flexible Lead-Free Perovskite Oxide Multilayer Film Capacitor Based on $(Na_{0.8}K_{0.2})_{0.5}Bi_{0.5}TiO_3$/$Ba_{0.5}Sr_{0.5}(Ti_{0.97}Mn_{0.03})O_3$ for High-Performance Dielectric Energy Storage. *Adv. Energy Mater.* **2020**, *10*, 1904229. [CrossRef]

47. Yang, C.H.; Qian, J.; Lv, P.P.; Wu, H.T.; Lin, X.J.; Wang, K.; Ouyang, J.; Huang, S.F.; Cheng, X.; Cheng, Z.X. Flexible lead-free BFO-based dielectric capacitor with large energy density, superior thermal stability, and reliable bending endurance. *J. Mater.* **2020**, *6*, 200–208. [CrossRef]

48. Pan, H.; Zeng, Y.; Shen, Y.; Lin, Y.-H.; Ma, J.; Li, L.L.; Nan, C.-W. $BiFeO_3$-$SrTiO_3$ thin film as new lead-free relaxor-ferroelectric capacitor with ultrahigh energy storage performance. *J. Mater. Chem. A* **2017**, *5*, 5920. [CrossRef]

49. Shen, Z.B.; Wang, X.H.; Luo, B.C.; Li, L.T. Correction: $BaTiO_3$-$BiYbO_3$ perovskite materials for energy storage applications. *J. Mater. Chem. A* **2015**, *3*, 18146. [CrossRef]

50. Liang, Z.S.; Ma, C.R.; Shen, L.K.; Lu, L.; Lu, X.L.; Lou, X.J.; Liu, M.; Jia, C.-L. Flexible lead-free oxide film capacitors with ultrahigh energy storage performances in extremely wide operating temperature. *Nano Energy* **2019**, *57*, 519–527. [CrossRef]

51. Johnson, R.W.; Evans, J.L.; Jacobsen, P.; Thompson, J.R.; Christopher, M. The Changing Automotive Environment: High-Temperature Electronics. *IEEE Trans. Electron. Packag. Manuf.* **2005**, *27*, 164–176. [CrossRef]

52. Hengst, S.; Luong-Van, D.M.; Everett, J.R.; Lawrence, J.S.; Ashley, M.C.B.; Castel, D.; Storey, J.W.V. A small, high-efficiency diesel generator for high-altitude use in Antarctica. *Int. J. Energy Res.* **2010**, *34*, 827–838. [CrossRef]

53. Watson, J.; Castro, G. High-Temperature Electronics Pose Design and Reliability Challenges. *Analog Dialog.* **2012**, *46*, 1–9.

54. Han, S.-T.; Zhou, Y.; Roy, V.A.L. Towards the development of flexible non-volatile memories. *Adv. Mater.* **2013**, *25*, 5425–5449. [CrossRef] [PubMed]

55. Zhang, Y.; Shen, L.K.; Liu, M.; Li, X.; Lu, X.L.; Lu, L.; Ma, C.R.; You, C.Y.; Chen, A.P.; Huang, C.W.; et al. Flexible quasi-two-dimensional $CoFe_2O_4$ epitaxial thin films for continuous strain tuning of magnetic properties. *ACS Nano* **2017**, *11*, 8002–8009. [CrossRef]

56. Fu, H.X.; Cohen, R.E. Polarization rotation mechanism for ultrahigh electromechanical response in single-crystal piezoelectrics. *Nature* **2000**, *403*, 281–283. [CrossRef] [PubMed]

 *nanomaterials*

MDPI

Article

Excellent Energy Storage Performance in Bi(Fe$_{0.93}$Mn$_{0.05}$Ti$_{0.02}$)O$_3$ Modified CaBi$_4$Ti$_4$O$_{15}$ Thin Film by Adjusting Annealing Temperature

Tong Liu [1,2,†], Wenwen Wang [1,†], Jin Qian [1], Qiqi Li [1], Mengjia Fan [1], Changhong Yang [1,*], Shifeng Huang [1] and Lingchao Lu [1]

[1] Shandong Provincial Key Laboratory of Preparation and Measurement of Building Materials, University of Jinan, Jinan 250022, China; shandongmems@163.com (T.L.); wangwenwen_0717@163.com (W.W.); j28_qian@163.com (J.Q.); q1635376692@163.com (Q.L.); fan15176103198@163.com (M.F.); mse_huangsf@ujn.edu.cn (S.H.); mse_lulc@ujn.edu.cn (L.L.)

[2] MEMS Institute of Zibo National High-Tech Development Zone, Zibo 255000, China

* Correspondence: mse_yangch@ujn.edu.cn

† These authors contributed equally to this work.

Citation: Liu, T.; Wang, W.; Qian, J.; Li, Q.; Fan, M.; Yang, C.; Huang, S.; Lu, L. Excellent Energy Storage Performance in Bi(Fe$_{0.93}$Mn$_{0.05}$Ti$_{0.02}$)O$_3$ Modified CaBi$_4$Ti$_4$O$_{15}$ Thin Film by Adjusting Annealing Temperature. *Nanomaterials* **2022**, *12*, 730. https://doi.org/10.3390/nano12050730

Academic Editors: Dong-Joo Kim and Alain Pignolet

Received: 29 January 2022
Accepted: 18 February 2022
Published: 22 February 2022

Publisher's Note: MDPI stays neutral with regard to jurisdictional claims in published maps and institutional affiliations.

Abstract: Dielectric capacitors with ultrahigh power density are highly desired in modern electrical and electronic systems. However, their comprehensive performances still need to be further improved for application, such as recoverable energy storage density, efficiency and temperature stability. In this work, new lead-free bismuth layer-structured ferroelectric thin films of CaBi$_4$Ti$_4$O$_{15}$-Bi(Fe$_{0.93}$Mn$_{0.05}$Ti$_{0.02}$)O$_3$ (CBTi-BFO) were prepared via chemical solution deposition. The CBTi-BFO film has a small crystallization temperature window and exhibits a polycrystalline bismuth layered structure with no secondary phases at annealing temperatures of 500–550 °C. The effects of annealing temperature on the energy storage performances of a series of thin films were investigated. The lower the annealing temperature of CBTi-BFO, the smaller the carrier concentration and the fewer defects, resulting in a higher intrinsic breakdown field strength of the corresponding film. Especially, the CBTi-BFO film annealed at 500 °C shows a high recoverable energy density of 82.8 J·cm^{-3} and efficiency of 78.3%, which can be attributed to the very slim hysteresis loop and a relatively high electric breakdown strength. Meanwhile, the optimized CBTi-BFO film capacitor exhibits superior fatigue endurance after 10^7 charge–discharge cycles, a preeminent thermal stability up to 200 °C, and an outstanding frequency stability in the range of 500 Hz–20 kHz. All these excellent performances indicate that the CBTi-BFO film can be used in high energy density storage applications.

Keywords: CBTi-BFO; fine grain; electric breakdown strength; recoverable energy storage

1. Introduction

At present, energy and environmental issues are the focus of social attention. The vigorous development of green and clean energy (such as wind and solar energy) is one of the future trends. The instability and intermittency of green energy put forward higher requirements for energy storage technology [1–5]. Dielectric capacitors typically display ultrafast charge–discharge rates and long life-time, temperature/frequency stability, fatigue resistance, which play key roles in various modern electrical and electronic systems, such as hybrid electric vehicle, aircraft and military [6–9]. Film capacitors offer a smaller size and higher energy storage density, making them easier to integrate into circuits than other devices such as ceramic capacitors [10]. Currently, most commercial dielectrics are mainly made of organic polymers, such as biaxially oriented polypropylene (BOPP), which have been widely used as the dielectric layer in power inverter capacitor systems, making the storage system bulky due to the low energy density (<5 J·cm^3). Furthermore, the operating temperature of BOPP cannot be higher than 80 °C, which increases the difficulty of structure design due to the need for an extra cooling system [11,12]. By contrast, inorganic dielectric film capacitors have the advantages of relatively high energy densities, better thermal stability in wider operating temperature ranges, and long-term endurance. Among this, inorganic ferroelectric film capacitors (TFFCs) are considered as good candidates for energy storage due to their large polarization and high temperature resistance [13,14]. However, low energy storage density and efficiency limit its further development in energy storage applications; thus, further improvements are needed.

For film capacitors, two important energy storage parameters, the recoverable energy storage density (W_{rec}) and energy storage efficiency (η), can be calculated from the measured hysteresis loops adopting the following equations [15,16]:

$$W_{rec} = \int_{P_r}^{P_m} E\, dP \tag{1}$$

$$W_t = \int_{0}^{P_m} E\, dP \tag{2}$$

$$n = \frac{W_{rec}}{W_t} \times 100\% \tag{3}$$

where E, W_t, P_m and P_r are the applied electric field, total energy storage density, the maximum polarization and remanent polarization during the discharge process, respectively. Therefore, W_{rec} can be improved by increasing the difference between P_m and P_r, and the electric breakdown strength (E_b). It is well known that the E_b of dielectric materials is mainly contingent on its microstructure, such as grain size and degree of densification. Therefore, increasing E_b by reducing grain size is an effective way to improve energy storage performance [17]. Wang et al. sort out the relationship between grain size and electric breakdown strength, confirming the optimization effect of energy storage via grain size-engineering [6]. As is well known, the annealing temperature has an immense impact on the quality of films prepared by chemical solution deposition (CSD). For instance, Wang et al. unveil a large value of W_{rec} up to 91.3 J·cm^{-3} at 4993 kV·cm^{-1} for Pb$_{0.88}$Ca$_{0.12}$ZrO$_3$ (PCZ) antiferroelectric thin films by designing a nanocrystalline structure of the pyrochlore phase by optimizing the annealing temperature to 550 °C [18]. However, the negative effect caused by the application of lead-containing dielectrics to human health and environmental sustainability cannot be ignored, and the exploration of lead-free energy storage materials is raised in the agenda. For example, Zuo et al. investigate that a high W_{rec} of 8.12 J·cm^{-3} and a great η of ~90% are obtained simultaneously in BiFeO$_3$-BaTiO$_3$-NaNbO$_3$ ceramics, which can be attributed to the significantly enhanced E_b of BiFeO$_3$-based ternary solid solutions originating from the increased resistivity and refined grain size [19].

Bismuth layer-structured ferroelectric (BLSF) compounds, such as SrBi$_2$Nb$_2$O$_9$ (SBN), SrBi$_2$Ta$_2$O$_9$ (SBT), Bi$_4$Ti$_3$O$_{12}$ (BIT), CaBi$_4$Ti$_4$O$_{15}$ (CBTi), belong to a large category of ferroelectric materials [13,20–22]. They have the advantages of excellent anti-fatigue property, large dielectric constant and small dielectric loss, high resistivity and low leakage current density, high ferroelectric Curie transition temperature, and so on [23–26]. Those traits show a good application prospect in the field of dielectric energy storage, but there is little research on BLSF compounds in this field [27,28]. This is mainly due to their intrinsic shortcomings, namely, relatively low polarization and high coercive field, which lead to lower energy density and higher losses in energy storage applications [29]. Recently, Pan et al. presented a composition modification method in ferroelectric Aurivillius Bi$_{3.25}$La$_{0.75}$Ti$_3$O$_{12}$ by introducing BiFeO$_3$ to increase the polarization value and optimize hysteresis loops, in which W_{rec} (113 J·cm^{-3}) and η (80.4%) are observed. Yang et al. prepared a series of 0.6BaTiO$_3$-0.4Bi$_{3.25}$La$_{0.75}$Ti$_3$O$_{12}$ thin films, and the modified thin film also shows higher dielectric breakdown strength and polarization. CBTi is also a representative BLSF compound, which exhibits distinct advantages including being lead-free and fatigue-free. Meanwhile, it possesses a high Curie point of about 790 °C to be used in relatively high temperature applications [30]. However, it also faces troubles of low spontaneous polarization.

In this work, we select $Bi(Fe_{0.93}Mn_{0.05}Ti_{0.02})O_3$ introduced into CBTi, namely, CBTi-BFO, to reduce leakage current and enhance breakdown field strength. In order to further optimize the energy storage performance of CBTi-BFO thin films, the effect of annealing temperature on their energy storage capacity has been studied in detail. We found that the microstructures of the CBTi-BFO thin films can be dominated by adjusting the annealing temperature. The CBTi-BFO film annealing at 500 °C possesses an excellent W_{rec} of 82.8 J·cm^{-3} and η of 78.3%, simultaneously, due to the obviously enhanced E_b of 3596 kV·cm^{-1}. Meanwhile, the film shows outstanding temperature/frequency stability up to 150 °C and superior fatigue stability after 10^7 switch cycles. The findings overcome the shortcomings of organic thin films in energy storage, including low energy storage density and low application temperature, unveiling an effective way towards high performance lead-free and eco-friendly ferroelectric materials for energy storage applications.

2. Materials and Methods

Fabrication: CBTi-BFO films were synthesized on Pt/Ti/SiO$_2$/Si substrates by chemical solution deposition. Bismuth nitrate pentahydrate [Bi(NO$_3$)$_3$·5H$_2$O], calcium nitrate tetrahydrate [Ca(NO$_3$)$_2$·4H$_2$O], iron nitrate nonahydrate [Fe(NO$_3$)$_2$·9H$_2$O], manganese acetate tetrahydrate [C$_4$H$_6$MnO$_4$·4H$_2$O] as raw materials were dissolved in ethylene glycol and acetic acid. Here, 10 mol% excess Bi was added to compensate for elements volatilization. After that, the tetrabutyl titanate and acetylacetone were added into the mixed clarified salt solution. The final concentration of the precursor solution was 0.1 M. After 24 h of aging, the precursor solution was spin coated on Pt/TiO$_2$/SiO$_2$/Si substrates with a speed of 4000 rpm for 30 s. After that, the as-prepared CBTi-BFO films were pyrolyzed at 350 °C for 120 s and annealed at 450, 500, 550, 600 °C for 10 min in a rapid thermal annealing procedure, respectively. The spin coating and annealing process procedures were duplicated up till the desired thickness of 300~700 nm was obtained. Circular Pt top electrodes, ~200 μm in diameter, were sputtered through a shadow mask on the films for the next electrical measurements.

Characterization and Measurements: The crystalline structure of CBTi-BFO films was characterized by an X-ray diffractometer (XRD) with Cu Kα radiation (XRD, D8 ADVANCE, Karlsruhe, Germany). The cross-sectional microstructure and surface morphology were characterized by a field emission scanning electron microscope (FESEM, ZEISS Gemini300, Oberkochen, Germany). The polarization-electric field (*P-E*) loops and insulating characteristic were acquired from a standard ferroelectric tester (aixACCT TF3000, Aachen, Germany). The frequency-dependent dielectric properties and impedance data were measured using impedance analyzer (HP4294A, Agilent, Palo Alto, CA, US). Impedance data were analyzed by a Z-view software. The temperature-dependent electrical performance tests were completed with the help of a temperature-controlled probe station (Linkam-HFS600E-PB2, London, UK).

3. Results

Figure 1 displays the X-ray diffraction (XRD) patterns of the CBTi-BFO films annealing at four different temperatures and standard JCPDS cards of the target phases. From Figure 1a, as the temperature is at a lower value of 450 °C, most of the diffraction peak of Aurivillius phases have not appeared yet or just a bump, indicating that some amorphous phases formed due to insufficient heat energy. Note that the CBTi-BFO films annealed at 500 and 550 °C show the (119) and (200) diffraction peaks accompanied with another peak with low intensity, which are consistent with the reported results of other CBTi films with Aurivillius phase, demonstrating a polycrystalline bismuth layered structure without any second phase [31–33]. As the annealing temperature continued to rise to 600 °C, apart from the Aurivilius phase, two diffraction peaks that do not belong to the characteristic of the bismuth layered ferroelectric appeared near 15° and 30°, which is due to the formation of the pyrochlore phase of Bi$_2$Ti$_2$O$_7$ [34–37]. Generally, the pyrochlore phase can be formed frequently due to the bismuth element volatilize as the annealing temperature increases, resulting in the ratio of bismuth ions to titanium ions approaching 1:1. As shown in Figure 1b, the diffraction peak of (119) slightly shifts to a large angle with the increase of the annealing temperature, which may be caused by the release of surface residual stress [38,39].

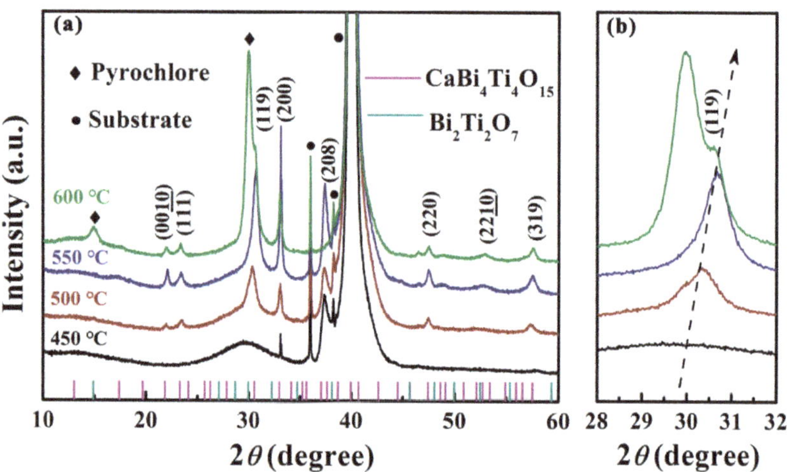

Figure 1. (**a**) X-ray diffraction (XRD) patterns in the 2θ range of 10–60° of the CBTi-BFO films annealed at various temperatures. (**b**) Enlarged XRD patterns of the diffraction peaks of 2θ at around 30°.

The SEM images clearly display the surface and the cross-sectional morphologies of the CBTi-BFO thin films annealing at different temperatures. In Figure 2, the inset images reveal that the thickness of all samples is approximately 550 nm. Meanwhile, all CBTi-BFO thin films present compact and pore-free surface, which is favorable to energy storage performance. As shown in Figure 2a, when the film is annealing at 450 °C, relatively uniform fine grains can be noticed on the surface. As the temperature rises to 500 °C, the grains absorb heat energy and thus, increase uniformly (Figure 2b). In sharp contrast, the CBTi-BFO film annealing at 550 °C possesses different grains with a wide range of grain size from ~17 to ~70 nm in Figure 2c. As the annealing temperature rises up to 600 °C, the grain size further increases in the range of ~20 to ~85 nm (Figure 2d). The phenomenon can be ascribed to: (i) different grain shapes corresponding to different orientations [40]; (ii) an inhomogeneous nucleation and grain-growth at a high annealing temperature [41].

It is well known that thin film with a high dielectric constant (ε_r) typically achieves tremendous recoverable energy density and energy efficiency [27,42]. Figure 3a presents the room-temperature frequency dependence of dielectric constant (ε_r) and the dissipation factor (tanδ) of the CBTi-BFO thin films annealing at different temperatures. The ε_r value of each film slightly decreases with the frequency raising, and increases obviously with the annealing temperature increasing. The values of ε_r for the films annealing at 450 and 500 °C have slightly changed with the frequency increases, indicating that the samples annealed at these two temperatures have better frequency stability. Generally, the dielectric properties of ferroelectric film contain the intrinsic and extrinsic contributions, which could be influenced by different factors, such as the grain size, preferred orientation, and so on [43,44]. The reason that in the changes of ε_r with increasing annealing temperature may be due to the fact that the increased annealing temperature resulted in the increased grain size and reduced grain boundaries, leading to the enhancement of ε_r [27]. Moreover, the dielectric loss (tanδ) gradually increases with increasing frequency in all samples. Besides, all samples possess smaller loss (tanδ < 0.08) at 10 kHz.

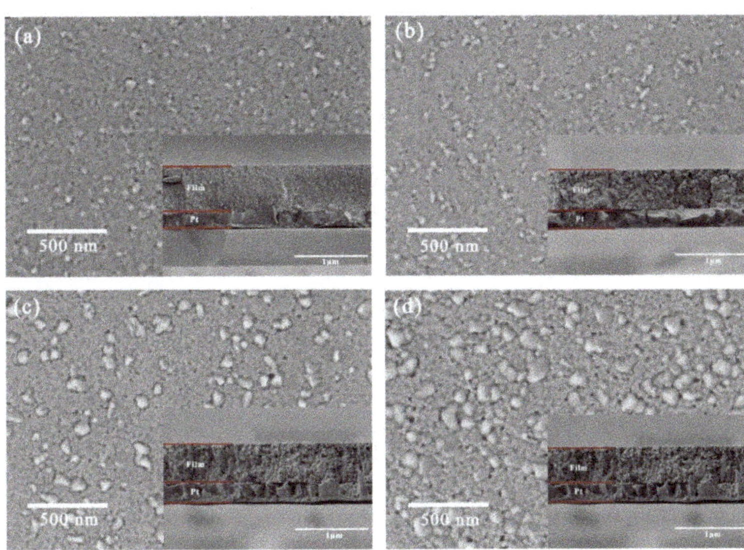

Figure 2. The scanning electron microscopic (SEM) images of the CBTi-BFO thin films annealed at (**a**) 450 °C, (**b**) 500 °C, (**c**) 550 °C and (**d**) 600 °C, and the inset shows their corresponding cross-sectional micrographs.

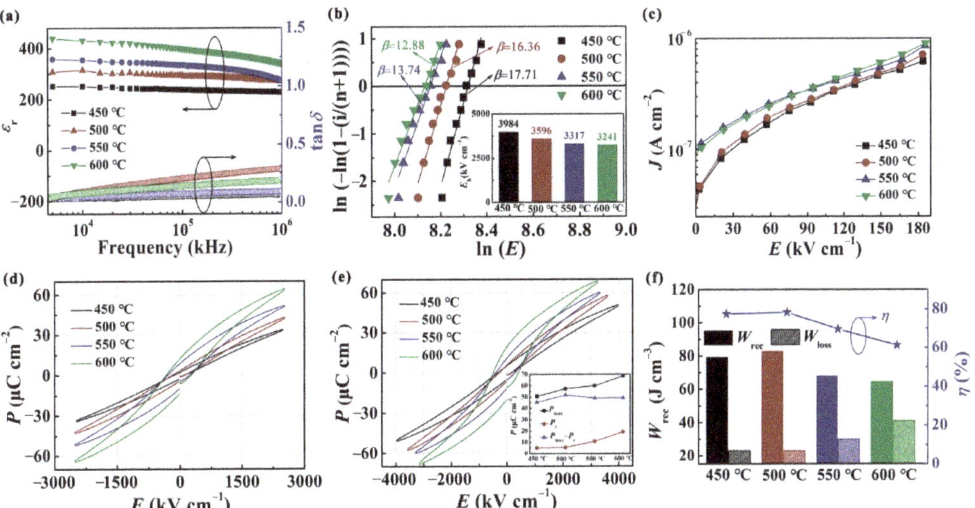

Figure 3. CBTi-BFO films annealed at 450, 500, 550 and 600 °C: (**a**) Frequency dependence of dielectric properties. (**b**) Weibull distributions dielectric breakdown strengths. (**c**) Leakage current density as a function of applied electric filed. (**d**) The bipolar *P-E* loops at 2500 kV·cm^{-1}. (**e**) The bipolar *P-E* loops around E_b, The inset shows the variation of P_{max}, P_r, and P_{max}-P_r as a function of annealing temperature. (**f**) The variations of W_{rec}, W_{loss} and η values.

Generally, E_b is analyzed by two parameter Weibull statistics, which is closely affected the energy storage performance of dielectric materials [45], as displayed in Figure 3b. Meanwhile, the E_b endurance of four films with different annealing temperatures are demonstrated in the inset of Figure 3b. The Weibull distribution of E_b can be expressed by the following formulas:

$$X_i = Ln(E_i) \tag{4}$$

$$Y_i = Ln(-Ln(1 - \frac{i}{n+1})) \tag{5}$$

where E_i is the breakdown electric field for each sample, i signifies the number of test samples and n denotes the total number of test samples. Based on the Weibull distribution function, the mean E_b for each film can be obtained from the intersection of the fitted lines with the horizontal axis at $Y_i = 0$. β represents Weibull shape parameter. It can be observed that E_b increases rapidly with the annealing temperature decreasing, as shown in the inset of Figure 3b. The average breakdown strength of the dielectric film increases from 3241 kV·cm^{-1} to 3984 kV·cm^{-1}, with the annealing temperature decrease from 600 to 450 °C. The enhancement of dielectric breakdown strength can be ascribed to the following factor. It is well-known that electric breakdown field strength is inversely proportional to the grain size (G), which can be manifested by the following formula:

$$E_b \propto (G)^{-a} \tag{6}$$

where a is the exponent values, being in the range of 0.2–0.4 [6,46]. It can be seen that the dielectric breakdown strength increases with the decreased grain size, which is aligned with the results in Figures 2 and 3b. That is owing to grain boundaries producing depletion regions similar to Schottky barriers located in semiconductor interfaces. Then, the grain boundaries depletion layers establish important barriers for the cross-transport of ionic and electronic charges [47]. Therefore, the E_b for CBTi-BFO film annealing at 450 °C is superior to the films that have higher annealing temperature. The slope parameter β, related to the scatter of E_b data, increases from 12.88 to 17.71 with the annealing temperature decrease from 600 °C to 450 °C, indicating an enhancement in dielectric reliability by annealing temperature decreasing. At the same time, the β of all films based on the linear fitting is higher than 12, suggestive of all samples possessing high reliability.

Figure 3c represents the leakage current of all films, which are measured by applying 0–185 kV·cm^{-1} to the electrodes. The CBTi-BFO films annealed at 450 and 500 °C illustrate well insulating properties with leakage current densities $< 7 \times 10^{-7}$ A cm^{-2} under an applied electric field of 185 kV·cm^{-1}. The CBTi-BFO films annealed at 550 and 600 °C exhibit higher leakage currents, which is ascribed to the greater grain size and the simultaneously declined number of the grain boundary.

Presented in Figure 3d are bipolar polarization-electric filed (*P-E*) loops of the CBTi-BFO films annealing at different temperatures applying the electric field of 2500 kV·cm^{-1} at frequency of 10 kHz. With the annealing temperature decreasing, a monotonous decrease of polarization (P_{max}) is observed, from 33 μC cm^{-2} of 450 °C to 63 μC cm^{-2} of 600 °C.

Figure 3e shows *P-E* loops of CBTi-BFO films near their respective E_b annealing at different temperatures. It can be seen that the CBTi-BFO thin film annealing at 600 °C presents a practically saturated *P-E* loop with a remnant polarization P_r of 20 μC/cm^2, which is lower than previous reports [32,33], suggesting that the derived CBTi-BFO thin films have somewhat discrepant ferroelectric properties. For the CBTi-BFO thin films annealing at lower temperatures, the obtained *P-E* hysteresis loops exhibit slim shape, which can be ascribed to the function of lower leakage current and small crystallite size [48–50]. Figure 3f presents the corresponding energy storage parameters of W_{rec}, W_{loss} and η determined from *P-E* loops about all CBTi-BFO films. The value of W_{rec} and η for the CBTi film annealing at 500 °C, respectively, reaches 82.8 J·cm^{-3} and 78.3% due to an integration of remarkable E_b of 3596 kV·cm^{-1} and a large polarization disparity of 52.3 μC cm^{-2} (inset of Figure 3e). It is well known that the dielectric film can induce outstanding energy performance due to high P_{max}, low P_r and high electric breakdown strength. The lower annealing temperature leads to more slender electric hysteresis loops and lower W_{loss}. Meanwhile, E_b is enhanced with the decrease of the annealing temperature.

Figure 4a–d show the Nyquist plots of films measured in the frequency range of 100 Hz to 1 MHz at a series of temperatures (225–300 °C). As is known to all, a low-frequency arc means the dielectric response of grain boundaries, while a high-frequency arc means the dielectric response of grains. One semicircular arc is observed for the samples. As the grain boundary dielectric relaxation of the material mainly contributes to the total impedance, only the semi-circular arc corresponding to the grain boundary response is observed. The intercept of impedance semicircular arcs on the Z'-axis can represent the total resistance (R_b) values of the film. Clearly, the value of R_b gradually decreased with the increasing measured temperature in each film, showing a negative temperature characteristic. This phenomenon can be attributed to the fact that the mobility of the space charge becomes easier, and more charge carriers will accumulate at the grain boundaries with the increasing measured temperature, thus resulting in increased electrical conductivity and decreased grain boundary resistance [51,52].

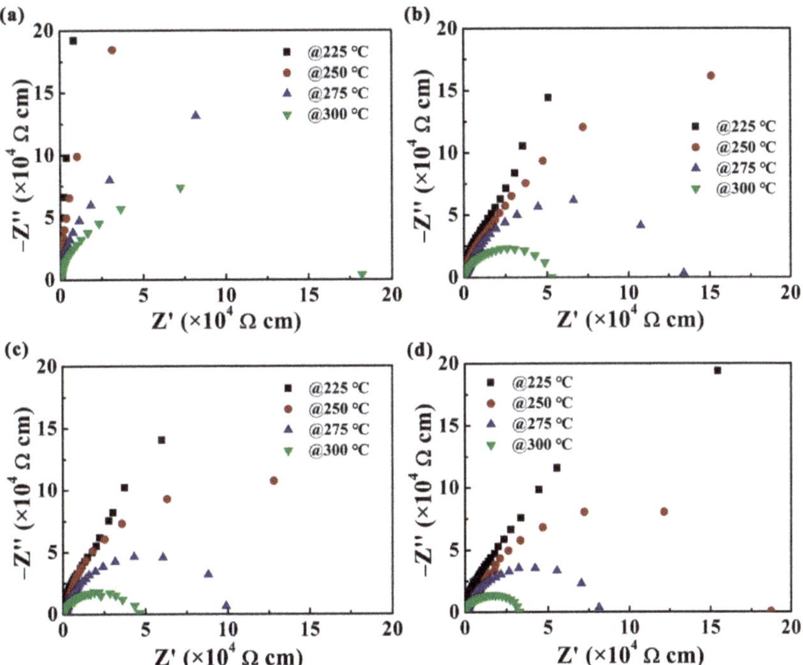

Figure 4. Nyquist diagrams of complex-plane of the CBTi-BFO films with annealing temperature of (**a**) 450, (**b**) 500, (**c**) 550, (**d**) 600 °C measured at 20 V under four measured temperatures.

In order to compare the impedance differences of the four samples in more detail, we separately take out the impedance diagrams tested at 300 °C and showed them in Figure 5a. It can be seen that R_b of CBTi-BFO films gradually increases with a decreasing annealing temperature, indicating that the charge carrier concentration decreases in low-temperature annealed samples [53].

Generally, the energy required for carriers to cross the energy barrier is called conductive activation energy (E_a), which can be calculated by Arrhenius formula:

$$\sigma = \sigma_0 \exp(-E_a/k_B T) \tag{7}$$

where T, σ_0, E_a and k_B are the measuring temperature on the Kelvin scale, preexponential factor, activation energy and Boltzmann constant, respectively. The plots of $\ln(\sigma)$ as a function of 1000/T for the films and linear fittings are shown in Figure 5b. The activation energies (E_a) obtained from the slopes of fitting lines for films annealing at 450, 500, 550, 600 °C are estimated to be 1.21, 0.84, 0.68, 0.66 eV, respectively. Generally, the higher value of activation energy means fewer defects exist in the film [54]. Thus, we can deduce that the sample with the lowest annealing temperature has the least defects. Combining the above two points, it can be seen that CBTi-BFO with lowest annealing temperature possesses the smallest charge carrier concentration and fewer defects, which contribute greatly to the observed highest intrinsic breakdown field in this film.

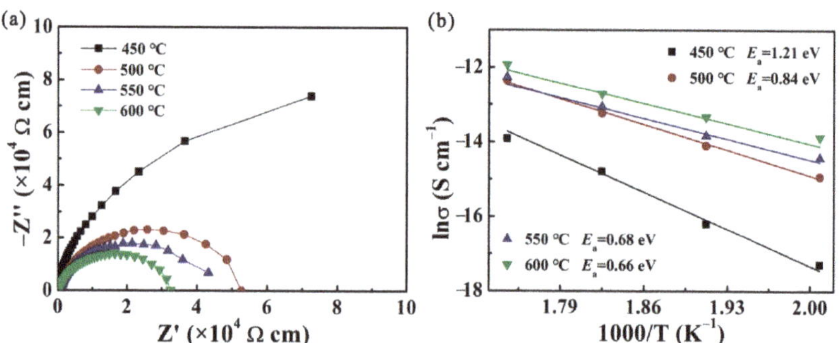

Figure 5. (**a**) The impedance and fit result of CBTi-BFO films measured at 300 °C. (**b**) Arrhenius plots of CBTi-BFO films annealing at different temperatures.

For practical application, the requirement of reliability and stability is emphasized, such as under high/low temperature, high/low frequency and long-term working environments. Thus, in view of the CBTi-BFO film annealing at 500 °C possessing large W_{rec} and η among all samples, further investigations are carried out on it under an electric field of 2500 kV·cm^{-1}. Firstly, the temperature stability of the CBTi-BFO film annealing at 500 °C is measured. The bipolar *P-E* loops of the film measured from relatively low temperature of −25 °C to ultrahigh temperature of 200 °C are displayed in Figure 6a. For the purpose of expressing the variation of polarization more clearly, the unipolar hysteresis loop diagrams' dependence on temperature is drawn in Figure 6d. It can be seen that the P_{max} value varies slightly from 42.62 to 45.89 µC cm^{-2}. Correspondingly, W_{rec} values are slightly increased by 2% from 44.88 to 45.84 J·cm^{-3} and η is reduced by 6% from 80.04 to 75.19% with temperature increasing, as shown in Figure 6g. The aforesaid energy storage performance of the film shows good temperature stability, which is adequate to meet the demands of applications in extreme environments (capacitors used in underground industrial instruments need to work at temperatures higher than 150 °C and the inverter must work at about 140 °C in HEV) [55–57]. Besides, frequency dependence of the energy storage behavior is also researched at 2500 kV·cm^{-1}. Figure 6b, e are the bipolar and unipolar hysteresis loops measured at the frequency range of 500 Hz to 20 kHz, respectively; and the corresponding energy performance W_{rec} and η are shown in Figure 6h. Clearly, the *P-E* loop can maintain a slim feature and no discernible decline in energy storage performance can be discovered. Even though the frequency rises from 500 Hz to 20 kHz, the W_{rec} and η values slightly drop from 48.71 to 43.80 J·cm^{-3} and 83.94 to 77.47%, respectively, indicating that a good frequency stability can be realized. Furthermore, to assess the long-term charging–discharging stability of dielectric capacitors, the fatigue endurance should be investigated. The *P-E* loops of CBTi-BFO film annealing at 500 °C over 10^7 charge–discharge cycles are exhibited in Figure 6c, f. It can be seen that there is no obvious change in the hysteresis loops. The corresponding W_{rec} and an excellent η present a negligible degradation with 1% and 0.3% as shown in Figure 6i. The observed superior antifatigue feature is closely related to the $(Bi_2O_2)^{2+}$ layer in the Aurivillius phase, which has a good insulating effect inhibiting the flow and generation of leakage charges during the repeated polarization process. Thus, the Aurivillius phase can more effectively suppress the electrical breakdown during the fatigue process and improve the anti-fatigue performance.

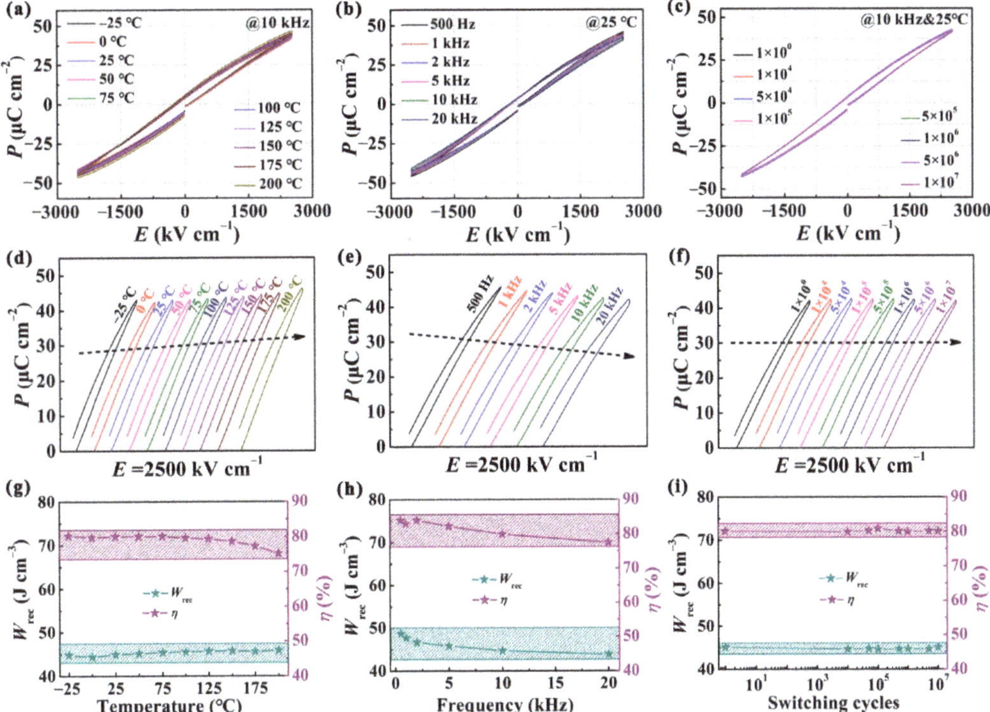

Figure 6. CBTi-BFO thin film annealed at 500 °C: (**a–c**) bipolar *P-E* hysteresis loops measured at different temperatures/frequencies/switching cycles; (**d–f**) unipolar *P-E* hysteresis loops of CBTi-BFO films at different temperatures/frequencies/switching cycles; (**g–i**) the changes of W_{rec} and η depending on temperature/frequencies/switching cycles.

Figure 7 summarizes a serious of results on the energy storage performance of Aurivillius ferroelectric films [12,13,27,28,58–60]. In this study, the CBTi-BFO thin film annealing at 500 °C possesses a relatively high η (~78%) superior than $BaLa_{0.2}Bi_{3.8}Ti_4O_{15}$ (BLBT~60%) and $Bi_{3.25}La_{0.75}Ti_3O_{12}/BiFeO_3/Bi_{3.25}La_{0.75}Ti_3O_{12}$ (BLT/BFO/BLT~74%), but inferior to $Ba_2Bi_4Ti_5O_{18}$ (BBT~92%), $0.6BaTiO_3$-$0.4Bi_{3.25}La_{0.75}Ti_4O_{12}$ (0.6BT-0.4BLT~84%), $CaBi_2Nb_2O_9$ (CBNO~82%), $Sr_2Bi_4Ti_5O_{18}$ (SBT~81%), $Bi_{3.25}La_{0.75}Ti_3O_{12}$-$BiFeO_3$ (BLT-BFO~80%). In contrast, its W_{rec} clearly outperforms the surveyed dielectric systems with a high η. That is, the CBTi-BFO thin film annealing at 500 °C has excellent comprehensive properties, namely, a good balance between W_{rec} and η. Meanwhile, the film has a great E_b (~3596 kV·cm^{-1}) at a high level. Taken together, the CBTi-BFO thin film annealing at 500 °C is a good candidate for application in energy storage devices.

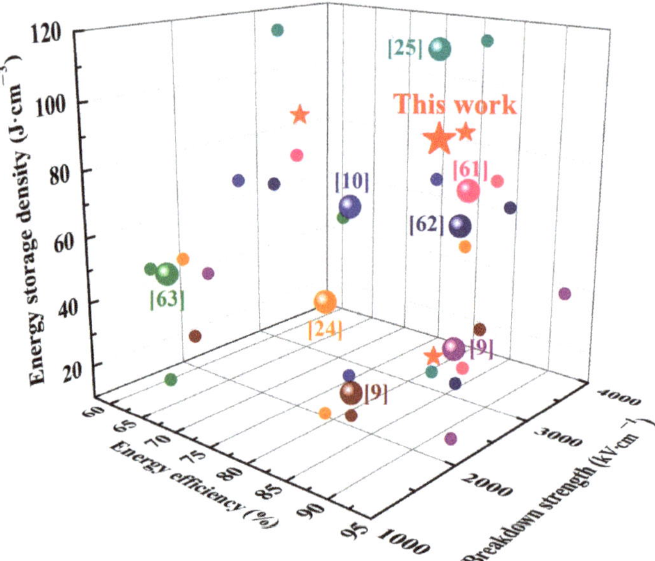

Figure 7. Summary of recently reported the core parameters of E_b, W_{rec} and η for energy storage properties of representative Aurivillius ferroelectrics film [12,13,27,28,58–60].

4. Conclusions

In this work, a series of CBTi-BFO thin films are prepared by chemical solution deposition under different annealing temperatures. We found that by optimizing the microstructures and crystalline structure of films via adjusting the annealing temperature, the electrical properties and energy storage performance can be greatly improved, especially in the film with a relatively lower annealing temperature. In the annealing temperature of 500–550 °C, CBTi-BFO thin films demonstrates a polycrystalline bismuth layered structure without any second phase. The CBTi-BFO with the lowest annealing temperature has the smallest carrier concentration and fewer defects, which greatly contribute to the highest intrinsic breakdown field of this film. Here, ultrahigh energy storage density of W_{rec} (~82.8 J·cm⁻³) and η (~78.3%) are achieved in the CBTi-BFO film that annealed at 500 °C. The excellent energy storage performance can be ascribed to its uniform fine grain size. The film also shows superior thermal stability (from −25 to 200 °C), frequency stability (from 500 Hz to 20 kHz) and fatigue endurance (after 10^7 switching cycles). In a word, annealing temperature plays an important part in performance tuning, which is a vital factor needed to be considered for preparing thin film capacitors with high energy storage characteristics.

Author Contributions: Conceptualization, C.Y., S.H. and L.L.; methodology, C.Y.; investigation, T.L., W.W., J.Q. and Q.L.; resources, C.Y.; data curation, T.L., W.W. and M.F.; writing—original draft preparation, T.L. and W.W.; writing—review and editing, C.Y. and S.H.; supervision, C.Y., S.H. and L.L. All authors have read and agreed to the published version of the manuscript.

Funding: This work was supported by the National Natural Science Foundation of China (Grant Nos. 51972144, and U1806221), the Shandong Provincial Natural Science Foundation (Grant No. ZR2020KA003), the Project of "20 Items of University" of Jinan (Grant Nos. T202009 and T201907), and the Synergetic Innovation Center Research Project (Grant No. WJGTT-XT2), the Introduction Program of Senior Foreign Experts (G2021024003L).

Data Availability Statement: Not applicable.

Conflicts of Interest: The authors declare no conflict of interest.

References

1. Yan, F.; Huang, K.W.; Jiang, T.; Zhou, X.F.; Shi, Y.J.; Ge, G.L.; Shen, B.; Zhai, J.W. Significantly enhanced energy storage density and efficiency of BNT-based perovskite ceramics via A-site defect engineering. *Energy Storage Mater.* **2020**, *30*, 392–400. [CrossRef]
2. Yao, Z.H.; Song, Z.; Hao, H.; Yu, Z.Y.; Cao, M.; Zhang, S.; Lanagan, M.T.; Liu, H.X. Homogeneous/Inhomogeneous-Structured Dielectrics and Their Energy-Storage Performances. *Adv. Mater.* **2017**, *29*, 1601727. [CrossRef] [PubMed]
3. Yan, F.; Shi, Y.J.; Zhou, X.F.; Zhu, K.; Shen, B.; Zhai, J.W. Optimization of polarization and electric field of bismuth ferrite-based ceramics for capacitor applications. *Chem. Eng. J.* **2021**, *417*, 127945. [CrossRef]
4. Silva, J.P.B.; Silva, J.M.B.; Oliveira, M.J.S.; Weingärtner, T.; Sekhar, K.C.; Pereira, M.; Gomes, M.J.M. High-Performance Ferroelectric-Dielectric Multilayered Thin Films for Energy Storage Capacitors. *Adv. Funct. Mater.* **2018**, *29*, 1807196. [CrossRef]
5. Wang, W.W.; Qian, J.; Geng, C.H.; Fan, M.J.; Yang, C.H.; Lu, L.C.; Cheng, Z.X. Flexible Lead-Free $Ba_{0.5}Sr_{0.5}TiO_3/0.4BiFeO_3$-$0.6SrTiO_3$ Dielectric Film Capacitor with High Energy Storage Performance. *Nanomaterials* **2021**, *11*, 3065. [CrossRef] [PubMed]
6. Wang, G.; Lu, Z.L.; Li, Y.; Li, L.H.; Ji, H.F.; Feteira, A.; Zhou, D.; Wang, D.W.; Zhang, S.J.; Reaney, I.M. Electroceramics for High-Energy Density Capacitors: Current Status and Future Perspectives. *Chem. Rev.* **2021**, *121*, 6124–6172. [CrossRef] [PubMed]
7. Palneedi, H.; Peddigari, M.; Hwang, G.-T.; Jeong, D.-Y.; Ryu, J. High-Performance Dielectric Ceramic Films for Energy Storage Capacitors: Progress and Outlook. *Adv. Funct. Mater.* **2018**, *28*, 1803665. [CrossRef]
8. Qian, J.; Han, Y.J.; Yang, C.H.; Lv, P.P.; Zhang, X.F.; Feng, C.; Lin, X.J.; Huang, S.F.; Cheng, X.; Cheng, Z.X. Energy storage performance of flexible NKBT/NKBT-ST multilayer film capacitor by interface engineering. *Nano Energy* **2020**, *74*, 104862. [CrossRef]
9. Chu, B.J.; Zhou, X.; Ren, K.L.; Neese, B.; Lin, M.; Wang, Q.; Bauer, F.; Zhang, Q.M. A Dielectric Polymer with High Electric Energy Density and Fast Discharge Speed. *Science* **2006**, *313*, 334–336. [CrossRef]
10. Fan, Z.J.; Li, L.L.; Mei, X.S.; Zhao, F.; Li, H.J.; Zhuo, X.S.; Zhang, X.F.; Lu, Y.; Zhang, L.; Liu, M. Multilayer ceramic film capacitors for high performance energy storage: Progress and outlook. *J. Mater. Chem. A* **2021**, *9*, 946. [CrossRef]
11. Dang, Z.M. *Dielectric Polymer Materials for High-Density Energy Storage*; William Andrew: Norwich, NY, USA, 2018.
12. Yang, B.B.; Guo, M.Y.; Tang, X.W.; Wei, R.H.; Hu, L.; Yang, J.; Song, W.H.; Dai, J.M.; Lou, X.J.; Zhu, X.B.; et al. Lead-free $A_2Bi_4Ti_5O_{18}$ thin film capacitors (A = Ba and Sr) with large energy storage density, high efficiency, and excellent thermal stability. *J. Mater. Chem. C* **2019**, *7*, 1888–1895. [CrossRef]
13. Yang, B.B.; Guo, M.Y.; Jin, L.H.; Tang, X.W.; Wei, R.H.; Hu, L.; Yang, J.; Song, W.H.; Dai, J.M.; Lou, X.J.; et al. Ultrahigh energy storage in lead-free $BiFeO_3/Bi_{3.25}La_{0.75}Ti_3O_{12}$ thin film capacitors by solution processing. *Appl. Phys. Lett.* **2018**, *112*, 033904. [CrossRef]
14. Chen, P.; Chu, B.J. Improvement of dielectric and energy storage properties in $Bi(Mg_{1/2}Ti_{1/2})O_3$-modified $(Na_{1/2}Bi_{1/2})_{0.92}Ba_{0.08}TiO_3$ ceramics. *J. Eur. Ceram. Soc.* **2016**, *36*, 81–88. [CrossRef]
15. Burn, I.; Smyth, D.M. Energy Storage in Ceramic Dielectrics. *J. Mater. Sci.* **1972**, *7*, 339–343. [CrossRef]
16. Lv, P.P.; Yang, C.H.; Qian, J.; Wu, H.T.; Huang, S.F.; Cheng, X.; Cheng, Z.X. Flexible Lead-Free Perovskite Oxide Multilayer Film Capacitor Based on $(Na_{0.8}K_{0.2})_{0.5}Bi_{0.5}TiO_3/Ba_{0.5}Sr_{0.5}(Ti_{0.97}Mn_{0.03})O_3$ for High-Performance Dielectric Energy Storage. *Adv. Energy Mater.* **2020**, *10*, 1904229. [CrossRef]
17. Zhao, L.; Liu, Q.; Gao, J.; Zhang, S.; Li, J.F. Lead-Free Antiferroelectric Silver Niobate Tantalate with High Energy Storage Performance. *Adv. Mater.* **2017**, *29*, 1701824. [CrossRef]
18. Li, Y.Z.; Lin, J.L.; Bai, Y.; Li, Y.X.; Zhang, Z.D.; Wang, Z.J. Ultrahigh-Energy Storage Properties of $(PbCa)ZrO_3$ Antiferroelectric Thin Films via Constructing a Pyrochlore Nanocrystalline Structure. *ACS Nano* **2020**, *14*, 6857–6865. [CrossRef]
19. Qi, H.; Xie, A.; Tian, A.; Zuo, R. Superior Energy-Storage Capacitors with Simultaneously Giant Energy Density and Efficiency Using Nanodomain Engineered $BiFeO_3$-$BaTiO_3$-$NaNbO_3$ Lead-Free Bulk Ferroelectrics. *Adv. Energy Mater.* **2020**, *10*, 1903338. [CrossRef]
20. Kato, K.; Zheng, C.; Finder, J.M.; Dey, S.K. Sol-Gel Route to Ferroelectric Layer-Structured Perovskite $SrBi_2Ta_2O_9$ and $SrBi_2Nb_2O_9$ Thin Films. *J. Am. Ceram. Soc.* **1998**, *81*, 1869–1875. [CrossRef]
21. Osaka, T.; Sakakibara, A.; Seki, T.; Ono, S.; Koiwa, I.; Hashimoto, A. Phase Transition in Ferroelectric $SrBi_2Ta_2O_9$ Thin Films with Change of Heat-treatment Temperature. *Jpn. J. Appl. Phys.* **1998**, *37*, 597–601. [CrossRef]
22. Cheng, C.-M.; Chen, K.-H.; Tsai, J.-H.; Wu, C.-L. Solution-based fabrication and electrical properties of $CaBi_4Ti_4O_{15}$ thin films. *Ceram. Int.* **2012**, *38*, S87–S90. [CrossRef]
23. Dubey, S.; Kurchania, R. Study of dielectric and ferroelectric properties of five-layer Aurivillius oxides: $A_2Bi_4Ti_5O_{18}$ (A = Ba, Pb and Sr) synthesized by solution combustion route. *Bull. Mater. Sci.* **2015**, *38*, 1881–1889. [CrossRef]
24. Park, B.H.; Kang, B.S.; Bu, S.D.; Noh, T.W.; Lee, J.; Jo, W. Lanthanum-substituted bismuth titanate for use in non-volatile memories. *Nature* **1999**, *401*, 682–684. [CrossRef]
25. Scott, J.F.; Ross, F.M.; Paz de Araujo, C.A.; Scott, M.C.; Huffman, M. Structure and Device Characteristics of $SrBi_2Ta_2O_9$-Based Nonvolatile Random-Access Memories. *MRS Bull.* **1996**, *21*, 33–39. [CrossRef]
26. Shet, T.; Bhimireddi, R.; Varma, K.B.R. Grain size-dependent dielectric, piezoelectric and ferroelectric properties of $Sr_2Bi_4Ti_5O_{18}$ ceramics. *J. Mater. Sci.* **2016**, *51*, 9253–9266. [CrossRef]
27. Yang, B.B.; Guo, M.Y.; Song, D.P.; Tang, X.W.; Wei, R.H.; Hu, L.; Yang, J.; Song, W.H.; Dai, J.M.; Lou, X.J.; et al. $Bi_{3.25}La_{0.75}Ti_3O_{12}$ thin film capacitors for energy storage applications. *Appl. Phys. Lett.* **2017**, *111*, 183903. [CrossRef]

28. Pan, Z.B.; Wang, P.; Hou, X.; Yao, L.M.; Zhang, G.Z.; Wang, J.; Liu, J.J.; Shen, M.; Zhang, Y.J.; Jiang, S.L.; et al. Fatigue-Free Aurivillius Phase Ferroelectric Thin Films with Ultrahigh Energy Storage Performance. *Adv. Energy Mater.* **2020**, *10*, 2001536. [CrossRef]

29. Tomar, M.S.; Melgarejo, R.E.; Hidalgo, A.; Mazumder, S.B.; Katiyar, R.S. Structural and ferroelectric studies of $Bi_{3.44}La_{0.56}Ti_3O_{12}$ films. *Appl. Phys. Lett.* **2003**, *83*, 341. [CrossRef]

30. Do, D.; Kim, S.S.; Kim, J.W.; Lee, Y.I.; Bhalla, A.S. Orientation Dependence of Ferroelectric and Dielectric Properties in $CaBi_4Ti_4O_{15}$ Thin Films. *Integr. Ferroelectr.* **2009**, *105*, 99–106. [CrossRef]

31. Kato, K.; Suzuki, K.; Nishizawa, K.; Miki, T. Ferroelectric properties of alkoxy-derived $CaBi_4Ti_4O_{15}$ thin films on Pt-passivated Si. *Appl. Phys. Lett.* **2001**, *78*, 1119. [CrossRef]

32. Kato, K.; Fu, D.; Suzuki, K.; Tanaka, K.; Nishizawa, K.; Miki, T. Ferro- and piezoelectric properties of polar-axis-oriented $CaBi_4Ti_4O_{15}$ films. *Appl. Phys. Lett.* **2004**, *84*, 3771. [CrossRef]

33. Kato, K.; Tanaka, K.; Suzuki, K.; Kimura, T.; Nishizawa, K.; Miki, T. Impact of oxygen ambient on ferroelectric properties of polar-axis-oriented $CaBi_4Ti_4O_{15}$ films. *Appl. Phys. Lett.* **2005**, *86*, 112901. [CrossRef]

34. Kim, S.S.; Kim, W.-J. Electrical properties of sol-gel derived pyrochlore-type bismuth magnesium niobate $Bi_2(Mg_{1/3}Nb_{2/3})_2O_7$ thin films. *J. Cryst. Growth* **2005**, *281*, 432–439. [CrossRef]

35. Kim, S.S.; Park, M.H.; Chung, J.K.; Kim, W.-J. Structural study of a sol-gel derived pyrochlore $Bi_2Ti_2O_7$ using a Rietveld analysis method based on neutron scattering studies. *J. Appl. Phys.* **2009**, *105*, 061641. [CrossRef]

36. Cho, S.Y.; Choi, G.P.; Jeon, D.H.; Johnson, T.A.; Lee, M.K.; Lee, G.J.; Bu, S.D. Effects of excess Bi_2O_3 on grain orientation and electrical properties of $CaBi_4Ti_4O_{15}$ ceramics. *Curr. Appl. Phys.* **2015**, *15*, 1332–1336. [CrossRef]

37. kato, k.; Suzuki, k.; tanaka, k.; Fu, D.; Nishizawa, k.; Miki, T. Effect of amorphous TiO_2 buffer layer on the phase formation of $CaBi_4Ti_4O_{15}$ ferroelectric thin films. *Appl. Phys. A* **2005**, *81*, 861–864. [CrossRef]

38. Jiao, L.L.; Liu, Z.M.; Hu, G.D.; Cui, S.G.; Jin, Z.J.; Wang, Q.; Wu, W.B.; Yang, C.H. Low-Temperature Fabrication and Enhanced Ferro- and Piezoelectric Properties of $Bi_{3.7}Nd_{0.3}Ti_3O_{12}$ Films on Indium Tin Oxide/Glass Substrates. *J. Am. Ceram. Soc.* **2009**, *92*, 1556–1559. [CrossRef]

39. Zheng, X.J.; Yang, Z.Y.; Zhou, Y.C. Residual stresses in $Pb(Zr_{0.52}Ti_{0.48})O_3$ thin films deposited by metal organic decomposition. *Scr. Mater.* **2003**, *49*, 71–76. [CrossRef]

40. Hu, G.D. Orientation dependence of ferroelectric and piezoelectric properties of $Bi_{3.15}Nd_{0.85}Ti_3O_{12}$ thin films on $Pt(100)/TiO_2/SiO_2/Si$ substrates. *J. Appl. Phys.* **2006**, *100*, 096109. [CrossRef]

41. Yan, J.; Hu, G.D.; Liu, Z.M.; Fan, S.H.; Zhou, Y.; Yang, C.H.; Wu, W.B. Enhanced ferroelectric properties of predominantly (100)-oriented $CaBi_4Ti_4O_{15}$ thin films on $Pt/Ti/SiO_2/Si$ substrates. *J. Appl. Phys.* **2008**, *103*, 056109. [CrossRef]

42. Hu, G.L.; Ma, C.R.; Wei, W.Z.; Sun, X.; Lu, L.; Mi, S.B.; Liu, M.; Ma, B.H.; Wu, J.; Jia, C.L. Enhanced energy density with a wide thermal stability in epitaxial $Pb_{0.92}La_{0.08}Zr_{0.52}Ti_{0.48}O_3$ thin films. *Appl. Phys. Lett.* **2016**, *109*, 193904. [CrossRef]

43. Xu, F.; Trolier-McKinstry, S.; Ren, W.; Xu, B.M.; Xie, Z.-L.; Hemker, K.J. Domain wall motion and its contribution to the dielectric and piezoelectric properties of lead zirconate titanate films. *J. Appl. Phys.* **2001**, *89*, 1336–1348. [CrossRef]

44. Cruz, J.P.D.L.; Joanni, E.; Vilarinho, P.M.; Kholkin, A.L. Thickness effect on the dielectric, ferroelectric, and piezoelectric properties of ferroelectric lead zirconate titanate thin films. *J. Appl. Phys.* **2010**, *108*, 114106. [CrossRef]

45. Weibull, W. A Statistical Distribution Function of Wide Applicability. *J. Appl. Mech.* **1951**, *73*, 293–297. [CrossRef]

46. Tunkasiri, T.; Rujijanagul, G. Dielectric Strength of Fine Grained Barium Titanate Ceramics. *J. Mater. Sci. Lett.* **1996**, *15*, 1767–1769. [CrossRef]

47. Waser, R. TrI4: The Role of Grain Boundaries in Conduction and Breakdown of Perovskite-Type Titanates. *Ferroelectrics* **1992**, *133*, 109–114. [CrossRef]

48. Frey, M.H.; Xu, Z.; Han, P.; Payne, D.A. The role of interfaces on an apparent grain size effect on the dielectric properties for ferroelectric barium titanate ceramics. *Ferroelectrics* **1998**, *206*, 337. [CrossRef]

49. Hoshina, T.; Furuta, T.; Kigoshi, Y.; Hatta, S.; Horiuchi, N.; Takeda, H.; Tsurumi, T. Size Effect of Nanograined $BaTiO_3$ Ceramics Fabricated by Aerosol Deposition Method. *Jpn. J. Appl. Phys.* **2010**, *49*, 09MC02. [CrossRef]

50. Deng, X.Y.; Wang, X.H.; Wen, H.; Chen, L.L.; Chen, L.; Li, L.T. Ferroelectric properties of nanocrystalline barium titanate ceramics. *Appl. Phys. Lett.* **2006**, *88*, 252905. [CrossRef]

51. Sarkar, S.; Jana, P.K.; Chaudhuri, B.K. Colossal internal barrier layer capacitance effect in polycrystalline copper (II) oxide. *Appl. Phys. Lett.* **2008**, *92*, 022905. [CrossRef]

52. Yang, C.H.; Feng, C.; Lv, P.P.; Qian, J.; Han, Y.J.; Lin, X.J.; Huang, S.F.; Cheng, X.; Cheng, Z.X. Coexistence of giant positive and large negative electrocaloric effects in lead-free ferroelectric thin film for continuous solid-state refrigeration. *Nano Energy* **2021**, *88*, 106222. [CrossRef]

53. Wang, X.Z.; Huan, Y.; Zhao, P.Y.; Liu, X.M.; Wei, T.; Zhang, Q.W.; Wang, X.H. Optimizing the grain size and grain boundary morphology of (K,Na) NbO_3-based ceramics: Paving the way for ultrahigh energy storage capacitors. *J. Mater.* **2021**, *7*, 780–789. [CrossRef]

54. Chao, X.L.; Ren, X.D.; Zhang, X.S.; Peng, Z.H.; Wang, J.J.; Liang, P.F.; Wu, D.; Yang, Z.P. Excellent optical transparency of potassium-sodium niobate-based lead-free relaxor ceramics induced by fine grains. *J. Eur. Ceram. Soc.* **2019**, *39*, 3684–3692. [CrossRef]

55. Watson, J.; Castro, G. High-Temperature Electronics Pose Design and Reliability Challenges. *Analog Dialog.* **2012**, *46*, 1–9.

56. Johnson, R.W.; Evans, J.L.; Jacobsen, P.; Thompson, J.R.; Christopher, M. The Changing Automotive Environment: High-Temperature Electronics. *IEEE Trans. Electron. Packag. Manuf.* **2005**, *27*, 164–176. [CrossRef]
57. Li, Q.; Chen, L.; Gadinski, M.R.; Zhang, S.H.; Zhang, G.Z.; Li, H.U.; Iagodkine, E.; Haque, A.; Chen, L.-Q.; Jackson, T.N.; et al. Flexible high-temperature dielectric materials from polymer nanocomposites. *Nature* **2015**, *523*, 576–579. [CrossRef]
58. Yan, J.; Wang, Y.L.; Wang, C.M.; Ouyang, J. Boosting energy storage performance of low-temperature sputtered $CaBi_2Nb_2O_9$ thin film capacitors via rapid thermal annealing. *J. Adv. Ceram.* **2021**, *10*, 627–635. [CrossRef]
59. Yang, B.B.; Guo, M.Y.; Song, D.P.; Tang, X.W.; Wei, R.H.; Hu, L.; Yang, J.; Song, W.H.; Dai, J.M.; Lou, X.J.; et al. Energy storage properties in $BaTiO_3$-$Bi_{3.25}La_{0.75}Ti_3O_{12}$ thin films. *Appl. Phys. Lett.* **2018**, *113*, 183902. [CrossRef]
60. Tang, Z.; Chen, J.Y.; Yang, B.; Zhao, S.F. Energy storage performances regulated by layer selection engineering for doping in multi-layered perovskite relaxor ferroelectric films. *Appl. Phys. Lett.* **2019**, *114*, 163901. [CrossRef]

nanomaterials

Article

Magnonic Metamaterials for Spin-Wave Control with Inhomogeneous Dzyaloshinskii–Moriya Interactions

Fengjun Zhuo [1,2], Hang Li [1,*], Zhenxiang Cheng [3,*] and Aurélien Manchon [4,*]

[1] School of Physics and Electronics, Henan University, Kaifeng 475004, China; fengjun.zhuo@kaust.edu.sa
[2] Physical Science and Engineering Division (PSE), King Abdullah University of Science and Technology (KAUST), Thuwal 23955-6900, Saudi Arabia
[3] Institute for Superconducting and Electronic Materials, Australian Institute of Innovative Materials, Innovation Campus, University of Wollongong, Squires Way, North Wollongong, NSW 2500, Australia
[4] Aix Marseille University, CNRS, CINAM, 13288 Marseille, France
* Correspondence: hang.li@vip.henu.edu.cn (H.L.); cheng@uow.edu.au (Z.C.); aurelien.manchon@univ-amu.fr (A.M.)

Abstract: A magnonic metamaterial in the presence of spatially modulated Dzyaloshinskii–Moriya interaction is theoretically proposed and demonstrated by micromagnetic simulations. By analogy to the fields of photonics, we first establish magnonic Snell's law for spin waves passing through an interface between two media with different dispersion relations due to different Dzyaloshinskii–Moriya interactions. Based on magnonic Snell's law, we find that spin waves can experience total internal reflection. The critical angle of total internal reflection is strongly dependent on the sign and strength of Dzyaloshinskii–Moriya interaction. Furthermore, spin-wave beam fiber and spin-wave lens are designed by utilizing the artificial magnonic metamaterials with inhomogeneous Dzyaloshinskii–Moriya interactions. Our findings open up a rich field of spin waves manipulation for prospective applications in magnonics.

Keywords: spin waves; Dzyaloshinskii–Moriya interaction; ferromagnetism; spintronics

Citation: Zhuo, F.; Li, H.; Cheng, Z.; Manchon, A. Magnonic Metamaterials for Spin-Wave Control with Inhomogeneous Dzyaloshinskii–Moriya Interactions. *Nanomaterials* **2022**, *12*, 1159. https://doi.org/10.3390/nano12071159

Academic Editor: José Antonio Sánchez-Gil

Received: 17 February 2022
Accepted: 28 March 2022
Published: 31 March 2022

Publisher's Note: MDPI stays neutral with regard to jurisdictional claims in published maps and institutional affiliations.

1. Introduction

Magnonics (or magnon spintronics) is an emerging field concentrating on the generation, detection and manipulation of magnons, the quanta of spin-wave, in ferromagnetic or antiferromagnetic metals and insulators [1–9]. As spin waves in magnetic insulators exhibit both low energy dissipation and long coherence length, these constitute a competitive alternative to electronic devices and are deemed to be a promising candidate as a high-quality information carrier [10–13]. Over the past decades, many properties of spin waves have been demonstrated experimentally, in analogy with electromagnetic waves: excitation and propagation [14–18], reflection and refraction [19–22], interference and diffraction [23–25] and tunneling and the Doppler effect [26–28].

Thus, far, based on recent progress in the fabrication of magnetic nanostructures, various device concepts have been proposed, such as spin-wave logic gates and circuits [10,29,30], waveguides [31,32], multiplexors [33], splitter [34] and diodes [35]. The implementations of those devices is usually achieved by the application of external local magnetic fields [26], spin current [28,36] and magnetic textures (for example, the chiral domain wall) [29,31,37] to control the dispersion relation of spin waves, thereby, steering the spin-wave propagation properties. Despite the soundness of the concepts, however, there are some inherent drawbacks and obstacles to applications. First, generating a local high-frequency magnetic field on micro-sized devices complicates the structure design, and the local field is often spatially inhomogeneous, which can inhibit the benefits of the device [38]. In addition, unstable magnetic textures under external excitation and at room temperature may give rise to poor reliability and high bit-error rates. Therefore, it is desirable to find a new method to manipulate the propagation of spin waves.

Recent discoveries in graded-index magnonics and magnonic metamaterials provide a new way to manipulate spin-wave propagation [39,40], which is inspired by the fields of graded-index photonics (or photonic metamaterials) [41–43]. The core idea of graded-index magnonics is to manipulate spin-wave propagation by designing a spatially varied magnonic refractive index. In magnetic thin films with in-plane magnetization, the spin-wave dispersion relation described by the Landau–Lifshitz–Gilbert (LLG) equation exhibits a much more complex structure compared to the isotropic dispersion relation of light. This offers extremely rich opportunities to modulate the magnonic refractive index.

Up to now, it has been shown that the graded magnonic refractive index can be created by modification of the material properties, such as non-uniform saturation magnetization or exchange constant [44–47], the magnetic anisotropy [19,20] or the internal magnetic field [37,48]. This index can be also achieved by utilizing a non-uniform external magnetic field [39,49–51], electric field (voltage) [52,53] or temperature [54,55]. Therefore, graded-index magnonics are expected to overcome the current limitation of magnonics and pave feasible routes for the implementation of spin-wave devices.

In this paper, we theoretically propose a magnonic metamaterial, in which we modulate the refractive index of spin waves with the inhomogeneous Dzyaloshinskii–Moriya interaction (DMI) to avoid a barely controllable local magnetic field and unstable magnetic textures. The DMI is an antisymmetric exchange interaction arising from the lack of structural inversion symmetry in magnetic films [56,57]. It has been found both for bulk materials [58–60] and magnetic interfaces [61].

Here, we focus on a spatial inhomogeneous interfacial DMI present in ferromagnet/heavy metal (FM/HM) bilayers realized by tuning the thickness of ferromagnetic layer or HM layer [62–65], the degree of hybridization between 3d-5d states [66] or utilizing a local gating [67]. We begin our work by rapidly deriving the spin-wave dispersion relation with spatially modulated DMI. Then, we further study spin-wave refraction and reflection at the interface between two magnetic media with different DMI and build a generalized Snell's law of spin waves, similar to Snell's law in optics.

According to the magnonic Snell's law, spin-wave can also experience total internal reflection (TIR) at the DMI step interface when their incident angle is larger than a critical value (i.e., the critical angle). Moreover, magnonic Snell's law and TIR are observed and confirmed by micromagnetic simulations. Utilizing the artificial magnonic metamaterials based on spatially modulated DMI, a spin-wave fiber owing to TIR (which can transmit spin waves over a long distance) and a spin-wave lens holding tremendous possibility to build spin-wave circuits are proposed as proofs of concept.

The paper is organized as follows. In Section 2, we introduce our theoretical model and method. Detailed results of micromagnetic simulations are presented in Section 3. Then, we discuss the realization of spin-wave fibers and lenses in Section 4. Finally, we end the paper with a summary in Section 5.

2. Analytical Model

2.1. Magnonic Snell's Law

We consider a thin magnetic film in the $x - y$ plane with the thickness much smaller than lateral dimensions of the film ($L_z \ll L_x, L_y$), whose initial magnetization is homogeneous along the $\hat{y}$ direction. The magnetization dynamics are governed by the LLG Equation [68],

$$\frac{\partial \mathbf{m}}{\partial t} = -\frac{\gamma}{M_s}\mathbf{m} \times \mathbf{H}_{eff} + \alpha \mathbf{m} \times \frac{\partial \mathbf{m}}{\partial t}, \tag{1}$$

where $\mathbf{m}$ is the unit direction of local magnetization $\mathbf{M} = M_s\mathbf{m}$ with a saturation magnetization M_s. α is the phenomenological Gilbert damping constant, and γ is the gyromagnetic ratio. Here, $\mathbf{H}_{eff} = A^*\nabla^2\mathbf{m} - D^*(x)(\hat{\mathbf{z}} \times \nabla) \times \mathbf{m} - K^*m_y\hat{\mathbf{y}}$ is the effective field [69], and $A^* = 2A/\mu_0 M_s$, $D^*(x) = 2D(x)/\mu_0 M_s$, $K^* = 2K/\mu_0 M_s$. A is the symmetric exchange constant, $D(x)$ is the interfacial antisymmetric DMI constant spatially inhomogeneous along the x direction, K is the in-plane anisotropy and μ_0 is the permeability of vacuum. Under the perturbative approximation, the small-amplitude spin waves propagating in the $x - y$ plane take the following form [70]:

$$\mathbf{m} = \hat{\mathbf{y}} + \delta\mathbf{m} \exp\left[i(\hat{\mathbf{k}} \cdot \hat{\mathbf{r}} - wt)\right], \tag{2}$$

where $\delta\mathbf{m} = (\delta m_x, 0, \delta m_z)$ is the spin-wave contribution to magnetization ($|\delta\mathbf{m}| \ll 1$). $\delta\mathbf{k} = (\mathbf{k}_x, \mathbf{k}_y, 0)$ is the spin-wave wavevector. Considering the system shown in Figure 1, we use a DMI step (i.e., $D = D_1$ in medium A and $D = D_2$ in medium B) to induce a difference in spin-wave dispersion relations between two magnetic domains. Inserting Equation (2) into Equation (1) and neglecting higher order terms, we obtain the spin-wave dispersion relation in each region [71,72],

$$\omega(\mathbf{k}_n) = \gamma\mu_0(K^* + A^*\mathbf{k}_n^2 - D_n^*\mathbf{k}_{n,y}), \tag{3}$$

with $D_n^* = 2D_n/\mu_0 M_s$ and $\mathbf{k}_n = \sqrt{\mathbf{k}_{n,x}^2 + \mathbf{k}_{n,y}^2}$. The spin-wave group velocity is $\mathbf{v}_{\mathbf{g,n}} = \partial w/\partial \mathbf{k}_n = 2A^*\mathbf{k}_{g,n}$, where $\mathbf{k}_{g,n} = \mathbf{k}_n - \delta_n\hat{\mathbf{y}}$ and $\delta_n = D_n^*/2A^*$. To simplify the model, we assume that the group velocity is parallel to the phase velocity at each point of the dispersion relation—that is to say, the dispersion relation is isotropic.

Equation (3) represents an isofrequency circle with radius $\mathbf{k}_{g,n}$ in momentum space, whose center deviates from the origin by δ_n in $-\hat{\mathbf{y}}$ direction as illustrated in Figure 2a. Nevertheless, in the magnetic films with in-plane magnetization, the spin-wave dispersion relation is anisotropic at low frequencies, where dipolar contribution dominates. When increasing frequency, the isofrequency contours smoothly transform through elliptical to almost circular. Consequently, the dispersion relation is isotropic as determined by the exchange interactions at high frequencies. In the following simulations, we use quite high frequency spin waves (100 GHz), and thus the spin-wave dynamic is determined by the exchange interactions.

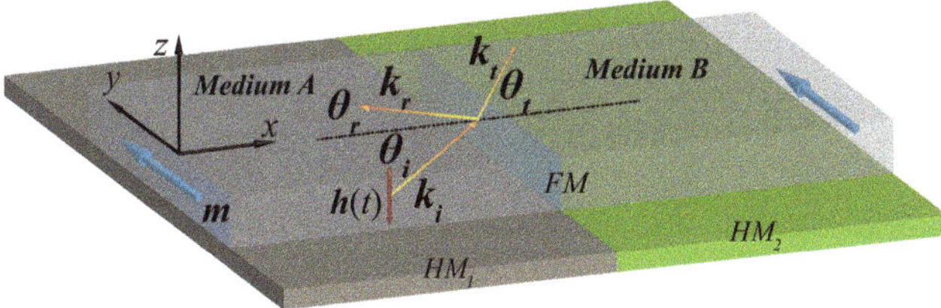

Figure 1. Schematic illustration of spin-wave transmission and reflection at an interface between media A and B with different interfacial DMI in a thin YIG film. The interfacial DMI step here is realized by utilizing two different HM layers (HM$_1$ and HM$_2$) below the YIG film. The blue arrows along the $\hat{y}$ direction denote the magnetization **m**. $\mathbf{k}_i$, $\mathbf{k}_t$ and $\mathbf{k}_r$ are the wave vectors of the incident, refracted and reflected spin-wave shown as the yellow and red arrows, respectively. $\theta_{i,t,r}$ denote their angles with respect to the interface normal. The red double-headed arrow shows the Gaussian distribution AC Magnetic field **h**(t) exciting the spin-wave.

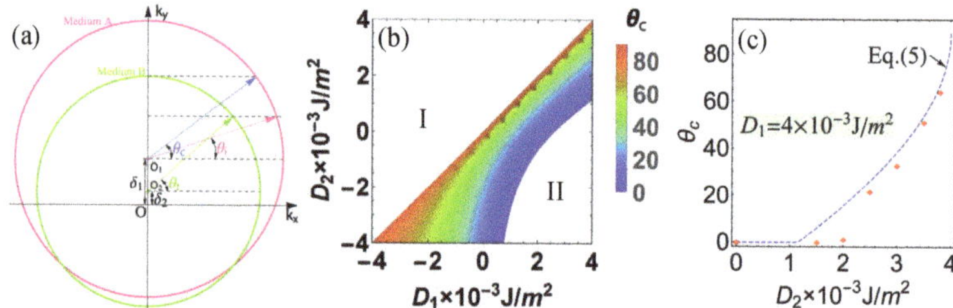

Figure 2. (**a**) Schematic illustrations of reflection and refraction of spin-wave at an interface between two different media in wave vector ($\mathbf{k}_x - \mathbf{k}_y$) space. The pink and green circles indicate the individual frequency contours of the allowed modes in the same-color-coded media A and B, respectively. The color-coded arrows denote the spin-wave vectors $\mathbf{k}$ propagating in each medium, as indicated by the incident (pink) and refracted (green) rays. The blue arrow denotes the critical angle. (**b**) Phase diagrams of critical angle θ_C in the $D_1 - D_2$ plane. No TIR exists in the white regions. (**c**) Critical angle θ_c as a function of DMI constants D_2 with a fixed DMI constant $D_1 = 4 \times 10^{-3}$ J/m^2. The symbols (red squares) are simulation data, and the solid curve represents the analytical results of Equation (5).

Based on translation symmetry considerations, the refraction angle obeys the generalized Snell's law, which guarantees continuity of the tangential components of the $\mathbf{k}$ vector across the DMI step interface along the $\hat{y}$ axis, such that $\mathbf{k}_{i,y} = \mathbf{k}_{t,y}$ [19,20,37]. Consequently, the generalized magnonic Snell's law based on modifying the dispersion relation with inhomogeneous DMI can be rewritten in the following form:

$$\mathbf{k}_{g,i} \sin\theta_i + \delta_i = \mathbf{k}_{g,t} \sin\theta_t + \delta_t, \tag{4}$$

where $\mathbf{k}_{g,n} = \sqrt{(\omega/\gamma\mu_0 - K^*)/A^* + \delta_n^2}$ is the value of $\mathbf{k}_{g,n}$. Here, the generalized Snell's law shown in Equation (4) is derived for an interface between two spin-wave media with different material parameters (interfacial DMI), which can be viewed as graded-index magnonic metamaterials. However, this is different from Snell's laws based on the interface inside magnetic textures, such as chiral domain walls (the interface formed by two opposite magnetic domains) [37].

2.2. Total Internal Reflection

Analogously to the case of electromagnetic waves in photonics or acoustic waves in phononics, spin waves are also expected to be completely reflected by the interface when a spin-wave travels from a *denser* medium with a higher refractive index to a *thinner* medium with a lower refractive index known as TIR. TIR occurs when the incident angle $\theta_i \geq \theta_c$, where θ_c is often called the critical angle. When $\theta_i = \theta_c$, the refracted spin-wave travels along the interface between the two media or the angle of refraction θ_t is $\pi/2$. According to the magnonic Snell's law in Equation (4), the critical angle can be expressed as

$$\theta_c = \arcsin\left(\frac{\mathbf{k}_{g,t} - \delta}{\mathbf{k}_{g,i}}\right), \tag{5}$$

where $\delta = \delta_i - \delta_t$. Specifically, Equation (5) shows that θ_c equals $\pi/2$ when the DMI is homogenous ($D_1 = D_2$), i.e., all incident spin waves are fully transmitted and no reflection occurs. Furthermore, when δ (the difference between DMI in two regions) is chosen to be large enough, a gap falls in between the two isofrequency circles and TIR occurs at all incident angles (i.e., $\theta_c = 0$). Equations (4) and (5) are the main analytical results in our paper.

3. Micromagnetic Simulations

To test the validity of these analytical findings in realistic situations, micromagnetic simulations have been proven to be an efficient tool for the investigation of spin-wave dynamics in various magnetic textures and geometries. The simulations here are performed in the GPU-accelerated micromagnetic simulations program MuMax3 [73], which solves the time-dependent LLG Equation (1) based on the finite difference method. In our simulations, we used typical magnetic parameters for YIG at zero temperature [74]: $M_s = 0.194 \times 10^5$ A/m, $A = 3.8$ pJ/m and $K = 10^4$ J/m^3.

All simulations presented here were performed for a thin film of size $L_x \times L_y \times L_z$, which discretized with cuboid meshes of dimensions $l_x \times l_y \times l_z$. The lateral dimensions of unit mesh (l_x and l_y) and the thickness of the film L_z are all smaller than the exchange length of YIG [3]. The simulations were implemented with the mesh size $2 \times 2 \times 2$ nm^3. The simulations were split into two stages: the static and dynamic stage. In the first stage, the static stage, the magnetic configuration is stabilized by minimization of the total energy starting from the random magnetic configuration with a high value of damping ($\alpha = 0.5$).

In the dynamic stage of the simulations, the equilibrium magnetic configuration was used to excite a spin-wave beam that propagates through the film with a small damping parameter ($\alpha = 0.0005$) to ensure long-distance propagation. During this step, a Gaussian type spin-wave beam was continuously generated by a harmonic dynamic external magnetic field following a Gaussian distribution function in a small rectangular region (red double-headed arrow shown in Figure 1). The detailed description of the Gaussian spin-wave beam generation procedure can be found in Ref. [75–77].

The Gaussian spin-wave beam is clearly visible and does not change with time after continuously exciting a sufficiently long time, which corresponds to a steady spin-wave propagation. Moreover, to avoid spin-wave reflection at the boundaries of the film, absorbing boundary conditions are applied on all boundaries by assigning a large damping constant ($\alpha = 1$) near the edges.

In order to verify the magnonic Snell's law in Equation (4) for the spin-wave propagation through a DMI step interface, we focus on a 4 µm $\times$ 4 µm $\times$ 2 nm nanowire. The spin-wave beams presented here are all exchange-dominated spin waves with 200 nm beam width and 30 nm wavelength generated by an external AC magnetic field with frequency $f = 100$ GHz. The phase diagram of the critical angle θ_c in the $D_1 - D_2$ plane is shown in Figure 2b. As $D_1 < D_2$ in the white region I, spin waves transmit from a thinner medium with a lower refractive index to a denser medium with a higher refractive index, and thus no TIR happens.

A gap falls in between the two isofrequency circles—in other words, TIR occurs in all incident angles when δ is chosen to be large enough as shown in the white region II. Figure 2c shows the critical angle as a function of the DMI constant D_2 in medium B, where the DMI constant of medium A is fixed at $D_1 = 4 \times 10^{-3}$ J/m^2. All incident angle spin waves are totally reflected at a small D_2 corresponding to Region II in Figure 2b. After that, the critical angle increases monotonically with D_2 and shows a good agreement with the analytical results.

In Figure 3a, we show the refracted angle θ_t as a function of the incident angle θ_i from micromagnetic simulations (red triangle) and the prediction from Equation (4) (blue curve) with DMI constants $D_1 = 4 \times 10^{-3}$ J/m^2 and $D_2 = 3.5 \times 10^{-3}$ J/m^2, respectively. The micromagnetic simulation for the five different incident angles, $\theta_i = 17°, 41.5°, 44°, 51.2°$ and $67°$, are displayed in Figure 3b–f. Figure 3b–d correspond to the refraction mode, and Figure 3e,f are the total reflection mode. Vertical dashed lines correspond to the interface at $x = 2000$ nm between medium A (left) and medium B (right).

The critical angle observed in our simulation is estimated to be $\theta_c = 51.2°$ as shown in Figure 3e. It is important to comment that the spin-wave propagation direction is not strictly perpendicular to the spin-wave wavefronts in our simulation. That is to say, it is easy to observe strong anisotropy in the propagation of spin waves. Typically, for in-plane magnetized films, spin waves dynamics are anisotropic. This means that iso-frequency dispersion relation lines (IFDRLs, slices of dispersion relations for particular frequencies) are not circular. Therefore, the group velocity and phase velocities (parallel to the wave-vector) are not parallel to each other, since the group velocity direction should be normal to the IFDRLs [78].

Such an intrinsic anisotropy called spin-wave collimation effect is common in ferromagnetic films with the magnetization fixed in the plane of the film by an external magnetic field or a strong in-plane anisotropy [75–77]. However, in the present case, the symmetry of the spin-wave dispersion relation is broken due to the presence of DMI, and the wave vector can be shifted from the direction perpendicular to spin-wave wavefronts [70]. This anisotropy decreases with the increasing frequency of the spin-wave but it is still present at the high frequency $f = 100$ GHz assumed in our simulations.

Moreover, a lateral shift $\triangle_{GH}$ of the spin-wave beam is observed at the interface between the reflected and the incident beams, which is called the Goos–Hänchen (GH) effect. The GH effect for spin-waves was reported in Refs. [75–78]. Furthermore, detailed investigations elucidating the role of inhomogeneous DMI on the GH shift in the reflection of the spin-wave at the interface were discussed in Ref. [22].

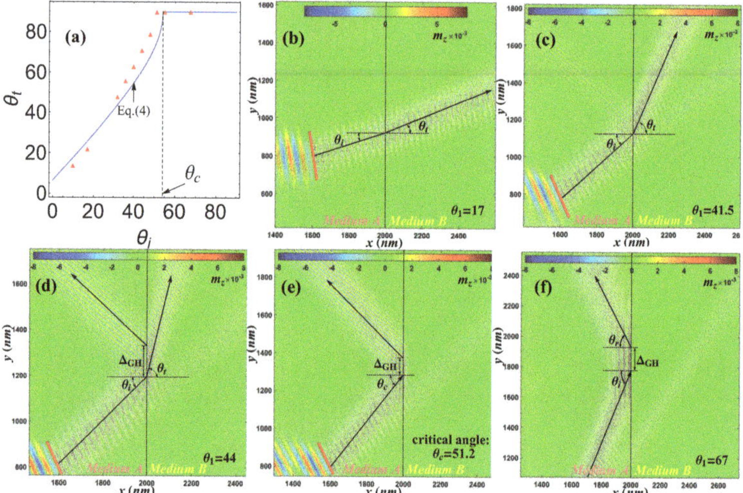

Figure 3. (a) The refracted angle as a function of the incident angle. Vertical dashed and solid lines correspond to the critical angle θ_c. (b–f) The micromagnetic simulations results for spin-wave beam reflection and refraction under different incident angles (b) $\theta_i = 17°$, (c) $\theta_i = 41.5°$, (d) $\theta_i = 44°$, (e) $\theta_i = 51.2°$ and (f) $\theta_i = 67°$. The DMI constants in medium A and B are $D_1 = 4 \times 10^{-3}$ J/m^2 and $D_2 = 3.5 \times 10^{-3}$ J/m^2, respectively. The color map shows the z component of the magnetization in the snapshot of micromagnetic simulations at some selected time. The black solid lines correspond to the rays of the incident and refractive beams. The red rectangular area is the excitation area of the spin-wave, and the exciting field frequency is $f = 100$ GHz.

4. Spin-Wave Fiber and Lens

We now turn to the realization of the spin-wave fiber and spin-wave lens, which are important to manipulate spin waves in spin-wave circuitry. Two kinds of spin-wave fiber have been proposed and designed, one based on the TIR by the magnetic domain wall [37] and the other based on the TIR in the medium with a uniform external magnetic field [51]. Here, utilizing the TIR at the interface with a DMI step, we propose a new type of spin-wave fiber as shown in Figure 4a of system size 12 μm × 1.6 μm × 2 nm. The DMI constant in the core (region II, $|x| \leq 400$ nm) is 0.5×10^{-3} J/m^2 surrounded by transparent cladding FM layers (region I, $|x| \geq 400$ nm) with a lower index of refraction ($D = -0.4 \times 10^{-3}$ J/m^2). Two DMI steps are formed with the critical angle $\theta_c = 48°$.

The upper one is located at $x = -400$ nm and the lower one is located at $x = 400$ nm. In Figure 4a, the spin-wave beam born at the middle of the nanowire (blue bar) propagates inside the core with an incident angle of $52°$ greater than θ_c. This is different from the unidirectional spin-wave fiber based on domain walls [37]. The fiber here is fully bidirectional for both right/left-moving spin-wave beams when the incident angle is greater than the critical angle.

More interestingly, a bound spin-wave mode propagates a long distance inside the DMI step interface as illustrated in the inset of Figure 4a. Similar to the bound spin wave mode inside a domain wall, which acts as a local potential well for spin waves [31,79], a DMI step also creates an imaginary potential well for the bound spin wave mode [80]. The details will be discussed in our future publications.

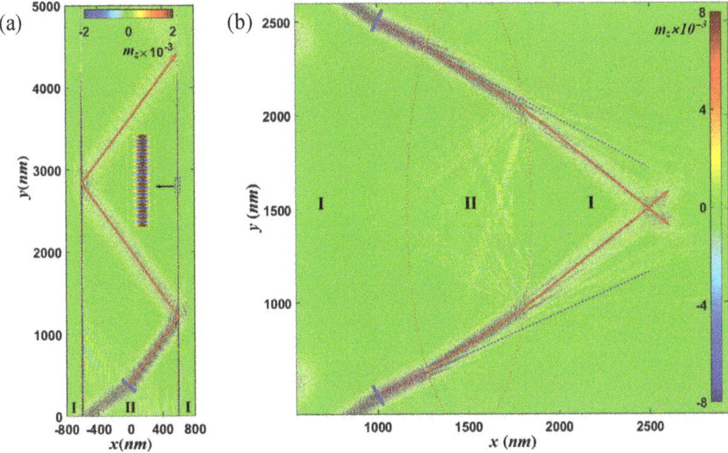

Figure 4. (**a**) Schematic illustration of a spin-wave fiber. The inset shows the enlarged figure at the interface. (**b**) Schematic illustration of a spin-wave convex lens. In all of the above figures, the color map shows z component of the magnetization in the snapshot of micromagnetic simulations at some selected time. The spin-wave trajectories are represented by solid red lines with an arrow. The simulated propagation of the spin wave excited by a AC source in blue bars with an exciting frequency $f = 100$ GHz.

A fundamental building block in spin-wave circuitry is a spin-wave lens that can focus or diverge spin-wave beams. Since the dispersion relation strongly depends on DMI constant, we propose a spin-wave lens by tuning the DMI distribution in the film. Figure 4b illustrates an example of a spin-wave convex lens (region I inside red dotted lines) with a DMI constant inside/outside the lens $D = 0.5/-0.4 \times 10^{-3}$ J/m^2, respectively. The size of the sample presented here is 6 μm × 6 μm × 2 nm.

Comparing the solid blue lines along the incident spin-wave beam propagation direction and the spin-wave trajectory (solid red lines with an arrow) passing through the lens, it is easy to observe focusing in the propagation of spin waves. Furthermore, a concave spin-wave lens can be obtained by reversing the DMI constants of regions I and II, which can be used to spit the spin-wave beams. Consequently, we believe that the inhomogeneous DMI can be a good playground to study spin-wave beam propagation [81].

5. Conclusions

In conclusion, we both theoretically and numerically studied spin-wave beam propagation in a two-dimensional ferromagnetic film with an inhomogeneous interfacial DMI. Utilizing a spatially varied magnonic refractive index introduced by the variation of DMI, a magnonic metamaterial or graded-index magnonic material can be realized. Snell's law and TIR for spin waves were predicted with a DMI step interface. Moreover, we designed and studied spin-wave fibers and spin-wave lenses via micromagnetic simulations. We believe that our findings shall open up alternative directions for building reconfigurable, stabilized and scalable spin-wave circuitry in magnon introspection devices.

However, the parameters that we adopted in our simulations to investigate spin-wave propagation in the presence of spatially modulated DMI are not meant to represent a specific material but rather to explore the physical conditions under which the spin-wave total reflection occurs. From the materials standpoint, we acknowledge that the dual requirements of low damping and large DMI may seem incompatible since spin–orbit coupling originating from the adjacent heavy metal layer is detrimental to the former but central to the latter.

The excitation of short-wavelength propagating spin waves with a wavelength of 45 nm in a YIG thin film covered by $Co_{25}Fe_{75}$ nanowires was reported in a recent experiment [82], where the effective damping was only enhanced to about 10^{-3}. Recent progress in materials science has proven that certain magnetic insulators do possess sizable DMIs either in their bulk [83–85] or at the interface [86–88]. Although these values remain small (typically $\sim 10^{-3}$–$10^{-2} \mathrm{mJ/m^2}$), these results open interesting perspectives for the achievement of large DMIs in magnetic insulators.

Author Contributions: Conceptualization, F.Z. and A.M.; methodology, F.Z.; software, F.Z.; validation, F.Z. and A.M.; formal analysis, F.Z., H.L., Z.C. and A.M.; investigation, F.Z.; resources, H.L. and A.M.; data curation, F.Z.; writing—original draft preparation, F.Z.; writing—review and editing, H.L., Z.C. and A.M.; visualization, H.L., Z.C. and A.M.; supervision, A.M.; project administration, A.M.; funding acquisition, H.L., Z.C. and A.M. All authors have read and agreed to the published version of the manuscript.

Funding: F.Z. and H.L. acknowledge the support from National Key R&D Program of China (No. 2018YFB0407600), Henan University (No. CJ3050A0240050) and National Natural Science Foundation of China (No. 11804078). F.Z. was supported by King Abdullah University of Science and Technology (KAUST). A.M. acknowledges support from the Excellence Initiative of Aix-Marseille Université— A*Midex, a French "Investissements d'Avenir" program. Z.C. acknowledges the support from Grant No. ARC (DP190100150).

Institutional Review Board Statement: Not applicable.

Informed Consent Statement: Not applicable.

Data Availability Statement: The data that support the findings of this study are available upon reasonable request from the authors.

Conflicts of Interest: The authors declare no conflict of interest.

References

1.   Kruglyak, V.V.; Hicken, R.J. Magnonics: Experiment to prove the concept. *J. Magn. Magn. Mater.* **2006**, *306*, 191. [CrossRef]
2.   Kruglyak, V.V.; Demokritov, S.O.; Grundler, D. Magnonics. *J. Phys. D Appl. Phys.* **2010**, *43*, 264001. [CrossRef]
3.   Serga, A.A.; Chumak, A.V.; Hillebrands, B. YIG magnonics. *J. Phys. D Appl. Phys.* **2010**, *43*, 264002. [CrossRef]
4.   Lenk, B.; Ulrichs, H.; Garbs, F.; Münzenberg, M. The Building Blocks of Magnonics. *Phys. Rep.* **2011**, *507*, 107. [CrossRef]
5.   Demokritov, S.O.; Slavin, A.N. *Magnonics: From Fundamentals to Applications*, 1st ed.; Springer: New York, NY, USA, 2013; p. 125.
6.   Chumak, A.V.; Vasyuchka, V.I.; Serga, A.A.; Hillebrands, B. Magnon spintronics. *Nat. Phys.* **2015**, *11*, 453. [CrossRef]
7.   Baltz, V.; Manchon, A.; Tsoi, M.; Moriyama, T.; Ono, T.; Tserkovnyak, Y. Antiferromagnetic spintronics. *Rev. Mod. Phys.* **2018**, *90*, 015005. [CrossRef]
8.   Barman, A.; Gubbiotti, G.; Ladak, S.; Adeyeye, A.O.; Krawczyk, M.; Gräfe, J.; Adelmann, C.; Cotofana, S.; Naeemi, A.; Vasyuchka, V.I.; et al. The 2021 Magnonics Roadmap. *J. Phys. Condens. Matter* **2021**, *33*, 413001. [CrossRef] [PubMed]
9.   Bonbien, V.; Zhuo, F.; Salimath, A.; Ly, O.; About, A.; Manchon, A. Topological aspects of antiferromagnets. *J. Phys. D Appl. Phys.* **2022**, *55*, 103002. [CrossRef]
10. Khitun, A.; Bao, M.; Wang, K.L. Magnonic logic circuits. *J. Phys. D Appl. Phys.* **2010**, *43*, 264005. [CrossRef]
11. Chumak, A.V.; Serga, A.A.; Hillebrands, B. Magnon transistor for all-magnon data processing. *Nat. Commun.* **2014**, *5*, 4700. [CrossRef]
12. Zhuo, F.; Li, H.; Manchon, A. Topological phase transition and thermal Hall effect in kagome ferromagnets. *Phys. Rev. B* **2021**, *104*, 144422. [CrossRef]

13. Zhuo, F.; Li, H.; Manchon, A. Topological thermal Hall effect and magnonic edge states in kagome ferromagnets with bond anisotropy. *New J. Phys.* **2022**, *24*, 023033. [CrossRef]
14. Liu, Z.; Giesen, F.; Zhu, X.; Sydora, R.D.; Freeman, M.R. Spin Wave Dynamics and the Determination of Intrinsic Damping in Locally Excited Permalloy Thin Films. *Phys. Rev. Lett.* **2007**, *98*, 087201. [CrossRef] [PubMed]
15. Serga, A.A.; Demokritov, S.O.; Hillebrands, B.; Slavin, A.N. Self-Generation of Two-Dimensional Spin-Wave Bullets. *Phys. Rev. Lett.* **2014**, *92*, 117203. [CrossRef] [PubMed]
16. Covington, M.; Crawford, T.M.; Parker, G.J. Time-Resolved Measurement of Propagating Spin Waves in Ferromagnetic Thin Films. *Phys. Rev. Lett.* **2002**, *89*, 237202. [CrossRef] [PubMed]
17. Demidov, V.E.; Jersch, J.; Demokritov, S.O.; Rott, K.; Krzysteczko, P.; Reiss, G. Transformation of propagating spin-wave modes in microscopic waveguides with variable width. *Phys. Rev. B* **2009**, *79*, 054417. [CrossRef]
18. Demidov, V.E.; Kostylev, M.P.; Rott, K.; Münchenberger, J.; Reiss, G.; Demokritov, S.O. Excitation of short-wavelength spin waves in magnonic waveguides. *Appl. Phys. Lett.* **2011**, *99*, 082507. [CrossRef]
19. Stigloher, J. Snell's Law for Spin Waves. *Phys. Rev. Lett.* **2016**, *117*, 037204. [CrossRef]
20. Kim, S.K.; Choi, S.; Lee, K.S.; Han, D.S.; Jung, D.E.; Choi, Y.S. Negative refraction of dipole-exchange spin waves through a magnetic twin interface in restricted geometry. *Appl. Phys. Lett.* **2008**, *92*, 212501. [CrossRef]
21. Wang, Z.; Zhang, B.; Cao, Y.; Yan, P. Probing the Dzyaloshinskii–Moriya Interaction via the Propagation of Spin Waves in Ferromagnetic Thin Films. *Phys. Rev. Applied* **2018**, *10*, 054018. [CrossRef]
22. Wang, Z.; Cao, Y.; Yan, P. Goos–Hänchen effect of spin waves at heterochiral interfaces. *Phys. Rev. B* **2019**, *100*, 064421. [CrossRef]
23. Choi, S.K.; Lee, K.S.; Kim, S.K. Spin-wave interference. *Appl. Phys. Lett.* **2006**, *89*, 062501. [CrossRef]
24. Perzlmaier, K.; Woltersdorf, G.; Back, C.H. Observation of the propagation and interference of spin waves in ferromagnetic thin films. *Phys. Rev. B* **2008**, *77*, 054425. [CrossRef]
25. Birt, D.R.; Gorman, B.O.; Tsoi, M.; Li, X.; Demidov, V.E.; Demokritov, S.O. Diffraction of spin waves from a submicrometer-size defect in a microwaveguide. *Appl. Phys. Lett.* **2009**, *95*, 122510. [CrossRef]
26. Demokritov, S.O.; Serga, A.A.; André, A.; Demidov, V.E.; Kostylev, M.P.; Hillebrands, B.; Slavin, A.N. Tunneling of Dipolar Spin Waves through a Region of Inhomogeneous Magnetic Field. *Phys. Rev. Lett.* **2004**, *93*, 047201. [CrossRef]
27. Stancil, D.D.; Henty, B.E.; Cepni, A.G.; Van't Hof, J.P. Observation of an inverse Doppler shift from left-handed dipolar spin waves. *Phys. Rev. B* **2006**, *74*, 060404. [CrossRef]
28. Vlaminck, V.; Bailleul, M. Current-Induced Spin-Wave Doppler Shift. *Science* **2008**, *322*, 410. [CrossRef] [PubMed]
29. Hertel, R.; Wulfhekel, W.; Kirschner, J. Domain-Wall Induced Phase Shifts in Spin Waves. *Phys. Rev. Lett.* **2004**, *93*, 257202. [CrossRef]
30. Lee, K.S.; Kim, S.K. Conceptual design of spin wave logic gates based on a Mach–Zehnder-type spin wave interferometer for universal logic functions. *J. Appl. Phys.* **2008**, *104*, 053909. [CrossRef]
31. Sanchez, F.G.; Borys, P.; Soucaille, R.; Adam, J.P.; Stamps, R.L.; Kim, J.V. Narrow Magnonic Waveguides Based on Domain Walls. *Phys. Rev. Lett.* **2015**, *114*, 247206. [CrossRef]
32. Mulkers, J.; Waeyenberge, B.V.; Milošević, M.V. Effects of spatially engineered Dzyaloshinskii–Moriya interaction in ferromagnetic films. *Phys. Rev. B* **2017**, *95*, 144401. [CrossRef]
33. Vogt, K.; Fradin, F.Y.; Pearson, J.E.; Sebastian, T.; Bader, S.D.; Hillebrands, B.; Hoffmann, A.; Schultheiss, H. Realization of a spin-wave multiplexer. *Nat. Commun.* **2014**, *5*, 3727. [CrossRef] [PubMed]
34. Sadovnikov, A.V.; Davies, C.S.; Grishin, S.V.; Kruglyak, V.V.; Romanenko, D.V.; Sharaevskii, Y.P.; Nikitov, S.A. Magnonic beam splitter: The building block of parallel magnonic circuitry. *Appl. Phys. Lett.* **2015**, *106*, 192406. [CrossRef]
35. Lan, J.; Yu, W.; Wu, R.; Xiao, J. Spin-Wave Diode. *Phys. Rev. X* **2015**, *5*, 041049. [CrossRef]
36. Seo, S.M.; Lee, K.J.; Yang, H.; Ono, T. Current-Induced Control of Spin-Wave Attenuation. *Phys. Rev. Lett.* **2009**, *102*, 147202. [CrossRef] [PubMed]
37. Yu, W.; Lan, J.; Wu, R.; Xiao, J. Magnetic Snell's law and spin-wave fiber with Dzyaloshinskii–Moriya interaction. *Phys. Rev. B* **2016**, *94*, 140410. [CrossRef]
38. Jamali, M.; Kwon, J.H.; Seo, S.M.; Lee, K.J.; Yang, H. Spin wave nonreciprocity for logic device applications. *Sci. Rep.* **2013**, *3*, 3160. [CrossRef] [PubMed]
39. Davies, C.S.; Francis, A.; Sadovnikov, A.V.; Chertopalov, S.V.; Bryan, M.T.; Grishin, S.V.; Allwood, D.A.; Sharaevskii, Y.P.; Nikitov, S.A.; Kruglyak, V.V. Towards graded-index magnonics: Steering spin waves in magnonic networks. *Phys. Rev. B* **2015**, *92*, 020408. [CrossRef]
40. Davies, C.S.; Kruglyak, V.V. Graded-index magnonics. *Low Temp. Phys.* **2015**, *41*, 760. [CrossRef]
41. Marchand, E.W. *Gradient Index Optics*, 1st ed.; ScienceDirect: London, UK, 1978; p. 125.
42. Chen, H.; Chan, C.T.; Sheng, P. Transformation optics and metamaterials. *Nat. Mater.* **2010**, *9*, 387. [CrossRef] [PubMed]
43. Pendry, J.B.; Domínguez, A.I.F.; Luo, Y.; Zhao, R. Capturing photons with transformation optics. *Nat. Phys.* **2013**, *9*, 518. [CrossRef]
44. Dadoenkova, Y.S.; Dadoenkova, N.N.; Lyubchanskii, I.L.; Sokolovskyy, M.L.; Kłos, J.W.; Romero-Vivas, J.; Krawczyk, M. Huge Goos–Hänchen effect for spin waves: A promising tool for study magnetic properties at interfaces. *Appl. Phys. Lett.* **2012**, *101*, 042404. [CrossRef]
45. Xi, H.; Xue, S. Spin-wave propagation and transmission along magnetic nanowires in long wavelength regime. *J. Appl. Phys.* **2007**, *101*, 123905. [CrossRef]

46. Xi, H.; Wang, X.; Zheng, Y.; Ryan, P.J. Spin wave propagation and coupling in magnonic waveguides. *J. Appl. Phys.* **2008**, *104*, 063921. [CrossRef]

47. Vogel, M.; Aßmann, R.; Pirro, P.; Chumak, A.V.; Hillebrands, B.; Freymann, G.V. Control of Spin-Wave Propagation using Magnetisation Gradients. *Sci. Rep.* **2018**, *8*, 11099. [CrossRef]

48. Xing, X.; Zhou, Y. Fiber optics for spin waves. *NPG Asia Mater.* **2016**, *8*, e246. [CrossRef]

49. Perez, N.; Diaz, L.L. Magnetic field induced spin-wave energy focusing. *Phys. Rev. B* **2015**, *92*, 014408. [CrossRef]

50. Houshang, A.; Iacocca, E.; Dürrenfeld, P.; Sani, S.; Akerman, J.; Dumas, R. Spin-wave-beam driven synchronization of nanocontact spin-torque oscillators. *Nat. Nanotechnol.* **2016**, *11*, 280. [CrossRef] [PubMed]

51. Gruszecki, P.; Krawczyk, M. Spin wave beam propagation in ferromagnetic thin film with graded refractive index: Mirage effect and prospective applications. *Phys. Rev. B* **2018**, *97*, 094424.

52. Wang, S.; Guan, X.; Cheng, X.; Lian, C.; Huang, T.; Miao, X. Spin-wave propagation steered by electric field modulated exchange interaction. *Sci. Rep.* **2016**, *6*, 31783. [CrossRef]

53. Kakizakai, H.; Yamada, K.; Ando, F.; Kawaguchi, M.; Koyama, T.; Kim, S.; Moriyama, T.; Chiba, D.; Ono, T. Influence of sloped electric field on magnetic-field-induced domain wall creep in a perpendicularly magnetized Co wire. *Jpn. J. Appl. Phys.* **2017**, *56*, 050305. [CrossRef]

54. Vogel, M.; Chumak, A.V.; Waller, E.H.; Langner, T.; Vasyuchka, V.I.; Hillebrands, B.; Freymann, G.V. Optically reconfigurable magnetic materials. *Nat. Phys.* **2015**, *11*, 487. [CrossRef]

55. Busse, F.; Mansurova, M.; Lenk, B.; Ehe, M.; Münzenberg, M. A scenario for magnonic spin-wave traps. *Sci. Rep.* **2015**, *5*, 12824. [CrossRef] [PubMed]

56. Dzyaloshinskii, I.E. A thermodynamic theory of "weak" ferromagnetism of antiferromagnetics. *J. Phys. Chem. Solids* **1958**, *4*, 241. [CrossRef]

57. Moriya, T. Anisotropic Superexchange Interaction and Weak Ferromagnetism. *Phys. Rev.* **1960**, *120*, 91. [CrossRef]

58. Mühlbauer, S.; Binz, B.; Jonietz, F.; Pfleiderer, C.; Rosch, A.; Neubauer, A.; Georgii, R.; Böni, P. Skyrmion Lattice in a Chiral Magnet. *Science* **2009**, *323*, 915. [CrossRef]

59. Huang, S.X.; Chien, C.L. Extended Skyrmion Phase in Epitaxial FeGe(111) Thin Films. *Phys. Rev. Lett.* **2012**, *108*, 267201. [CrossRef] [PubMed]

60. Zhuo, F.; Sun, Z.Z. Field-driven Domain Wall Motion in Ferromagnetic Nanowires with Bulk Dzyaloshinskii–Moriya Interaction. *Sci. Rep.* **2012**, *6*, 25122. [CrossRef] [PubMed]

61. Fert, A.; Cros, V.; Sampaio, J. Skyrmions on the track. *Nat. Nanotechnol.* **2013**, *8*, 152. [CrossRef] [PubMed]

62. Chen, G.; Zhu, J.; Quesada, A.; Li, J.; N'Diaye, A.T.; Huo, Y.; Ma, T.P.; Chen, Y.; Kwon, H.Y.; Won, C.; et al. Novel Chiral Magnetic Domain Wall Structure in Fe/Ni/Cu(001) Films. *Phys. Rev. Lett.* **2013**, *110*, 177204. [CrossRef] [PubMed]

63. Chen, G.; Ma, T.; N'Diaye, A.T.; Kwon, H.; Won, C.; Wu, Y.; Schmid, A.K. Tailoring the chirality of magnetic domain walls by interface engineering. *Nat. Commun.* **2013**, *4*, 2671. [CrossRef]

64. Torrejon, J.; Kim, J.; Sinha, J.; Mitani, S.; Hayashi, M.; Yamanouchi, M.; Ohno, H. Interface control of the magnetic chirality in CoFeB/MgO heterostructures with heavy-metal underlayers. *Nat. Commun.* **2014**, *5*, 4655. [CrossRef]

65. Tacchi, S.; Troncoso, R.E.; Ahlberg, M.; Gubbiotti, G.; Madami, M.; Akerman, J.; Landeros, P. Interfacial Dzyaloshinskii–Moriya Interaction in Pt/CoFeB Films: Effect of the Heavy-Metal Thickness. *Phys. Rev. Lett.* **2017**, *118*, 147201. [CrossRef] [PubMed]

66. Belabbes, A.; Bihlmayer, G.; Bechstedt, F.; Blügel, S.; Manchon, A. Hund's Rule-Driven Dzyaloshinskii–Moriya Interaction at 3d-5d Interfaces. *Phys. Rev. Lett.* **2016**, *117*, 247202. [CrossRef] [PubMed]

67. Nawaoka, K.; Miwa, S.; Shiota, Y.; Mizuochi, N.; Suzuki, Y. Voltage induction of interfacial Dzyaloshinskii–Moriya interaction in Au/Fe/MgO artificial multilayer. *Appl. Phys. Express* **2015**, *8*, 063004. [CrossRef]

68. Gilbert, T.L. A phenomenological theory of damping in ferromagnetic materials. *IEEE Trans. Magn.* **2004**, *40*, 3443–3449. [CrossRef]

69. Bogdanov, A.N.; Rößler, U.K. Chiral Symmetry Breaking in Magnetic Thin Films and Multilayers. *Phys. Rev. Lett.* **2001**, *87*, 037203. [CrossRef]

70. Moon, J.H.; Seo, S.M.; Lee, K.J.; Kim, K.W.; Ryu, J.; Lee, H.W.; McMichael, R.D.; Stiles, M.D. Spin-wave propagation in the presence of interfacial Dzyaloshinskii–Moriya interaction. *Phys. Rev. B* **2013**, *88*, 184404. [CrossRef]

71. Di, K.; Zhang, V.L.; Lim, H.S.; Ng, S.C.; Kuok, M.H.; Qiu, X.; Yang, H. Asymmetric spin-wave dispersion due to Dzyaloshinskii–Moriya interaction in an ultrathin Pt/CoFeB film. *Appl. Phys. Lett.* **2015**, *106*, 052403. [CrossRef]

72. Manchon, A.; Ndiaye, P.; Moon, J.H.; Lee, H.W.; Lee, K.J. Magnon-mediated Dzyaloshinskii–Moriya torque in homogeneous ferromagnets. *Phys. Rev. B* **2014**, *90*, 224403. [CrossRef]

73. Vansteenkiste, A.; Leliaert, J.; Dvornik, M.; Helsen, M.; Garcia-Sanchez, F.; Van Waeyenberge, B. The design and verification of MuMax3. *AIP Adv.* **2014**, *4*, 107133. [CrossRef]

74. Klingler, S.; Chumak, A.V.; Mewes, T.; Khodadadi, B.; Mewes, C.; Dubs, C.; Surzhenko, O.; Hillebrands, B.; Conca, A. Measurements of the exchange stiffness of YIG films using broadband ferromagnetic resonance techniques. *J. Phys. D Appl. Phys.* **2015**, *48*, 015001. [CrossRef]

75. Gruszecki, P.; Romero-Vivas, J.; Dadoenkova, Y.S.; Dadoenkova, N.N.; LyubchanskiiI, L.; Krawczyk, M. Goos–Hänchen effect and bending of spin wave beams in thin magnetic films. *Appl. Phys. Lett.* **2014**, *105*, 242406. [CrossRef]

76. Gruszecki, P.; Dadoenkova, Y.S.; Dadoenkova, N.N.; Lyubchanskii, I.L.; Vivas, J.R.; Guslienko, K.Y.; Krawczyk, M. Influence of magnetic surface anisotropy on spin wave reflection from the edge of ferromagnetic film. *Phys. Rev. B* **2015**, *92*, 054427. [CrossRef]

77. Gruszecki, P.; Mailyan, M.; Gorobets, O.; Krawczyk, M. Goos–Hänchen shift of a spin-wave beam transmitted through anisotropic interface between two ferromagnets. *Phys. Rev. B* **2017**, *95*, 014421. [CrossRef]

78. Klos, J.W.; Gruszecki, P.; Serebryannikov, A.E.; Krawczyk, M. All-Angle Collimation for Spin Waves. *IEEE Magn. Lett.* **2015**, *6*, 3500804. [CrossRef]

79. Sanchez, F.G.; Borys, P.; Vansteenkiste, A.; Kim, J.V.; Stamps, R.L. Nonreciprocal spin-wave channeling along textures driven by the Dzyaloshinskii–Moriya interaction. *Phys. Rev. B* **2014**, *89*, 224408. [CrossRef]

80. Lee, S.J.; Moon, J.H.; Lee, H.W.; Lee, K.J. Spin-wave propagation in the presence of inhomogeneous Dzyaloshinskii–Moriya interactions. *Phys. Rev. B* **2017**, *96*, 184433. [CrossRef]

81. Korner, H.S.; Stigloher, J.; Back, C.H. Excitation and tailoring of diffractive spin-wave beams in NiFe using nonuniform microwave antennas. *Phys. Rev. B* **2017**, *96*, 100401. [CrossRef]

82. Wang, H.; Flacke, L.; Wei, W.; Liu, S.; Jia, H.; Chen, J.; Sheng, L.; Zhang, J.; Zhao, M.; Guo, C.; et al. Sub-50 nm wavelength spin waves excited by low-damping $Co_{25}Fe_{75}$ nanowires. *Appl. Phys. Lett.* **2021**, *119*, 152402. [CrossRef]

83. Janson, O.; Rousochatzakis, I.; Tsirlin, A.A.; Belesi, M.; Leonov, A.A.; Rößler, U.K.; van den Brink, J.; Rosner, H. The quantum nature of skyrmions and half-skyrmions in Cu_2OSeO_3. *Nat. Commun.* **2014**, *5*, 5376. [CrossRef] [PubMed]

84. Qian, F.; Bannenberg, L.; Wilhelm, H.; Chaboussant, G.; Debeer-Schmitt, L.M.; Schmidt, M.P.; Aqeel, A.; Palstra, T.T.M.; Brück, E.; Lefering, A.J.E.; et al. New magnetic phase of the chiral skyrmion material Cu_2OSeO_3. *Sci. Adv.* **2018**, *4*, eaat7323. [CrossRef] [PubMed]

85. Deng, L.; Wu, H.C.; Litvinchuka, A.P.; Yuan, N.F.Q.; Lee, J.-J.; Dahal, R.; Berger, H.; Yang, H.-D.; Chu, C.-W. Room-temperature skyrmion phase in bulk Cu_2OSeO_3 under high pressures. *Proc. Natl. Acad. Sci. USA* **2020**, *117*, 8783–8787. [CrossRef] [PubMed]

86. Avci, C.O.; Rosenberg, E.; Caretta, L.; Büttner, F.; Mann, M.; Marcus, C.; Bono, D.; Ross, C.A.; Beach, G.S.D. Interface-driven chiral magnetism and current-driven domain walls in insulating magnetic garnets. *Nat. Nanotechnol.* **2019**, *14*, 561–566. [CrossRef] [PubMed]

87. Caretta, L.; Rosenberg, E.; Büttner, F.; Fakhrul, T.; Gargiani, P.; Valvidares, M.; Chen, Z.; Reddy, P.; Muller, D.A.; Ross, C.A.; et al. Interfacial Dzyaloshinskii–Moriya interaction arising from rare-earth orbital magnetism in insulating magnetic oxides. *Nat. Commun.* **2020**, *11*, 1090. [CrossRef] [PubMed]

88. Ding, S.; Baldrati, L.; Ross, A.; Ren, Z.; Wu, R.; Becker, S.; Yang, J.; Jakob, G.; Brataas, A.; Kläui, M. Identifying the origin of the nonmonotonic thickness dependence of spin–orbit torque and interfacial Dzyaloshinskii–Moriya interaction in a ferrimagnetic insulator heterostructure. *Phys. Rev. B* **2020**, *102*, 054425. [CrossRef]

nanomaterials

MDPI

Article

High-Magnetic-Sensitivity Magnetoelectric Coupling Origins in a Combination of Anisotropy and Exchange Striction

Zhuo Zeng [1,†], Xiong He [1,†], Yujie Song [1], Haoyu Niu [1], Dequan Jiang [1], Xiaoxing Zhang [1], Meng Wei [1], Youyuan Liang [1], Hao Huang [1], Zhongwen Ouyang [1], Zhenxiang Cheng [2,*] and Zhengcai Xia [1,*]

[1] Wuhan National High Magnetic Field Center & School of Physics, Huazhong University of Science and Technology, Wuhan 430074, China
[2] Institute for Superconducting and Electronic Materials, Australia Institute for Innovation Materials, Innovation Campus, University of Wollongong, Squires Way, North Wollongong, NSW 2500, Australia
*   Correspondence: cheng@uow.edu.au (Z.C.); xia9020@hust.edu.cn (Z.X.)
†   These authors contributed equally to this work.

Abstract: Magnetoelectric (ME) coupling is highly desirable for sensors and memory devices. Herein, the polarization (P) and magnetization (M) of the $DyFeO_3$ single crystal were measured in pulsed magnetic fields, in which the ME behavior is modulated by multi-magnetic order parameters and has high magnetic-field sensitivity. Below the ordering temperature of the Dy^{3+}-sublattice, when the magnetic field is along the c-axis, the P (corresponding to a large critical field of 3 T) is generated due to the exchange striction mechanism. Interestingly, when the magnetic field is in the ab-plane, ME coupling with smaller critical fields of 0.8 T (a-axis) and 0.5 T (b-axis) is triggered. We assume that the high magnetic-field sensitivity results from the combination of the magnetic anisotropy of the Dy^{3+} spin and the exchange striction between the Fe^{3+} and Dy^{3+} spins. This work may help to search for single-phase multiferroic materials with high magnetic-field sensitivity.

Keywords: multiferroic materials; anisotropy; $DyFeO_3$; magnetoelectric coupling; pulsed high magnetic field

Citation: Zeng, Z.; He, X.; Song, Y.; Niu, H.; Jiang, D.; Zhang, X.; Wei, M.; Liang, Y.; Huang, H.; Ouyang, Z.; et al. High-Magnetic-Sensitivity Magnetoelectric Coupling Origins in a Combination of Anisotropy and Exchange Striction. *Nanomaterials* **2022**, *12*, 3092. https://doi.org/10.3390/nano12183092

Academic Editor: Sam Lofland

Received: 1 August 2022
Accepted: 2 September 2022
Published: 6 September 2022

Publisher's Note: MDPI stays neutral with regard to jurisdictional claims in published maps and institutional affiliations.

1. Introduction

Multiferroic materials [1–7], with the coupling of two or more ferroic orders, have been attracting much attention due to their intriguing physics and great application potential. The *Pbnm* structured orthoferrites $RFeO_3$ (R = rare earth element) have great potential value for application as magnetoelectric (ME) devices based on the mutual control of magnetization (M) and electric polarization (P) [8–11]. For example, at lower temperatures, the application of a large critical magnetic field along the c-axis induces a multiferroic (weakly ferromagnetic of Fe^{3+}-sublattice and ferroelectric) state in $DyFeO_3$, and the magnetic field induced P results from the movement of the Dy^{3+} ions toward the Fe^{3+} ions and backward, corresponding to the exchange striction [12]. The ME coupling and P are decided by the spin configurations of both Fe^{3+} and Dy^{3+} ions. In $DyFeO_3$, Fe^{3+} ions ($S = 5/2$) exhibit the $G_xA_yF_z$ (magnetic configuration in Bertaut's notation below its Néel temperature of the Fe^{3+}-sublattice $T_N(Fe) = 650$ K) [13–16], in which the main component of the magnetic moment of Fe^{3+} ions lies along the a-axis, and due to the Dzyaloshinskii–Moriya interaction (DMI), a small fraction of the moment is canted along the c-axis, causing weak ferromagnetism (wFM) in the material [17]. With decreasing temperature, the spin-reorientation (Morin) transition occurs at T_{SR} (in the range of about 35~70 K [18–21]), where the magnetic configuration of Fe^{3+} ions changes from $G_xA_yF_z$ to $A_xG_yC_z$, and then the wFM disappears. Below the antiferromagnetic (AFM) ordering temperature of the Dy^{3+}-sublattice $T_N(Dy) = 4.2$ K, the Dy^{3+} magnetic configuration is G_xA_y [22,23] with the Ising axis deviation of about 33° from the b-axis. When a magnetic field (higher than about 3 T) is applied along the c-axis below T_{SR}, the spin configuration of the Fe^{3+}-sublattice is driven to $G_xA_yF_z$ again [4].

For spin-driven ferroelectricity, there are mainly three types of microscopic mechanism models, i.e., the inverse Dzyaloshinskii–Moriya (IDM) mechanism [24], spin-dependent p-d hybridization model [25], and exchange striction model [26,27], for explaining ME behaviors. According to these models, the emergence of ferroelectricity is hardly understood only by the local spin arrangement, since the symmetry of the crystal structure needs to be considered. In $DyFeO_3$, below $T_N(Dy)$, the P_c (the direction of P is parallel to the c-axis) can be induced when the magnetic fields (higher than 2.4 T at 3.0 K [4]) are applied along the c-axis. However, the ME behaviors in $DyFeO_3$, i.e., the P_c is induced with the magnetic fields along other crystal axes, remain to be further understood. Moreover, some abnormal behaviors have been observed. For example, a small P drop was observed on the P-H curve at $B_C(Fe)$ [4]. Abnormal heat transport was measured, and a Fe_{III} state (a metastable phase) was speculated [18]. These abnormal behaviors indicate that there may be complex and delicate magnetic interactions in ME behaviors, such as the competition between the anisotropic energy of the Dy^{3+}-sublattice, the coupling energy between the Dy^{3+} and Fe^{3+}-sublattices, and Zeeman energy [28].

In $DyFeO_3$, when the magnetic field is in the ab plane, ME coupling with smaller critical fields of 0.8 T (a-axis) and 0.5 T (b-axis) is triggered. Inspired by these lower critical magnetic fields, we revisited the structure of $DyFeO_3$, in which two AFM sublattices (the Fe^{3+}- and Dy^{3+}-sublattices) are nesting with each other (see Figure 1a,b). The Fe^{3+}-sublattice (blue balls) has strong AFM coupling (G_xA_y) and wFM (F_z), and the Dy^{3+}-sublattice (red balls) has weak AFM coupling and strong magnetic anisotropy (the AFM vector is localized in the ab plane). Under a lower magnetic field in the easy plane (ab plane), the direction of the magnetic anisotropy of the Dy^{3+}-sublattice might be disturbed or changed, which leads to the change in exchange striction between the Fe^{3+}- and Dy^{3+}-sublattices and triggers P_c. Thus, we believe that single-phase materials with nested AFM lattices (as shown in Figure 1c,d) can be designed or found, where the A-sublattice (blue spheres) has strong AFM coupling in the plane and weak FM outside the plane, while the B-sublattice (red spheres) has weak AFM coupling, and the B-site ions have strong magnetic anisotropy. Such AFM systems are expected to achieve highly magnetically sensitive ME coupling induced by in-plane magnetic fields. Although the observed magnetoelectric effects mainly occurred at low temperatures (below $T_N(Dy)$), which may be difficult to apply directly in the traditional industry, our work deepens the understanding of the ME coupling in $DyFeO_3$. On the other hand, the ME systems controlled by a combination of multiple parameters (such as magnetic anisotropy and exchange striction) may have high sensitivity to the external magnetic field.

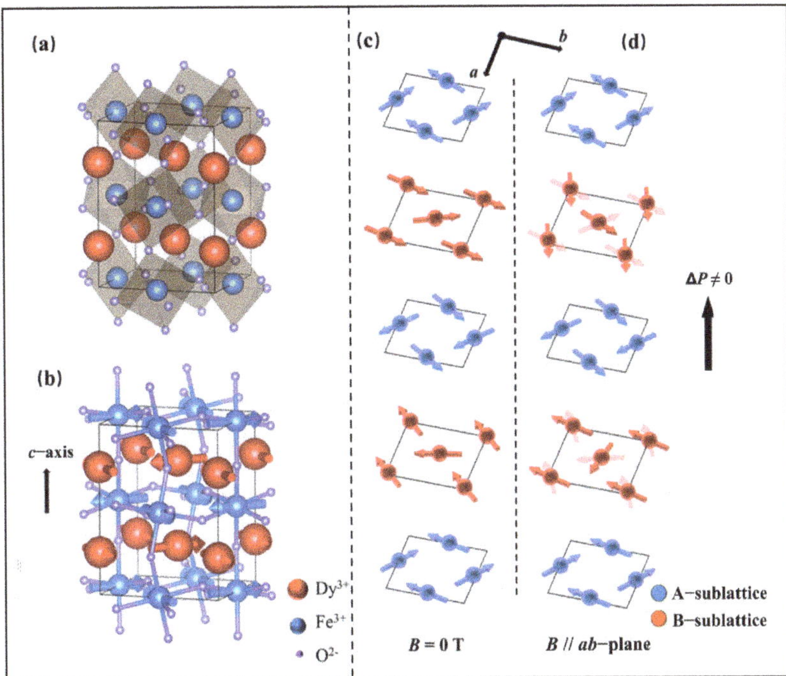

Figure 1. (**a**) The crystal structure and (**b**) magnetic configuration of $DyFeO_3$ below the Dy^{3+} ordering temperature. Two AFM sublattices nesting with each other at (**c**) zero field and (**d**) applied magnetic field.

2. Experiment

The $DyFeO_3$ polycrystalline sample was synthesized by the solid-phase synthesis method from Dy_2O_3 (99.99%) and Fe_2O_3 (99.99%), and the $DyFeO_3$ single crystal was grown by using a Four Mirror Optical Floating Zone Furnace (Crystal Systems Corp., Salem, MA, USA). The crystal structure and purity of both the $DyFeO_3$ powder (crushed single crystal) and the single-crystal samples were measured with an X-ray diffractometer (XRD, X'Pert MPD Powder-DY3734, PANalytical B.V., Almelo, NL) using Cu Ka radiation (λ = 1.5406 Å). The scan ranged from 10 to 90°, the step size was 0.013°, and the scan rate was 0.042°/s. The directions of three principal axes were determined according to the Laue X-ray diffractometer measurement results. The low magnetic-field magnetization was measured by using a superconducting quantum interference device (SQUID VSM, Quantum Design, San Diego, CA, USA). Dynamic behaviors of both the M and the P were measured under a pulsed high magnetic field at the Wuhan National High Magnetic Field Center. The pulsed-high-magnetic-field M was detected by the standard inductive method employing two concentric pick-up coils connected in series with opposite polarity [7]. The electrical polarization P_c measurement schematic is shown in Figure S1 of the supplementary material. The silver electrodes are evenly distributed on the upper and lower surfaces of the sample. When the magnetic field is applied to the sample, the current signal (corresponding to the change in charge density induced by magnetic fields) in the sample is converted into an electrical signal (V) at the reference resistor (R), and then the V is obtained after being processed by a preamplifier. Finally, the change in the electric polarization induced by the pulsed magnetic fields is obtained.

3. Results and Discussion

The XRD pattern of the $DyFeO_3$ powder and its fitting by the general structure analysis system (GSAS) are shown in Figure 2a. All the diffraction peaks are well indexed by a distorted orthorhombic structure with *Pbnm*. No impurity peaks are observed within the diffraction resolution, indicating the single-phase nature of the sample. The lattice parameters a = 5.3031 Å, b = 5.5983 Å, and c = 7.6228 Å and the detailed crystal parameters are listed in the Table S1 of the supplementary material, which are close to the values in the Inorganic Crystal Structure Database (ICSD 27280). In the diffraction pattern of the single-crystal sample, only the (002), (004), and (006) diffraction peaks are observed, which confirms the high quality and accurate c-axis orientation of the $DyFeO_3$ single-crystal sample.

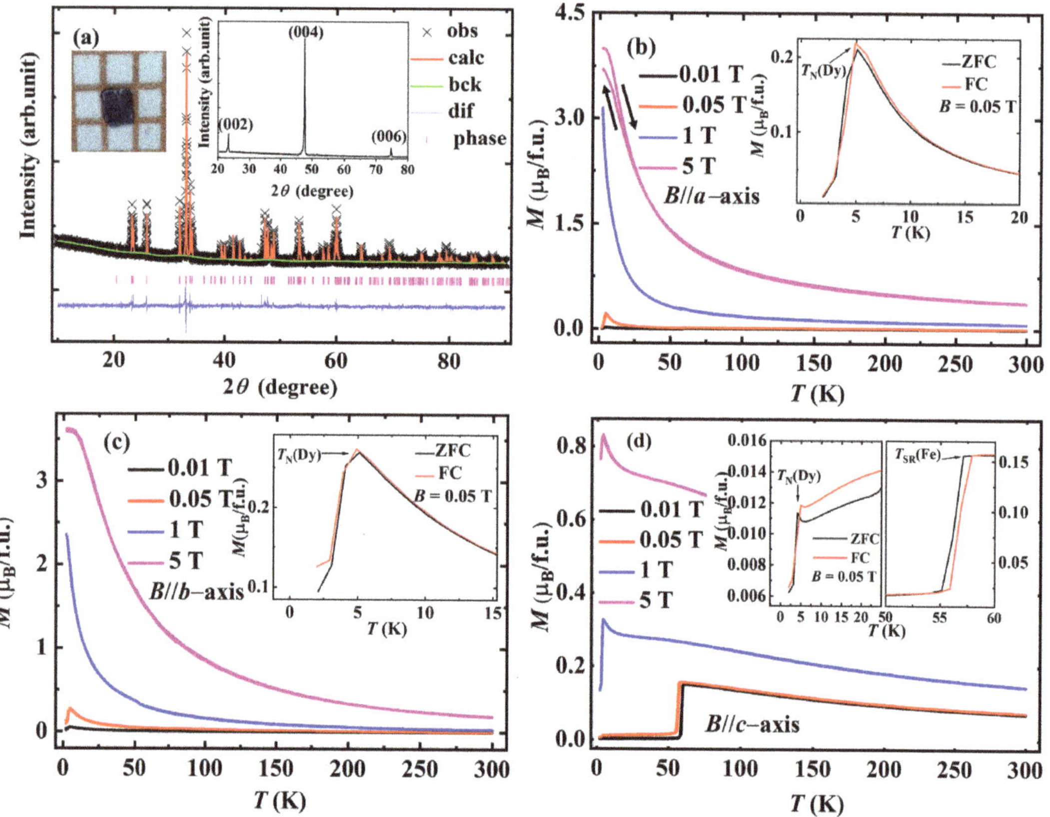

Figure 2. (**a**) The XRD patterns of $DyFeO_3$ powder and single crystal (right inset) and the $DyFeO_3$ single-crystal sample morphology (left inset). (**b**–**d**) The temperature dependence of the magnetization of the $DyFeO_3$ single crystal with magnetic fields along the a-, b-, and c-axes, respectively. The partial magnification near the transition temperatures of ZFC and FC curves measured at 0.05 T are shown in the corresponding insets.

The temperature dependence of magnetization measured in various magnetic fields (0.01 T, 0.05 T, 1 T, and 5 T, respectively) is shown in Figure 2b–d. In the lower-temperature region (below ~60 K), the zero-field-cooled (ZFC) and field-cooled (FC) magnetization curves have slight deviation, and the difference is presented in the insets of Figure 2b–d. In the higher-temperature region (above ~60 K), the difference between ZFC and FC becomes indiscernible. When the magnetic field is applied along the *c*-axis (see Figure 2d), there is an obvious transition with the magnetization jumps of ~0.13 μ_B/f.u. at T_{SR}(Fe) ~57 K, which is related to the spin-flop transition of the Fe^{3+}-sublattice. As the temperature decreases to T_N(Dy), an obvious drop can be observed (see the inset of Figure 2d), which indicates that the Dy^{3+}-sublattice undergoes a transition from a paramagnetic state to an AFM state $G_x A_y$. Since the magnetic moment of the Ising Dy^{3+} ions is localized in the *ab* plane, it is difficult to disturb the magnetic field (0.05 T, along the *c*-axis) or change the direction of the magnetic anisotropy (or the anisotropy energy) of the Dy^{3+} spin. With the magnetic field increasing, T_{SR}(Fe) moves to the low-temperature region. However, with the magnetic field applied along the *c*-axis, no obvious movement of T_N(Dy) is observed, which confirms the strong magnetic anisotropy and localization in the *ab* plane of the Dy^{3+} spins. The temperature and magnetic field dependence of the transitions are shown in the magnetic phase diagram of Figure S2 in the supplementary material. In $DyFeO_3$, there is the magnetic anisotropy of Dy^{3+} ions, field-induced spin flop of Fe^{3+} ions, temperature-driven spin reorientation of Fe^{3+} ions, AFM interaction between Dy^{3+} and Fe^{3+} ions, and thermal fluctuation [28], and these lead to the complex dependence of both T_{SR}(Fe) and T_N(Dy) on temperatures and magnetic fields.

The magnetization curves as functions of different magnetic fields along three principal axes are shown in Figure 3. Below T_N(Dy) (taken 2 K as an example), when the magnetic field is applied along the *a*-axis (see Figure 3a), a transition is observed at B_C(Dy) ~0.8 T (labeled with a red arrow), which is attributed to the spin-flop transition of the Dy^{3+}-sublattice (as shown in the inserted cartoon). With the magnetic field increasing, the magnetic field drives the spins of the Dy^{3+}-sublattice toward the *a*-axis as much as possible, resulting in a sharp increase in *M*. With the magnetic field further increasing, the AFM coupling of Dy^{3+}-sublattices may be partially broken. For the magnetic field parallel to the *b*-axis, similarly, a slightly smaller critical magnetic field of 0.5 T can be observed due to the smaller deviation (the angle ~33°) of the Ising vector of the Dy^{3+} spins from the *b*-axis. Above 0.5 T, a saturated magnetic moment of ~8.3 μ_B/f.u. is obtained, which indicates that the Dy^{3+} moment was almost magnetized to saturation by the magnetic field (as shown in the inserted cartoon in Figure 3b). The saturated magnetization shows that the Ising Dy^{3+} moment is mainly localized in the *ab* plane with the *b*-axis as the easy axis [18]. In the case with the magnetic field along the *c*-axis, a magnetization jump (ΔM) of ~0.13 μ_B/f.u. is observed around 3 T. The value of the ΔM is the same as the value of the magnetization jump in the *M-T* curves shown in Figure 2d, which confirms that the transition is mainly associated with the spin-flop transition of the Fe^{3+}-sublattice; the result is also consistent with previous studies [4]. In Figure 3c, two transitions are observed around 3 T and 3.2 T (labeled as black and red triangles, respectively). The lower one may be related to the spin-flop transition of the Fe^{3+} moment, while the higher one may be caused by the destruction of the AFM coupling between the FM components of the Dy^{3+} moment (induced by the magnetic field) and the Fe^{3+} moment (canted AFM) [29]. The magnetization behaviors measured in a pulsed higher magnetic field are shown in Figure 3d–f. Besides the low-field transitions, no additional transitions are induced up to 45 T, and the slopes of the *M-B* curves are almost constant, indicating that there is a strong AFM interaction in the Fe^{3+}-sublattice [4].

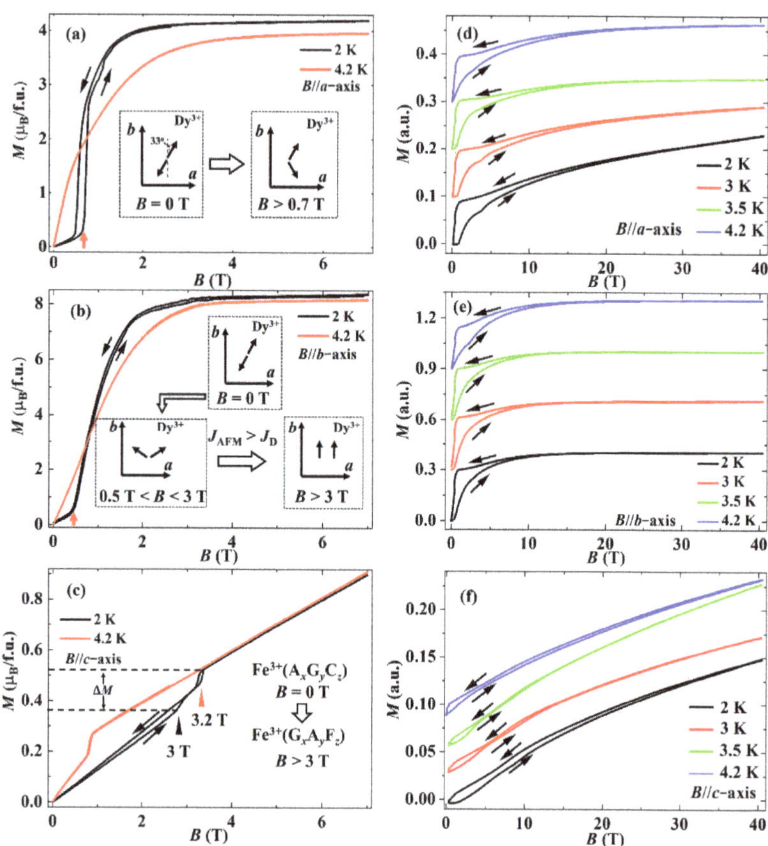

Figure 3. (**a–c**) Magnetization as a function of the static magnetic field along *a*-, *b*-, and *c*-axes, respectively. (**d–f**) The magnetization curves measured along the *a*-, *b*-, and *c*-axes under pulsed high magnetic field, respectively. The curves in the (**b,e,f**) are offset for clarity. J_{AFM} and J_D are the AFM interaction strength and anisotropy energy of the Dy^{3+}-sublattice, respectively.

In order to further investigate magnetic-field-induced transitions and the dynamic magnetization behavior of $DyFeO_3$, the magnetization results were investigated by pulsed magnetic fields. In this work, the waveform of the pulsed magnetic field is shown in Figure 4a, which is a full-wave pulsed magnetic field (including four quadrants, Q_A, Q_B, Q_C, and Q_D) with a maximum field sweep rate of about 10^4 T/s. When the magnetic field is along the *a*-axis and the temperature is below 4.2 K (see Figure 4b), in the field-increasing branch (quadrant Q_A), the transitions resulting from the Dy^{3+}-sublattice (marked with red arrows) and the field-induced spin flop in the Fe^{3+}-sublattice (marked with black arrows) are observed. When the temperature ranges from 4.2 K to 50 K, only the transition, corresponding to the spin flop in the Fe^{3+}-sublattice, could be observed, and its critical magnetic field decreases drastically with increasing temperature. Above 50 K, no obvious transition is observed (not shown). In the field-decreasing branch (quadrant Q_B), the transition field moves to the low-magnetic-field region (lower than 1 T). In the field-increasing and field-decreasing branches of the negative magnetic field (quadrants Q_C and Q_D), the magnetization behaviors are similar to those in Q_A and Q_B. As shown in Figure 4c, the magnetization behaviors of the magnetic field along the *b*-axis are similar to those along the *a*-axis.

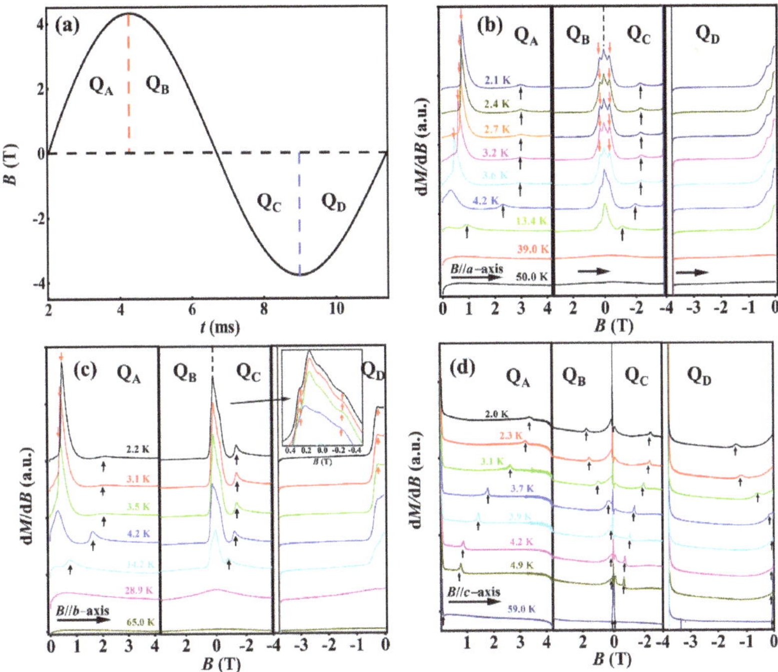

Figure 4. (**a**) The waveform of the pulsed magnetic field. (**b–d**) Magnetization derivative (dM/dB) as a function of the magnetic field measured along the three principal axes. The inset in (**c**) is the enhancement of the dM/dB around the zero-magnetic-field regions. The curves of (**b–d**) are offset for clarity. Q_A, Q_B, Q_C, and Q_D are the four quadrants of the pulsed magnetic field; all the transitions are labeled with arrows.

In Figure 4b,c, the temperature dependence of the transition field of the spin flop in the Fe^{3+}-sublattice is different above and below 4.2 K. Below 4.2 K, the critical field (corresponding to the spin flop of the Fe^{3+} moment) is essentially unchanged with the temperature increasing. Above 4.2 K, the transition field (labeled with arrows) has a strong dependency on temperature. These results suggest the pinning of the Dy^{3+}-sublattice on the spin flop of Fe^{3+} ions. With the temperature increasing, the pinning gradually becomes weaker due to the destruction of the long-range order of the Dy^{3+}-sublattice. The transition field (labeled with arrows) of the Dy^{3+}-sublattice has a strong temperature dependence, which indicates that the direction of the magnetic anisotropy (or anisotropy energy) of Dy^{3+} spins may also be disturbed by the magnetic field in the *ab* plane. The critical field moves toward the lower-magnetic-field region with the temperature increasing. To present the moving trend more clearly, the phase diagrams with the magnetic field parallel to the *a*- and *b*-axes are shown in Figure S2a,b of the supplementary material, respectively. When the magnetic field is applied along the *c*-axis, only the transitions related to the spin flop of Fe^{3+} ions are observed (see Figure 4d). Since the Dy^{3+} ion has strong Ising behavior and is localized in the *ab* plane, no obvious transition related to the Dy^{3+}-sublattice was observed for lower magnetic fields. The critical field (corresponding to the spin-flop of the Dy^{3+} moment) moves toward the lower-magnetic-field region with the temperature increasing, and the temperature and magnetic field dependence of the critical behaviors are shown in Figure S2c of the supplementary material.

In the measurement of the electric polarization, ΔP_c is a relative value, and it shows the change in P_c induced by magnetic fields, and the applied magnetic field includes four quadrants: Q_A, Q_B, Q_C, and Q_D. As shown in Figure 5a, both the magnetic field and the electric field are parallel to the c-axis; with the magnetic field increasing (quadrant Q_A), ΔP_c jumps (labeled with red pentacles) are observed at $B_P(\text{Fe})$ and a temperature below $T_N(\text{Dy})$. At 2 K, the transition is observed in $B_P(\text{Fe}) \sim 3$ T (labeled with a red pentacle), and the critical magnetic field is coincident with the result of magnetization measurement in the field-increasing branch (quadrant Q_A). As the temperature increases, the critical field $B_P(\text{Fe})$ moves to the low-field region, and a similar temperature dependence of the transition field is also observed in the magnetization curves (see Figure 4d). In the field-decreasing branch (quadrant Q_B), ΔP_c becomes zero at $B'_P(\text{Fe})$ (marked with black triangles). In quadrant Q_C (the field-increasing branch of the negative magnetic field), the transitions are also observed at $-B_P(\text{Fe})$ (marked with black diamonds). In the field-decreasing branches of the negative-magnetic-field region (quadrant Q_D), a transition is observed at $-B'_P(\text{Fe})$ (marked with crosses). Particularly, a metastable state (indicated by solid circles) and ΔP_c reversal (marked with a cross) are observed around 3.1 K in the negative-magnetic-field region. The transitions are affected by magnetic fields and temperatures, which may originate from the complicated interactions between the anisotropy energy of the Dy^{3+}-sublattice, the coupling energy between the Dy^{3+} and Fe^{3+}-sublattices, and Zeeman energy.

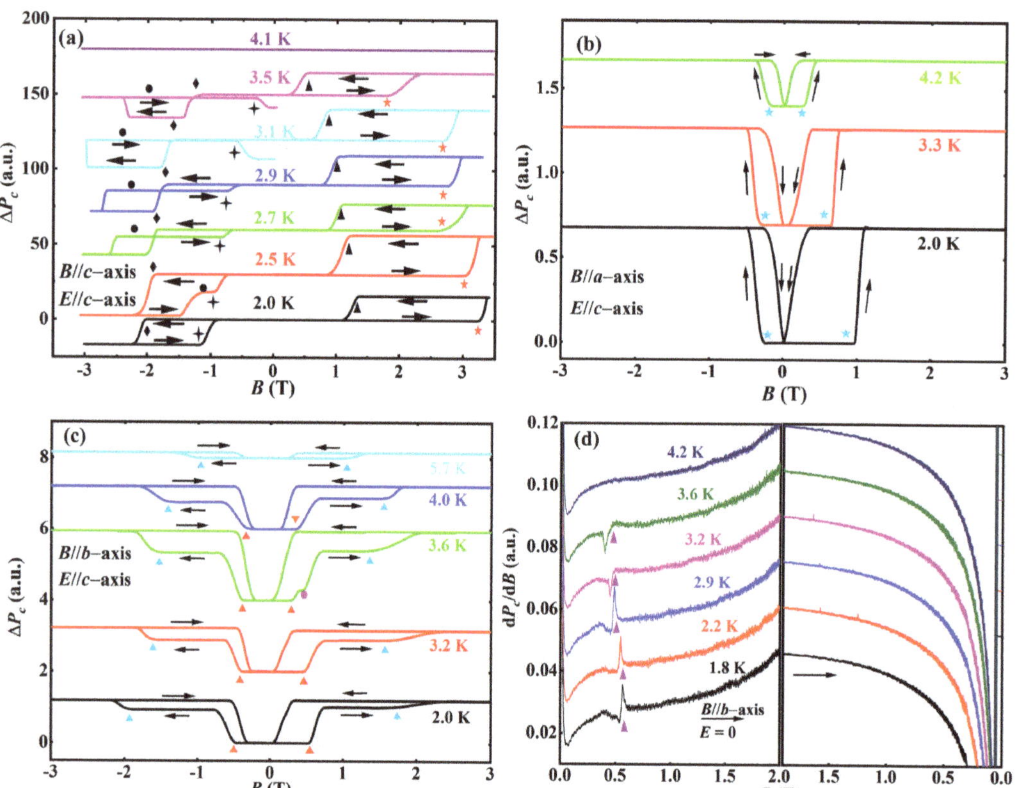

Figure 5. (**a–c**) Electric polarization as a function of pulsed magnetic fields, measured under the pulsed magnetic field along c-axis (**a**), a-axis (**b**), and b-axis (**c**), where the electric fields ($E = 1.5$ kV/cm) are along c-axis. (**d**) The magnetic field dependence of dP_c/dB measured at various temperatures with the applied pulsed magnetic field along b-axis and $E = 0$. The curves are offset for clarity. The various symbols of red pentacle, black triangle, black diamond, cross, solid circle, blue pentacle, red triangle, and purple triangle represent the transitions in the curves. The sweep directions of the magnetic field are labeled by black arrows.

According to the exchange striction model, P_c is related to the spin flop of both the Fe^{3+}-sublattice and the Dy^{3+}-sublattice [18]; to reverse the P_c, it is necessary to change the phase (0 or π) of the magnetic vector of either the Fe^{3+} or the Dy^{3+} ions. The magnetic vector of the Fe^{3+} ions is directly connected to the direction of the wFM of the Fe^{3+}-sublattice. Thus, the field-induced ΔP_c is observed when a large magnetic field is antiparallel to the c-axis. With the magnetic field decreasing and the temperature increasing, the interaction between the Dy^{3+} and Fe^{3+} ions, as well as the Zeeman energy, becomes weaker, and the magnetic anisotropy energies of the Dy^{3+}-sublattices gradually dominate, which leads to the ΔP_c reversal and metastable polarization states in the negative field (quadrants Q_C and Q_D of Figure 5a). On the other hand, the strong magnetic anisotropy of the Dy^{3+} ions becomes dominant, which drives the Dy^{3+} spins to its easy axis and leads to the change in the exchange striction and ΔP_c reversal to a lower value. For the observed metastable state (marked with solid circles in Figure 5a), we assume that this is due to the spin-pinning effect of Dy^{3+} ions on the change in the wFM of the Fe^{3+}-sublattice. The fact that metastable polarization behavior is more obvious when the temperature approaches $T_N(Dy)$ indicates that magnetic anisotropic Dy^{3+} is more easily magnetized by the magnetic field when the temperature approaches $T_N(Dy)$ than at lower temperatures.

For the pulsed magnetic field within the ab plane (as shown in Figure 5b–d), the change in the Dy^{3+} spins induced by the magnetic field also causes the change in exchange striction. The ΔP_c-B curves with magnetic fields along the a- and b-axes were measured (where the electrical polarization is along the c-axis). As shown in Figure 5b (the magnetic field along the a-axis) and Figure 5c (the magnetic field along the b-axis), the sign of ΔP_c is unchanged. This is totally different from the case with a magnetic field along the c-axis. The unchanged ΔP_c suggests the simultaneous flop of both the Fe^{3+} and Dy^{3+} spins when the magnetic field (in the ab plane) reverses. At 2 K, with the magnetic field along the a-axis and b-axis and the field increasing (quadrant Q_A), the field-induced ΔP_c are observed at $B_P(Dy)$ ~0.8 T (a-axis, marked with a blue pentacle) and 0.5 T (b-axis, marked with red triangles), respectively. Both the critical fields are lower than that along the c-axis, and with the increasing temperature, the transition fields move further to the lower magnetic fields. The effects of the temperature and magnetic field on the critical behaviors are shown in Figure S2 of the supplementary material.

In $DyFeO_3$, the AFM interaction in the Dy^{3+}-sublattice is weak and mainly localized within the ab plane. The lower magnetic field in the ab plane will disturb the direction of the magnetic anisotropy (or the anisotropy energy) of Dy^{3+} ions; that is, the Dy^{3+} moments are easily magnetized by the magnetic field, resulting in a higher magnetic field sensitivity. At zero fields, the magnetic vector (the Ising axis) of the Dy^{3+} ion deviates by about 33° from the b-axis (as shown in the inset of Figure 3a), which leads to the different critical field of the field-induced ΔP_c between the magnetic field along the a- and b-axes (see Figure 5b,c). In order to confirm the intrinsic effect of the magnetic field on polarization, the ΔP_c was also investigated with zero electric fields (see Figure 5d), and a weaker change of the electrical polarization was observed in the dP_c/dB-B curves, where the critical fields (marked with purple triangles) are coincident with those observed in the ΔP_c–B curves (see Figure 5c). With the temperature increasing, the transition peaks of the ΔP_c shift to the lower-field region. These experimental results indicate that the ΔP_c is an intrinsic behavior and can be induced by the magnetic field alone.

4. Conclusions

In this work, the combination of anisotropy and exchange striction in $DyFeO_3$ provides a guiding principle for designing high-sensitivity spin-driven multiferroicity. Especially in the ab plane, the direction of the magnetic anisotropy (or the anisotropy energy) of the Dy^{3+} ions can be modulated by a smaller magnetic field, which alters the exchange striction and leads to a ΔP_c sensitive to the external magnetic field. That is, the combination of the magnetic anisotropy of the Dy^{3+} spin and the exchange striction between the Fe^{3+} and Dy^{3+} spins leads to the ΔP_c. The ΔP_c exhibits high magnetic-field sensitivity. This work deepens the understanding of the effects of the multiple magnetic orders on the magnetoelectric coupling of multiferroic DyFeO$_3$ and will be beneficial in searching for novel multiferroic material systems with high magnetic-field sensitivity.

Supplementary Materials: The following supporting information can be downloaded at: https://www.mdpi.com/article/10.3390/nano12183092/s1, Table S1: The lattice parameters of the DyFeO$_3$ single crystal; Figure S1: The measurement schematic of the electric polarization (Pc) under pulsed magnetic fields; Figure S2: Magnetic phase diagrams determined with the magnetization and polarization measurements.

Author Contributions: Z.Z. and X.H. contributed equally to this work. Writing—original draft preparation, Z.Z. and X.H.; supervision, Y.S., H.N. and M.W.; investigation, D.J., X.Z., Y.L., H.H. and Z.O.; writing—review and editing, Z.C. and Z.X. All authors have read and agreed to the published version of the manuscript.

Funding: This research was funded by the National Natural Science Foundation of China (Grant Nos. 11674115, 51861135104, and 21875249). Z.C. thanks the Australian Research Council (Grant No. DP190100150) for support.

Institutional Review Board Statement: Not applicable.

Informed Consent Statement: Not applicable.

Data Availability Statement: Not applicable.

Conflicts of Interest: The authors declare no conflict of interest.

References

1. Holcomb, M.B.; Martin, L.W.; Scholl, A.; He, Q.; Yu, P.; Yang, C.-H.; Yang, S.Y.; Glans, P.-A.; Valvidares, M.; Huijben, M.; et al. Probing the evolution of antiferromagnetism in multiferroics. *Phys. Rev. B* **2010**, *81*, 134406. [CrossRef]
2. Weston, L.; Cui, X.Y.; Ringer, S.P.; Stampfl, C. Multiferroic crossover in perovskite oxides. *Phys. Rev. B* **2016**, *93*, 165210. [CrossRef]
3. Stein, J.; Baum, M.; Holbein, S.; Finger, T.; Cronert, T.; Tölzer, C.; Fröhlich, T.; Biesenkamp, S.; Schmalzl, K.; Steffens, P.; et al. Control of Chiral Magnetism Through Electric Fields in Multiferroic Compounds above the Long-Range Multiferroic Transition. *Phys. Rev. Lett.* **2017**, *119*, 177201. [CrossRef]
4. Tokunaga, Y.; Iguchi, S.; Arima, T.; Tokura, Y. Magnetic-Field-Induced Ferroelectric State in DyFeO$_3$. *Phys. Rev. Lett.* **2008**, *101*, 097205. [CrossRef] [PubMed]
5. Petrov, V.M.; Srinivasan, G. Theory of domain wall motion mediated magnetoelectric effects in a multiferroic composite. *Phys. Rev. B* **2014**, *90*, 144411. [CrossRef]
6. Prokhnenko, O.; Feyerherm, R.; Dudzik, E.; Landsgesell, S.; Aliouane, N.; Chapon, L.C.; Argyriou, D.N. Enhanced Ferroelectric Polarization by Induced Dy Spin Order in Multiferroic DyMnO$_3$. *Phys. Rev. Lett.* **2007**, *98*, 057206. [CrossRef] [PubMed]
7. Zhang, X.X.; Xia, Z.C.; Ke, Y.J.; Zhang, X.Q.; Cheng, Z.H.; Ouyang, Z.W.; Wang, J.F.; Huang, S.; Yang, F.; Song, Y.J.; et al. Magnetic behavior and complete high-field magnetic phase diagram of the orthoferrite ErFeO$_3$. *Phys. Rev. B* **2019**, *100*, 054418. [CrossRef]
8. Tokunaga, Y.; Taguchi, Y.; Arima, T.-H.; Tokura, Y. Electric-field-induced generation and reversal of ferromagnetic moment in ferrites. *Nat. Phys.* **2012**, *8*, 838. [CrossRef]
9. Kimura, T.; Goto, T.; Shintani, H.; Ishizaka, K.; Arima, T.; Tokura, Y. Magnetic control of ferroelectric polarization. *Nature* **2003**, *426*, 55. [CrossRef]
10. Eerenstein, W.; Mathur, N.D.; Scott, J.F. Multiferroic and magnetoelectric materials. *Nature* **2006**, *442*, 759. [CrossRef]
11. Feyerherm, R.; Dudzik, E.; Aliouane, N.; Argyriou, D.N. Commensurate Dy magnetic ordering associated with incommensurate lattice distortion in multiferroic DyMnO$_3$. *Phys. Rev. B* **2006**, *73*, 180401. [CrossRef]
12. Tokunaga, Y.; Furukawa, N.; Sakai, H.; Taguchi, Y.; Arima, T.-H.; Tokura, Y. Composite domain walls in a multiferroic perovskite ferrite. *Nat. Mater.* **2009**, *8*, 558. [CrossRef] [PubMed]
13. Das, M.; Roy, S.; Mandal, P. Giant reversible magnetocaloric effect in a multiferroic GdFeO$_3$ single crystal. *Phys. Rev. B* **2017**, *96*, 174405. [CrossRef]
14. Nikitin, S.E.; Wu, L.S.; Sefat, A.S.; Shaykhutdinov, K.A.; Lu, Z.; Meng, S.; Pomjakushina, E.V.; Conder, K.; Ehlers, G.; Lumsden, M.D.; et al. Decoupled spin dynamics in the rare-earth orthoferrite YbFeO$_3$: Evolution of magnetic excitations through the spin-reorientation transition. *Phys. Rev. B* **2018**, *98*, 064424. [CrossRef]
15. Chiang, F.-K.; Chu, M.-W.; Chou, F.C.; Jeng, H.T.; Sheu, H.S.; Chen, F.R.; Chen, C.H. Effect of Jahn-Teller distortion on magnetic ordering in Dy(Fe,Mn)O$_3$ perovskites. *Phys. Rev. B* **2011**, *83*, 245105. [CrossRef]
16. Ke, Y.J.; Zhang, X.Q.; Ge, H.; Ma, Y.; Cheng, Z.H. Low field induced giant anisotropic magnetocaloric effect in DyFeO$_3$ single crystal. *Chin. Phys. B* **2015**, *24*, 037501. [CrossRef]

17. Lee, J.-H.; Jeong, Y.K.; Park, J.H.; Oak, M.-A.; Jang, H.M.; Son, J.Y.; Scott, J.F. Spin-Canting-Induced Improper Ferroelectricity and Spontaneous Magnetization Reversal in SmFeO$_3$. *Phys. Rev. Lett.* **2011**, *107*, 117201. [CrossRef]
18. Zhao, Z.Y.; Zhao, X.; Zhou, H.D.; Zhang, F.B.; Li, Q.J.; Fan, C.; Sun, X.F.; Li, X.G. Ground state and magnetic phase transitions of orthoferrite DyFeO$_3$. *Phys. Rev. B* **2014**, *89*, 224405. [CrossRef]
19. Prelorendjo, L.A.; Johnson, C.E.; Thomas, M.F.; Wanklyn, B.M. Spin reorientation transitions in DyFeO$_3$ induced by magnetic fields. *J. Phys. C Solid State Phys.* **1980**, *13*, 2567. [CrossRef]
20. Du, Y.; Cheng, Z.X.; Wang, X.L.; Dou, S.X. Lanthanum doped multiferroic DyFeO$_3$: Structural and magnetic properties. *J. Appl. Phys.* **2010**, *107*, 09D908. [CrossRef]
21. Jaiswal, A.; Das, R.; Maity, T.; Poddar, P. Dielectric and spin relaxation behaviour in DyFeO$_3$ nanocrystals. *J. Appl. Phys.* **2011**, *110*, 124301. [CrossRef]
22. Holmes, L.M.; Van Uitert, L.G.; Hecker, R.R.; Hull, G.W. Magnetic Behavior of Metamagnetic DyAlO$_3$. *Phys. Rev. B* **1972**, *5*, 138. [CrossRef]
23. Stroppa, A.; Marsman, M.; Kresse, G.; Picozzi, S. The multiferroic phase of DyFeO$_3$: An ab initio study. *New J. Phys.* **2010**, *12*, 093026. [CrossRef]
24. Sergienko, I.A.; Dagotto, E. Role of the Dzyaloshinskii-Moriya interaction in multiferroic perovskites. *Phys. Rev. B* **2006**, *73*, 094434. [CrossRef]
25. Arima, T.-H. Ferroelectricity Induced by Proper-Screw Type Magnetic Order. *J. Phys. Soc. Jpn.* **2007**, *76*, 073702. [CrossRef]
26. Zhao, Z.Y.; Wang, X.M.; Fan, C.; Tao, W.; Liu, X.G.; Ke, W.P.; Zhang, F.B.; Zhao, X.; Sun, X.F. Magnetic phase transitions and magnetoelectric coupling of GdFeO$_3$ single crystals probed by low-temperature heat transport. *Phys. Rev. B* **2011**, *83*, 014414. [CrossRef]
27. Bousquet, E.; Cano, A. Non-collinear magnetism in multiferroic perovskites. *J. Phys. Condens. Matter* **2016**, *28*, 123001. [CrossRef] [PubMed]
28. Hoogeboom, G.R.; Kuschel, T.; Bauer, G.E.W.; Mostovoy, M.V.; Kimel, A.V.; van Wees, B.J. Magnetic order of Dy^{3+} and Fe^{3+} moments in antiferromagnetic DyFeO$_3$ probed by spin Hall magnetoresistance and spin Seebeck effect. *Phys. Rev. B* **2021**, *103*, 134406. [CrossRef]
29. Cao, S.X.; Chen, L.; Zhao, W.Y.; Xu, K.; Wang, G.H.; Yang, Y.L.; Kang, B.J.; Zhao, H.J.; Chen, P.; Stroppa, A.; et al. Tuning the Weak Ferromagnetic States in Dysprosium Orthoferrite. *Sci. Rep.* **2016**, *6*, 37529. [CrossRef] [PubMed]

 *nanomaterials*

Article

Presence of Induced Weak Ferromagnetism in Fe-Substituted $YFe_xCr_{1-x}O_3$ Crystalline Compounds

Roberto Salazar-Rodriguez [1,*], Domingo Aliaga Guerra [1], Jean-Marc Greneche [2], Keith M. Taddei [3], Noemi-Raquel Checca-Huaman [4], Edson C. Passamani [5] and Juan A. Ramos-Guivar [6]

[1] Facultad de Ciencias, Universidad Nacional de Ingeniería (UNI), Av. Túpac Amaru 210, Rímac, Lima 15333, Peru

[2] Institut des Molécules and Matériaux du Mans (IMMM UMR CNRS 6283), University Le Mans, Avenue Olivier Messiaen, Cedex 9, 72085 Le Mans, France

[3] Oak Ridge National Laboratory (ORNL), Oak Ridge, TN 37831, USA

[4] Centro Brasileiro de Pesquisas Físicas (CBPF), R. Xavier Sigaud, 150, Urca, Rio de Janeiro 22290-180, Brazil

[5] Departamento de Física, Universidade Federal do Espírito Santo, Vitória 29075-910, Brazil

[6] Grupo de Investigación de Nanotecnología Aplicada para Biorremediación Ambiental, Energía, Biomedicina y Agricultura (NANOTECH), Facultad de Ciencias Físicas, Universidad Nacional Mayor de San Marcos, Av. Venezuela Cdra 34 S/N, Ciudad Universitaria, Lima 15081, Peru

* Correspondence: rsalazarr@uni.edu.pe; Tel.: +51-19-9161-9615

Citation: Salazar-Rodriguez, R.; Aliaga Guerra, D.; Greneche, J.-M.; Taddei, K.M.; Checca-Huaman, N.-R.; Passamani, E.C.; Ramos-Guivar, J.A. Presence of Induced Weak Ferromagnetism in Fe-Substituted $YFe_xCr_{1-x}O_3$ Crystalline Compounds. *Nanomaterials* **2022**, *12*, 3516. https://doi.org/10.3390/nano12193516

Academic Editors: Jordi Sort, Zhenxiang Cheng, Chang-hong Yang and Chunchang Wang

Received: 5 September 2022
Accepted: 29 September 2022
Published: 8 October 2022

Publisher's Note: MDPI stays neutral with regard to jurisdictional claims in published maps and institutional affiliations.

Abstract: Fe-substituted $YFe_xCr_{1-x}O_3$ crystalline compounds show promising magnetic and multiferroic properties. Here we report the synthesis and characterization of several compositions from this series. Using the autocombustion route, various compositions (x = 0.25, 0.50, 0.6, 0.75, 0.9, and 1) were synthesized as high-quality crystalline powders. In order to obtain microscopic and atomic information about their structure and magnetism, characterization was performed using room temperature X-ray diffraction and energy dispersion analysis as well as temperature-dependent neutron diffraction, magnetometry, and ^{57}Fe Mössbauer spectrometry. Rietveld analysis of the diffraction data revealed a crystallite size of 84 (8) nm for $YFeO_3$, while energy dispersion analysis indicated compositions close to the nominal compositions. The magnetic results suggested an enhancement of the weak ferromagnetism for the $YFeO_3$ phase due to two contributions. First, a high magnetocrystalline anisotropy was associated with the crystalline character that favored a unique high canting angle of the antiferromagnetic phase (13°), as indicated by the neutron diffraction analysis. This was also evidenced by the high magnetic hysteresis curves up to 90 kOe by a remarkable high critical coercivity value of 46.7 kOe at room temperature. Second, the Dzyaloshinskii–Moriya interactions between homogenous and heterogeneous magnetic pairs resulted from the inhomogeneous distribution of Fe^{3+} and Cr^{3+} ions, as indicated by ^{57}Fe Mössbauer studies. Together, these results point to new methods of controlling the magnetic properties of these materials.

Keywords: DM interaction; crystalline $YFeO_3$; magnetic properties; enhanced weak ferromagnetism; exchange interactions

1. Introduction

Bulk orthoferrites and orthochromites have been the subject of various studies since the 1950s, but the interest of the scientific community has recently been renewed due to their possible technological applications in sensors, switching devices, and spintronics [1,2]. New research is focused on improving novel synthesis methods, consequently producing materials with different and improved magnetic and electrical properties. It is important first to point out that the first publications were carried out almost exclusively with single crystal samples, a synthesis process that requires specialized equipment not often found in ordinary laboratories, and that inherently prevents the production of large quantities of materials [3,4]. Currently, new synthesis methods have been explored to produce polycrystalline samples that have the advantage of rapid preparation, low cost, and the ability to produce relatively large masses of material [4]. On the other hand, developments in nanotechnology have also made it possible to obtain nanocrystalline orthoferrites with highly controlled stoichiometry using wet methods, such as the sol–gel approach, allowing continuous doping between the orthoferrite and orthochromite endmembers [5]. Such synthesis developments may be instrumental in studying the Fe–Cr phase diagram, which has been historically challenging due to the thermodynamic considerations of the Y_2O_3–Fe_2O_3 binary phase diagram, in which magnetite or ternary garnets can easily be obtained as parasitic (or secondary) phases. Therefore, the use of these new techniques to scale up the synthesis and carefully control the composition allows unprecedented access to the Fe–Cr phase diagram, potentially enabling the optimization and understanding of the interesting magnetic properties of these materials.

The $YFeO_3$ orthoferrite (YFO) and the $YCrO_3$ orthochromite (YCO) compounds crystallize in the Pnma centrosymmetric space group symmetry and are biferroic with high Néel temperatures (T_N). Specifically, these materials are antiferromagnetic with T_N = 644 K and 140 K for YFO and YCO, respectively, with the a-direction of the unit cell as the easy magnetization axis. Moreover, these materials exhibit weak ferromagnetism, which has been attributed to the canting of magnetic moments along the c-direction, leading to the convincing observation of ferromagnetic behavior [5]. As early as the 1960s, Treves [6] proposed an antisymmetric interaction mechanism (Dzyaloshinskii–Moriya (DM) exchange) for this type of ferromagnetism in orthoferrites based on torque measurements in single crystals of rare earth orthoferrites ($RFeO_3$, R = rare earth and Y). Considering these intriguing physical properties and the possibility to combine the two above materials (i.e., forming a Fe-substituted YCO phase), improved electrical and magnetic properties are expected. In particular, the case of the $YFe_{0.5}Cr_{0.5}O_3$ compound is very interesting because it has the largest magnetoelectric effect in the series [7] and may present a spin reorientation phenomenon according to the literature [8]. However, most of these studies have been performed on bulk polycrystalline samples, leaving the effects of another tuning parameter, namely, the crystallite size, unstudied. To optimize and understand the magnetoelectric effect in these materials, one could use the crystalline size as well as the application of large magnetic fields to try to stabilize different magnetic configurations or tune the canting angle and thus potentially improve the magnetoelectric effect.

In this work, the structural and magnetic properties of the Fe-substituted YCO phase (i.e., the $YFe_xCr_{1-x}O_3$ compounds) in the crystalline regime are studied in detail. The structural and composition features were obtained from Rietveld refinements and scanning electron microscopy (SEM), which showed (i) a single phase in the studied compounds and (ii) their nanoscale characteristics. The magnetic properties of the Fe-substituted compounds were studied by performing direct current magnetization measurements at and between 300 K (room temperature (RT)) and 5 K using both zero-field-cooling (ZFC) and field-cooling (FC) protocols, while the local magnetic properties were investigated using ^{57}Fe Mössbauer spectrometry performed at 300 K and 77 K and under an external magnetic field. The obtained results suggested an enhancement of WFM associated mainly with the sub-micrometric size character of the YFO phase (84 (8) nm), which has favored a relevant spin canting of 13°, i.e., a net spin contribution related to the sample finite-size effect. This property is not observed in the single and polycrystalline systems reported in the literature. In addition, the presence of the finite-size effect (spin-canting) in our sample was also reflected in the increase of the saturation magnetization that reached 0.79 emu/g at 3.5 kOe.

2. Materials and Methods

The whole series of $YFe_xCr_{1-x}O_3$ perovskites was synthesized by the combustion method by stoichiometrically mixing the following initial reactants: $Y(NO_3)\cdot6H_2O$, $Fe(NO_3)_3\cdot9H_2O$, $Cr(NO_3)_3\cdot9H_2O$, urea, and glycine. To improve the crystallization process, all powder samples were heated up to 1200 °C and annealed under ambient conditions. The synthesized samples were labeled as the RSx series, where RS1, RS2, RS3, RS4, RS5, RS6, and RS7 correspond to x = 0, 0.25, 0.50, 0.60, 0.75, 0.90, and 1.0, respectively.

Structural characterization was carried out using Bruker Advance D8 X- ray diffraction equipment (Bruker Corporation, Billerica, MA, USA), operating with a Cu–Kα radiation source (1.5418 Å wavelength), and the X-ray diffraction (XRD) diffractograms were recorded at RT with a 2θ from 15° to 65° in a step of 0.02° and with an accumulating time of 10 s. Neutron diffraction (NPD) of the RS3 ($YFe_{0.5}Cr_{0.5}O_3$) sample was performed on the HB-2A line of the High Flux Isotope Reactor (HFIR) at Oak Ridge National Laboratory (ORNL) [9]. X-ray and NPD data were analyzed using the FullProf suite (Gif sur Yvette Cedex, France, version January 2021). In all nuclear diffraction peak modeling, the previously reported orthorhombic crystal structure (space group Pnma) was found to account for all observed peak positions and intensities, identified using the software Match v3. As initial cell parameter values, we employed a = 5.59 Å, b = 7.59 Å, and c = 5.27 Å (Match entry 210–1387) and allowed the parameters to refine during the profile fitting for the different temperatures and compositions. The instrumental resolution function (IRF) of the X-ray diffractometer was obtained from the aluminum oxide (Al_2O_3) standard with Caglioti parameters: U = 0.0093, V = −0.0051, and W = 0.0013 [10]. The morphology, size, and composition of the powders were obtained using a TESCAN LYRA3 high-resolution scanning electron microscope (Tescan Brno s.r.o., Brno, Czech Republic) with an FEG type electron source coupled with an Oxford energy-dispersive X-ray spectroscopy (EDS) detector. Secondary electron imaging and atomic element mapping were acquired simultaneously using an accelerating voltage of 15 kV and a working distance of 9 mm.

Zero-field-cooling (ZFC) magnetic hysteresis loops (*M(H)* loops) were recorded at 300 K and 5 K using the vibrating sample magnetometer (VSM) option operating in a Dynacool (Quantum Design North America, San Diego, CA, USA) setup for a maximum applied field of 90 kOe. ZFC and warm-field-cooling (WFC) magnetization measurements, *M(T)*, were performed under two different probe fields: 50 Oe and 1000 Oe.

^{57}Fe Mössbauer spectra were obtained with WissEl equipment (WissEl—Wissenschaftliche Elektronik GmbH, Starnberg, Germany) in a transmission geometry using a ^{57}Co source diffused into an Rh matrix with an activity of about 1.5 GBq and mounted on a conventional constant acceleration vibrating electromagnetic transducer. The sample was in the form of a powder layer containing about 5 mg Fe/cm^2. Spectra were obtained at 300 K and at 77 K in a bath cryostat. A thin foil of α-Fe was used at 300 K for calibration of the spectrometry (isomer shift values are given relative to Fe at 300 K). The modeling of the hyperfine structures was performed using a homemade Mosfit program based on the least squares method, and magnetic and quadrupolar components were composed of Lorentzian peaks.

3. Results and Discussion

3.1. XRD and Rietveld Analysis

From the Rietveld refinement of the XRD and neutron diffraction powder (NPD) patterns measured at RT and 2 K, respectively, the results suggested for all samples the presence of only an orthorhombic Pnma (No. 62) crystal structure [11], i.e., no secondary phase was observed. In addition, both experiments showed similar behavior for the lattice parameters a, b, and c within their uncertainties. Thus, the refined values of the cell parameters are plotted in Figure 1a,b, which suggested: (i) a linear evolution of the cell parameters (Figure 1a) at RT and 2 K as a function of iron concentration (x), (ii) the c parameter changed more rapidly than the a or b parameters, and (iii) a continuous increase in cell volume with Fe concentration (Figure 1b) at both temperatures. For the size estimation of $YFeO_3$ crystallites (RS7 sample) (see refined diffractogram in Figure 1c), we used the modified Scherrer's formula that expresses the anisotropic size broadening as a linear combination of spherical harmonics (SHP) if the anisotropic size contribution belongs only to the Lorentzian component of the total Voigt function [10]. Therefore, the explicit formula for the SPH approach of size broadening is given by Equation (1) [12]:

$$\beta_h = \frac{\lambda}{D_h \cos\theta} = \frac{\lambda}{\cos\theta} \sum_{lmp} a_{lmp} y_{lmp}(\Theta_h, \Phi_h) \tag{1}$$

where h is assigned to the (hkl) indices, β_h is the size contribution to the integral width of reflection (hkl), $y_{lmp}(\Theta_h, \Phi_h)$ are the real components of spherical harmonics (arguments Θ_h and Φ_h are the polar and azimuthal angles of vector (hkl) with respect to a Cartesian crystallographic frame), and a_{lmp} are the refined coefficients, related to the Laue class [13]. For the RS7 sample, a -1 Laue class was used.

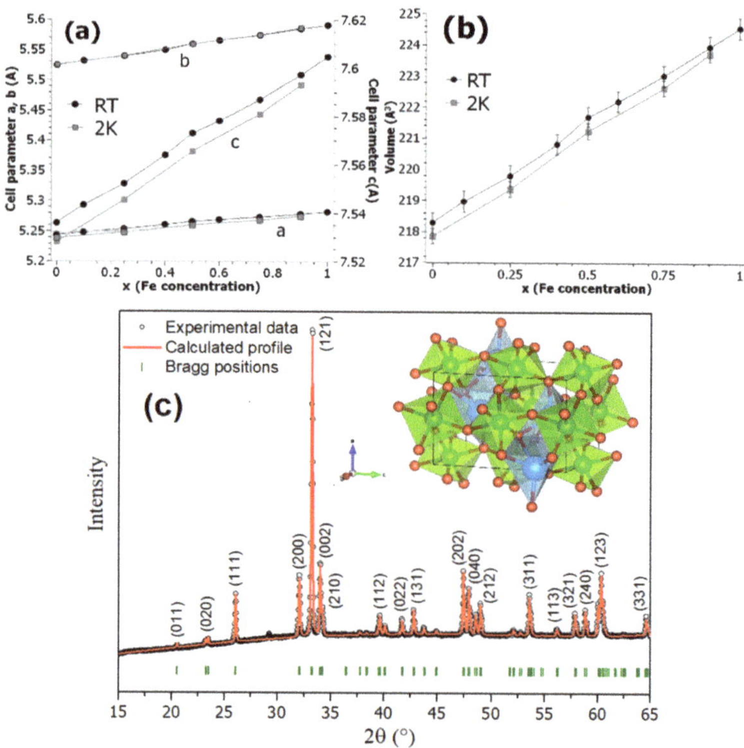

Figure 1. Cell parameter (**a**) and volume (**b**) variation with the x (Fe concentration) and Rietveld refinement XRD diffractogram (**c**) for the RS7 sample. The inset in (**c**) is the simulated structure obtained after refinement using VESTA (red spheres are oxygen atoms, white and blue are yttrium atoms, and green spheres are iron atoms). While blue, red, and green arrows indicate the a, b, and c crystallographic axes, respectively. Miller indexes are given between parentheses, black dots are the experimental data, the red line is the calculated diffractogram, and the vertical green lines are the Bragg's diffraction position. Solid lines in (**a,b**) are guides for visualization.

The obtained profile refinement gave acceptable statistical parameters for the reliability factor, R_p (%); weighted profile residual, Rw_p (%); expected profile residual, R_{exp}(%); and goodness of fit, χ^2; having quantitative values as follows: $R_p = 16.2\%$, $R_{wp} = 10.4\%$, $R_{exp} = 26.19\%$, and $\chi^2 = 0.94$, while the refined harmonic coefficients were found equal to $Y_{00} = -0.00574$, $Y_{20} = 0.01395$, $Y_{21}{}^+ = -0.06448$, $Y_{21}{}^- = 0.01518$, $Y_{22}{}^+ = 0.07557$, and $Y_{22}{}^- = -0.03690$, respectively. With these values, the anisotropic Lorentzian size broadening gave a mean crystallite value of 84 (8) nm. Hence, combining all above data, it can be inferred that the autocombustion method allowed nanocrystalline Fe-substituted $YCrO_3$ powders with a single phase, orthorhombic-like structure to be obtained.

3.2. SEM Analysis

For all samples, the autocombustion method with final annealing up to 1200 °C yielded a distribution of agglomerates about (~200–500 nm) that form a series of interconnected chains, as typical found in autocombustion synthesis [4]. In Figure 2a–q, the RSx series shows similar morphologies as those obtained by Zhang et al. [4,14]. In all of them, the notorious polycrystalline nature can be observed. The STEM images and elemental analyses given by yellow, red, blue, and green colors (Figure 2d–m) show the evolution of the systems when the Fe concentration increases. The systematic formation of the Fe-substituted $YCrO_3$ phase is noted in the EDS pattern given in Figure 2r, and the atomic percentage contribution is summarized in Table 1. We can roughly say that the particles produced by this combustion method have a similar morphology and similar dispersion in all tested concentrations. In addition, considering the uncertainties of the element contents in the samples, we can also affirm that Y and O are quite constant, while Fe increases and Cr decreases its contribution; this indicates that Fe enters the crystalline cell, due to the larger ionic radius of Fe^{3+} (0.645 Å) (compared to Cr^{3+} = 0.615 Å), which is consistent with the unit cell volume determined from X-ray and neutron diffractions, as previously discussed.

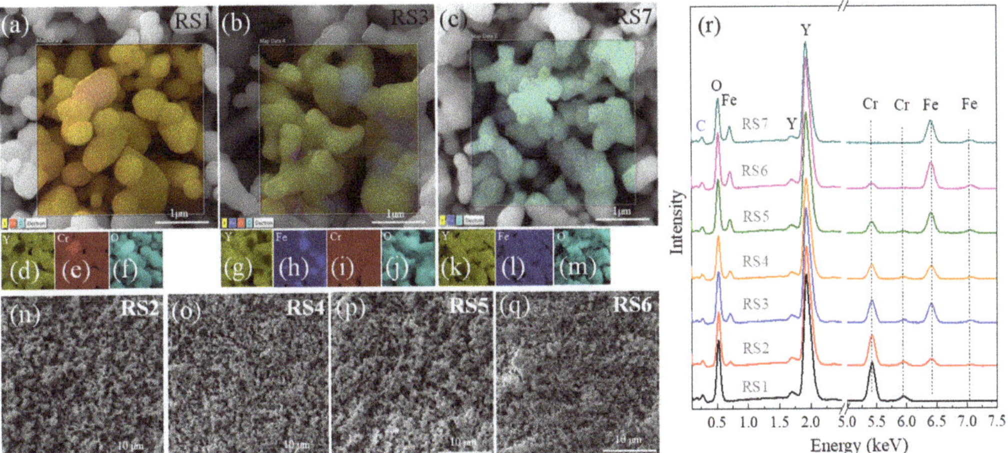

Figure 2. (**a–c**) SEM images for the RS1, RS3, and RS7 samples (bar length = 1 μm). The magnified area was performed in high resolution mode, and the elemental mapping area for each sample is given in (**d–m**) images. SEM images for the RS2, RS4, RS5, and RS6 samples (**n–q**) (bar length =10 μm). (**r**) The EDS for the RS1–RS7 samples.

Table 1. Weight percentage composition of the elements found in all samples.

Sample	Y (% wt) ± 2	Cr (% wt) ± 2	Fe (% wt) ± 1	O (% wt) ± 1
RS1	52	31	-	17
RS2	51	25	9	15
RS3	47	17	23	14
RS4	51	14	20	15
RS5	51	9	25	16
RS6	51	4	30	15
RS7	51	-	33	16

3.3. VSM Analysis

Figures 3 and 4 illustrate the ZFC ± 90 kOe *M(H)* curves of the YFe$_x$Cr$_{1-x}$O$_3$ series taken at 5 K (top) and 300 K (bottom), respectively. At RT, the samples with x = 0.25 and 0.50 behaved as ordinary paramagnets (see Figure 3d,f), while the samples with x = 0.60, 0.75, and 0.90 suggested an onset of a weak ferromagnetism (see Figure 4b,d,f). For the x = 1.0 sample (YFeO$_3$ compound, RS7), shown in Figure 5a, *M(H)*, curves recorded at different temperatures show the magnetic features of a weak ferromagnet with high magnetic anisotropy, i.e., with characteristics similar to that found in the pure YFeO$_3$ compound (set-like *M(H)* curve). Therefore, this sample (RS7) revealed an interesting and complex magnetic behavior that is mainly attributed in the literature to an exchange spring effect. In particular, the magnetic spring effect can be often observed by an exchange magnetic coupling between coexisting and interacting soft and hard magnets in a sample, as reported by Popkov et al. [5].

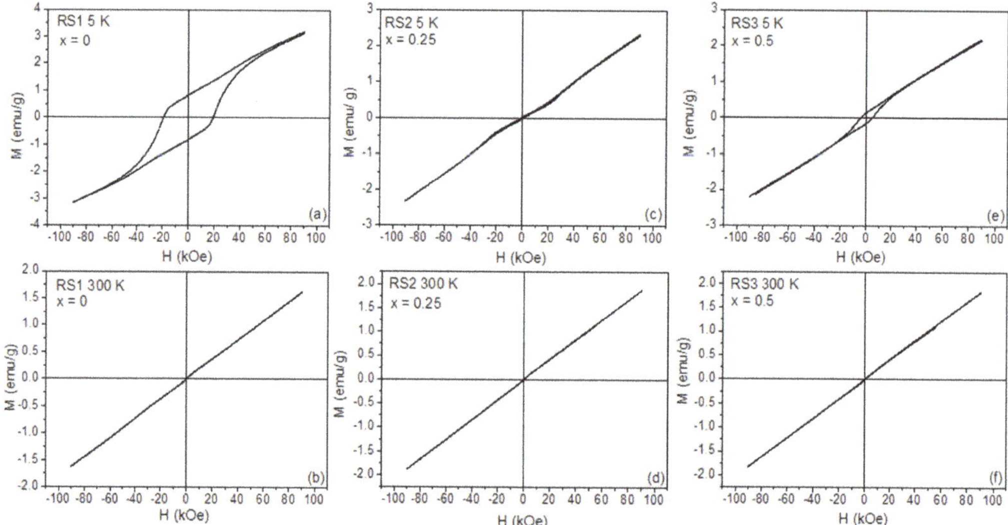

Figure 3. *M(H)* curves recorded for a maximum field of 90 kOe at 5 K (top) and 300 K (bottom) for the RSX samples. RS1 in (**a**,**b**), RS2 in (**c**,**d**), and RS3 in (**e**,**f**), respectively.

Thus, since the NDP, Rietveld, and SEM data of the SR7 sample suggested the presence of a single-phase structure, and the presence of two magnetic phases of two crystalline structures (as occurs in bilayer films) cannot be inferred as a reason for the observed phenomenon. However, the atomic disordering in the orthorhombic structure and the change in cell volume could lead to local different magnetic phases, which will be magnetically interacting and producing the observed *M(H)* behavior discussed above.

Looking at the 5 K *M(H)* loop for the x = 0 sample (see Figure 3a), we can observe the characteristic *M(H)* curve reported for the YCrO$_3$ compound [15]. Below T_N, the non-saturation regime of the *M(H)* curve occurred till values of +90 kOe, indicating a remarkable antiferromagnetic state, while above T_N, a paramagnetic-like behavior was regarded; see Figure 3b. On the other hand, the loss of hysteresis in the RS2 and RS6 samples (Figure 3c,e and Figure 4a,c,e) indicated the substitution of Cr by Fe atoms in the orthorhombic crystal configuration, as confirmed by our XRD data. Table 2 contains the remanence (M_r), coercivity (H_C), and saturation magnetization (σ_{sat}) values of the hysteresis ferromagnetic part. Using the slope of the *M(H)* curves, it was possible to subtract the antiferromagnetic contribution of the *M(H)* curves, thus leaving the purely ferromagnetic component, as shown in Figure 5b. Thus, these *M(H)* curves recorded at different temperatures really showed large H_C fields, but their value decreased when the temperature decreased, concomitantly with the increase of the saturation magnetization (the area inside the *M(H)* loop remained nearly constant). In addition, all *M(H)* curves show more clearly the step-like behavior near the zero-applied field region of the *M(H)* curve, a feature discussed above and attributed to a magnetic spring-like effect.

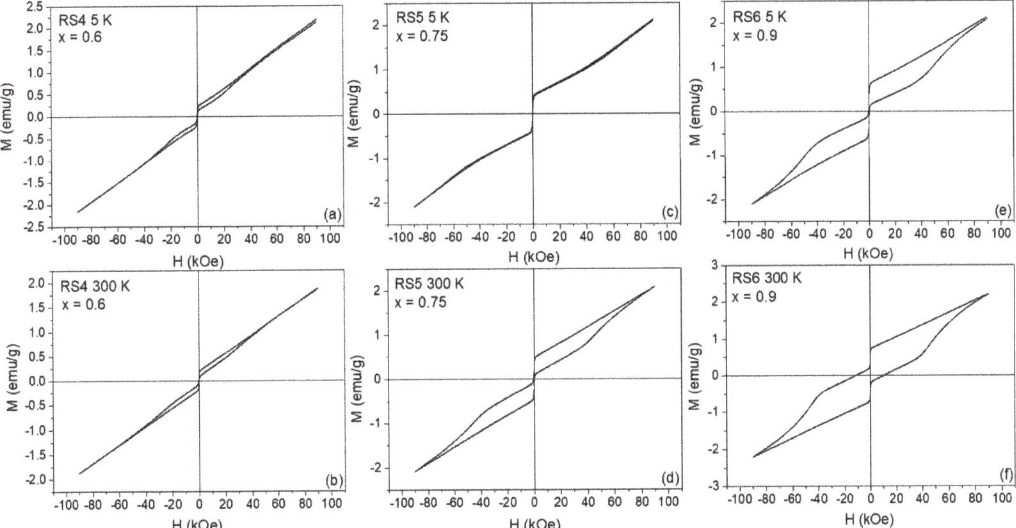

Figure 4. *M(H)* curves recorded for a maximum field of 90 kOe at 5 (top) and 300 K (bottom) for the RSX samples. RS4 in (**a,b**), RS5 in (**c,d**), and RS6 in (**e,f**), respectively.

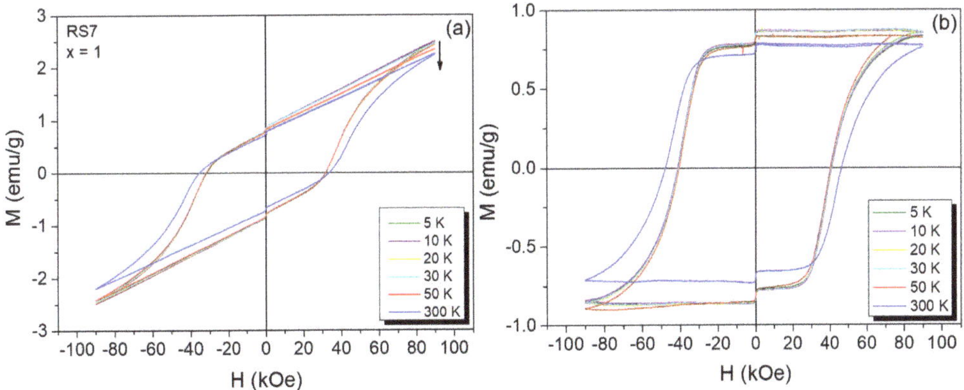

Figure 5. *M(H)* loops for the RS7 sample recorded at different temperatures (**a**). *M(H)* loops after the subtraction of the paramagnetic contribution of the AFM phase (**b**).

The magnetic parameters in Table 2 were plotted as a function of Fe concentration (x), as seen in Figure 6a,b,d. At 300 K, the H_C field dependence with x had two marked regions (I and II): (i) region-I can be interpreted as the magnetic domain reorientations (magnetization reversal) due to the increasing concentration of Fe atoms that are replacing Cr, forming the pure $YFeO_3$ crystalline phase; (ii) region-II has relatively high values of the H_C fields and that occur above x = 0.75, reaching a maximum value of 46.7 kOe for x = 1 (RS7 sample).

Table 2. Remanence (M_r), coercive field (H_C), and magnetic saturation (σ_{Sat}) of the ferromagnetic component, and susceptibility of the antiferromagnetic component to the RSx samples with x = 1, 0.9, 0.75, and 0.60. For the samples with x = 0, 0.25, and 0.50, the values correspond to the 'paramagnetic' state.

x	T (K)	M_r (emu/g) $\pm$ 0.05	H_C (kOe) $\pm$ 0.5	σ_{Sat} (emu/g) $\pm$ 0.05	Susc. (emu/g $\times$ Oe) $\pm$ 0.01
1.00	300	0.75	46.7	0.79	0.02
1.00	5	0.84	40.9	0.87	0.02
0.90	300	0.64	43.2	0.77	0.02
0.90	5	0.27	0.3	0.61	0.02
0.75	300	0.18	0.3	0.55	0.02
0.75	5	0.02	0.02 (5)	0.41	0.02
0.60	300	0.05	0.1	0.20	0.02
0.60	5	0.05	0.1	0.23	0.02
0.50	300	0	0	0	0.02
0.50	5	0	0	0	0.02
0.25	300	0	0	0	0.02
0.25	5	0	0	0	0.03
0.00	300	0	0	0	0.02
0.00	5	0.80	18.9	0.89	0.03

The remanence (M_r) is calculated from $M_r = (M_{R+} + M_{R-})/2$, where M_{R+} and M_{R-} are the values of the upper and lower magnetization, respectively, when the magnetic field is zero. The coercive field (H_C) is calculated from $H_C = (H_{C+} - H_{C-})/2$, where H_{C+} and H_{C-} are the values of the right and left fields when magnetization is zero.

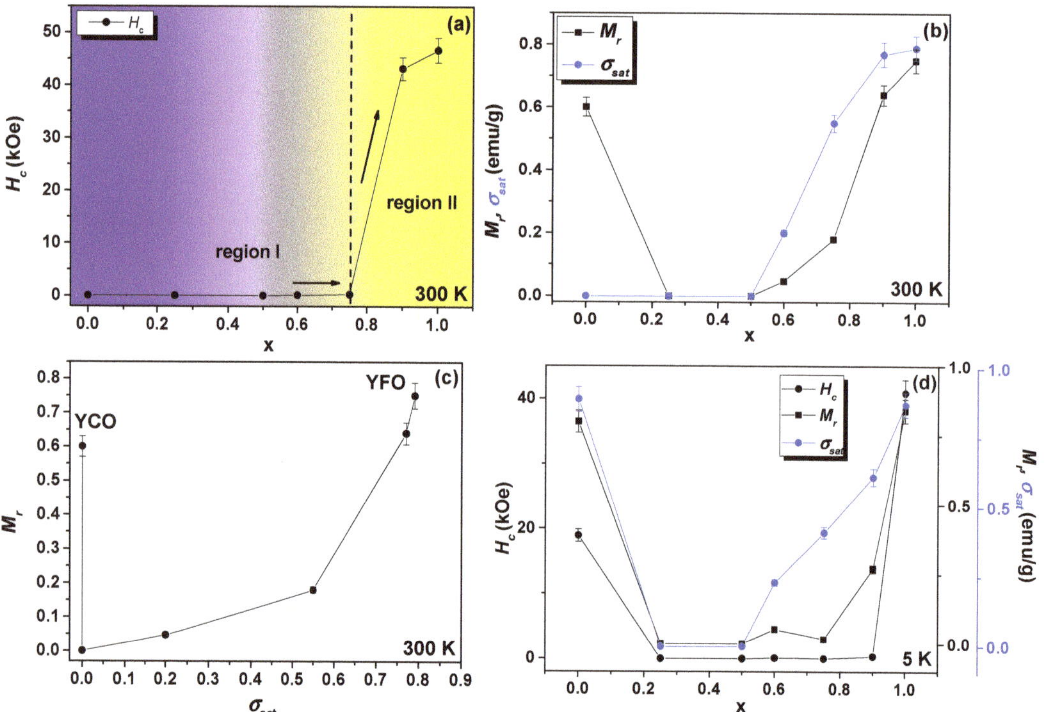

Figure 6. (**a**) Dependence of the H_C (kOe) vs. x (Fe concentration) at 300 K. (**b**) Dependence of the M_r and σ_{sat} vs. x (Fe concentration) at 300 K. (**c**) M_r vs. σ_{sat} graph at 300 K. (**d**) H_C, M_r, and σ_{sat} vs. (Fe concentration) at 5 K.

These values were larger than those reported by Popkov et al. [5] for four $YFeO_3$ crystalline samples synthesized by different routes. On the other hand, the M_r and σ_{sat} values had a similar dependence with Fe concentration at 300 K and 5 K, as can be seen in Figure 6b,d. The behavior of M_r vs. σ_{sat} is shown in Figure 6c. The M_r and σ_{sat} quantities could reach maximum values of 0.75 and 0.79 emu/g, respectively. This σ_{sat} value of 0.79 emu/g was consistent with others found in the literature for either powder or single crystals [3,4,16–19], as summarized in Figure 7. In particular, the value of 0.79 emu/g, obtained for a field of 3.5 kOe, was almost two times higher than the values reported by Zhang et al. [4] and four times higher than that obtained by Shen et al. [20] for a similar system. Therefore, the RS7 sample behaved as an ordinary single crystal of the YFO phase with a multidomain magnetic structure [4,20]. In addition, it is worth mentioning that the RS7 sample exhibited weak ferromagnetism enhanced at 90 kOe.

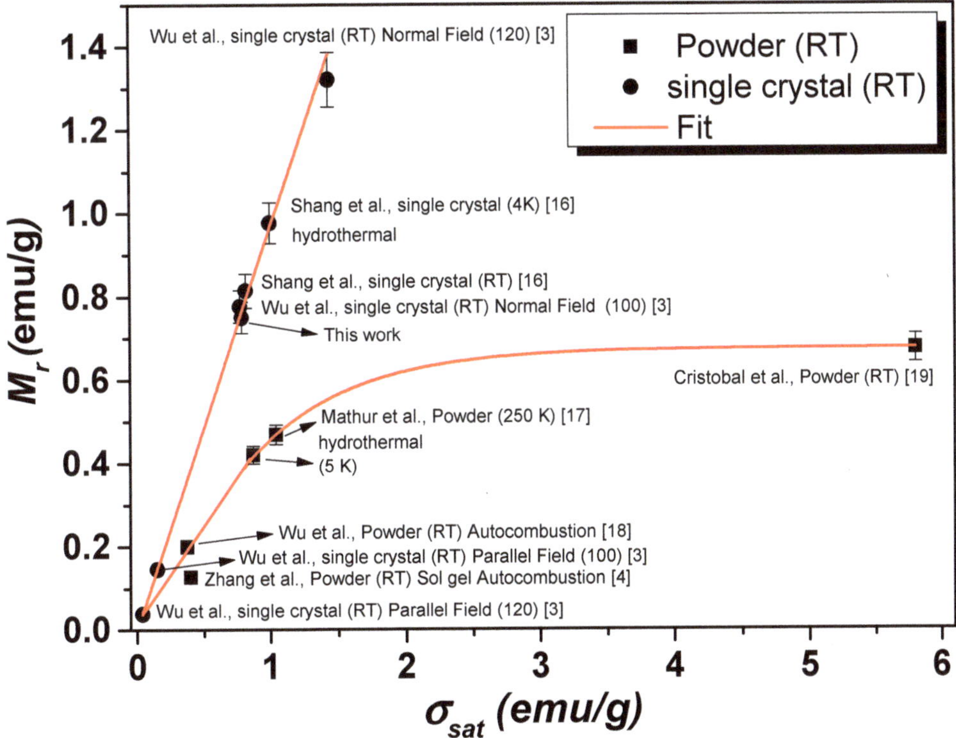

Figure 7. M_r vs. σ_{sat} relation built from data recorded at RT and reported in the literature for similar compounds. Comparison between powder and single crystal $YFe_xCr_{1-x}O_3$ [3,5,16–19].

The WFC and ZFC $M(T)$ measurements for all RSx samples were collected under two probe fields, namely, 50 Oe and 1000 Oe, and the results are shown in Figures 8–10. For the lowest applied field, the ZFC and WFC $M(T)$ curves, displayed in Figure 8a, clearly show the magnetization transition from the AFM to PM state of the $YCrO_3$ compound at 159 K, assigned to T_N. No other magnetic transition was observed in $M(T)$ curves, indicating that no secondary phase was formed during the auto combustion synthesis, in agreement with the XRD data. For the $YFe_{0.25}Cr_{0.75}O_3$ compound, the T_N value increased to 174 K (see Figure 8b), but a further increase of Fe content, for example, x = 0.50, led to a cancelation of total magnetization and a compensation temperature between the antiferromagnetic sub-lattices of 245 K. The zero-net magnetization was observed as an enhancement of the diamagnetism contribution, as shown in Figure 8c. At x = 0.60 and 0.75, see Figure 8d,e, a slight increase in the magnetization was observed, in agreement with the onset of WFM, as also seen in the *M(H)* curves. At x = 0.90 and 1.0 (Figures 8f and 9), a significant increase in the magnetization was observed with significant overlap between ZFC and WFC *M(T)* curves above 250 K.

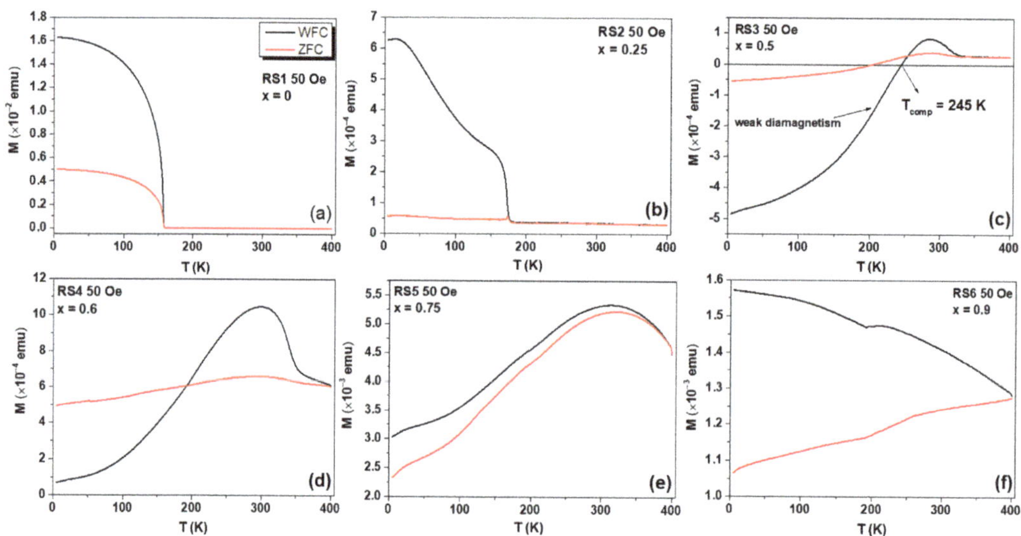

Figure 8. ZFC and WFC *M(T)* curves at a probe field of 50 Oe for the (**a**) RS1, (**b**) RS2, (**c**) RS3, (**d**) RS4, (**e**) RS5, and (**f**) RS6 samples.

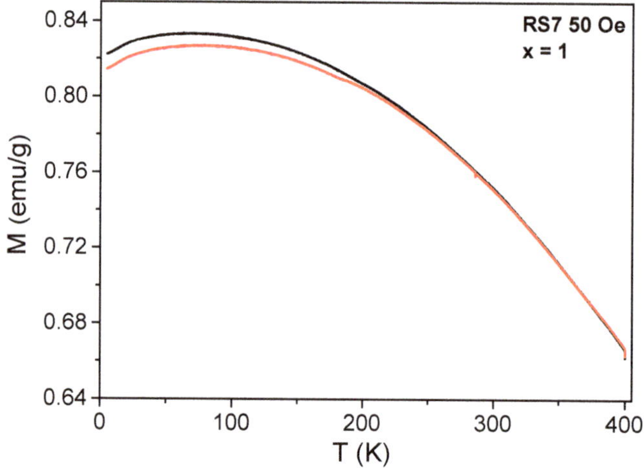

Figure 9. ZFC and WFC *M(T)* curves recorded for a probe field of 50 Oe for the RS7 sample. Black line indicates WFC and red line ZFC, respectively.

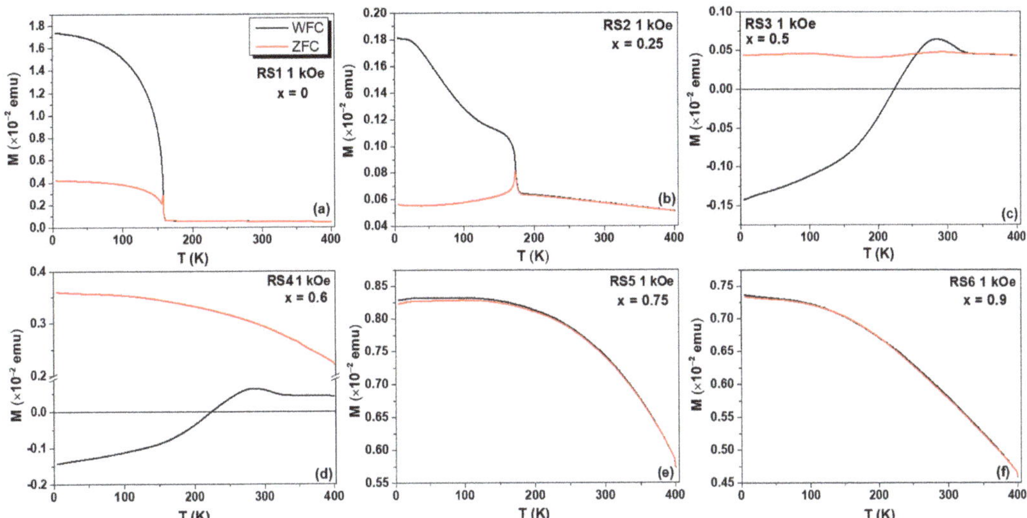

Figure 10. ZFC and WFC *M(T)* curves recorded at a probe field of 1000 Oe for the (**a**) RS1, (**b**) RS2, (**c**) RS3, (**d**) RS4, (**e**) RS5, and (**f**) RS6 samples.

The determination of the T_N of the Fe-substituted YCO compounds was done recording the ZFC and WFC *M(T)* curves at a higher field (1000 Oe). At this probe field, the magnetization of the samples with x = 0.75 and x = 0.90 showed a strong interaction with the external field, confirming the enhancement of the WFM. From ca. 5 K to higher temperatures, both ZFC and WFC *M(T)* curves coincided for the sample with x = 0.9.

Based on the above experimental results, it can be inferred that the anisotropic exchange-spring in crystalline compounds cause a significant increase in the coercive field of 46.7 kOe at 300 K. This interesting magnetic response has also been observed by Popkov et al. [5]. In our case, the hard and soft magnetic phases are intrinsically correlated to the same structure, but they are due to chemical disorders in the sites of the orthorhombic crystal nanostructure. Moreover, the hysteresis loop shape depends on the finite-size effects under an applied DC magnetic field (in our case, we use the highest value reported in the literature of 90 kOe). Hence, the observed ascending/descending hysteresis loops at several temperatures is explained due to spin reorientation of the antiferromagnetic vector in the *x–z* plane, reaching the *z*-axis at a critical magnetic field, as reported by Jacobs et al. [21], where a value of 74 kOe at 4.2 K was obtained for the $YFeO_3$ single crystal. According to Popkov et al. [5], in nanocrystalline materials, the typical WFM hysteresis cycle is observed only for the YFO phase when their grain sizes are equal and larger than 41 nm, i.e., the YFO material may exhibit WFM, and the exchange spring-like effect may occur due to its high magnetocrystalline anisotropy energy. Consequently, considering our experimental results that showed a grain size of 84 (8) nm, we can also expect the observed ascending/descending branch behaviors of the *M(H)* loops of the YFO sample. More precisely, the combined magnetic effects of the enhanced WFM and the presence of AFM interactions among the Fe ions of the different sites of the orthorhombic crystal structure gave rise to different local anisotropy contributions, producing high magnetocrystalline anisotropy due to the size effect and the enhancement of DM (Dzyaloshinskii–Moriya) interactions in the samples.

The two effects cannot be separated, and the improvement of WFM features can be explained assuming a canting angle of 13°, as demonstrated by previous neutron diffraction analysis [11]. The presence of AFM interactions in the Fe-substituted YCO compounds is also confirmed by the changes of T_N values as a function of Fe content, as displayed in Figure 11. Indeed, the T_N values increase nonlinearly with increasing Fe content, reaching the reference value for YFO [8]. Of course, the Fe substitution phenomenon is randomly changing locally the anisotropy by changing the lattice parameters, as shown by the XRD results. These modifications favor the spin reorientation and magnetization reversal phenomena.

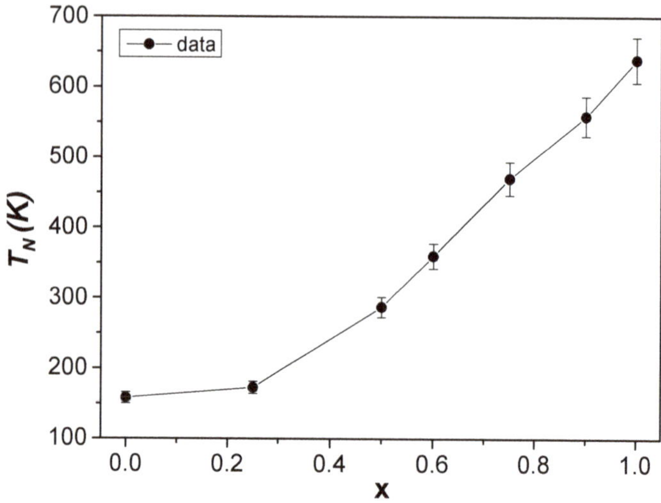

Figure 11. Néel temperature T_N dependence on x (Fe concentration) estimated from 1000 Oe *M(T)* curves. The solid line passing by experimental points is only a guide for viewing.

3.4. Mössbauer Analysis

3.4.1. Measurements at 300 K

In agreement with the magnetization data, the 300 K ^{57}Fe Mössbauer spectra recorded for samples with different Fe concentrations show, on the one hand, partial (x = 0.50) or total (x = 0.25) paramagnetic behavior (see Figure 12a). On the other hand, the 300 K ^{57}Fe Mössbauer spectra of the samples with x = 0.75 and 1.0 show six absorption lines due to the nuclear Zeeman interaction with a local magnetic hyperfine field (B_{hf}). The refined values of the corresponding hyperfine parameters are given in Table 3. The values of isomer shift are typical of the presence of Fe^{3+} ions. Another important feature that should be highlighted is that the line widths of the Mössbauer spectra are generally broader for samples with x = 0.25, 0.50, and 0.75 compared with those of the RS7 sample (x = 1.0), the latter being expected to show less atomic disorder. Therefore, the broadening effect of magnetic lines is probably caused by different iron environments, since in the orthorhombic crystal structure of these perovskites, a $3d^5$ Fe^{3+} ion is usually surrounded by 2, 3, 4, 5, or 6 Cr^{3+} ions in octahedral sites. The result of the chemical disorder is a hyperfine magnetic field distribution, i.e., a distribution of static sextets. In particular, the fit of the ^{57}Fe Mössbauer spectrum of the sample with x = 0.50 was done with two magnetic sextets and one quadrupolar doublet. The two sextets will represent the different local Fe environments of the orthorhombic crystal structure, while the doublet, best seen in the inset spectrum recorded in a low-velocity range, must be associated with Fe^{3+} ions in the paramagnetic state resulting from Cr^{3+}-rich environments ($T_N < 300$ K, e.g., for the x = 0.25, $T_N = 153$ K). The features discussed above tell us that the Fe substitution is not homogeneous, leading to an assembly of clusters with different compositions, i.e., a chemical disorder in the octahedral sites (B-sites) of the orthorhombic crystal structure. Thus, the largest magnetic component can be attributed to Fe^{3+} ions preferentially surrounded by Fe^{3+} ions ($T_N > 300$ K), while the quadrupolar doublet is associated with a neighborhood rich in Cr^{3+} ions, of course, with T_N values lower than 300 K, as shown by our magnetization data.

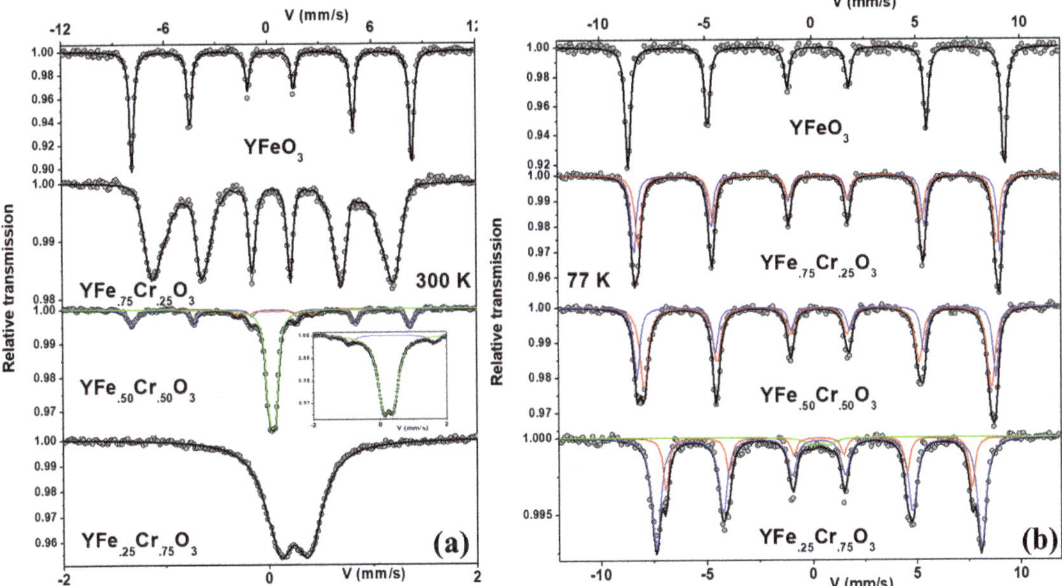

Figure 12. (**a**) RT Mössbauer measurements of the Fe-substituted $YFe_xCr_{1-x}O_3$ compounds. Samples with x = 1.0, 0.75, and 0.5 are measured with ±12 mm/s. The inset shows the same spectra for x = 0.5 but with a maximal velocity of ±2 mm/s, as in the case of the $YFe_{0.25}Cr_{0.75}O_3$ sample (last spectrum). (**b**) Same sample measured at 77 K. The subspectra used to fit these spectra are also shown. The two sextets represent the two iron configurations in octahedral sites of the orthorhombic crystal structure of the perovskite.

Table 3. Refined values of the hyperfine parameters at given temperatures.

x	T (K)	IS (mm/s) $\pm$ 0.01	2ε or Δ (mm/s) $\pm$ 0.01	B_{hf} (T) $\pm$ 0.5	Absorption Area Ratio % $\pm$ 2
1	300	0.37	0.00	50.1	100
	77	0.48	0.01	55.2	100
0.75	300	<0.38>	<0.04>	<40.7>	100
	77	0.47	0.03	53.8	49
		0.47	0.05	52.6	51
		<0.47>	<0.04>	<53.2>	
0.50	300	0.39	0.28		62
		0.40	−0.24	50.1	22
		0.46	−0.12	48.5	8
		0.35	0.15	13.3	8
	77	0.48	0.02	51.0	62
		0.48	−0.10	52.7	38
		<0.48>	<−0.02>	<51.7>	
0.25	300	0.36	0.28		100
	77	0.47	0.08	47.9	74
		0.47	0.08	45.0	23
		0.47	0.79		3

3.4.2. Measurements at 77 K

To better understand the local environment of the ^{57}Fe ions in the orthorhombic crystal structure, additional measurements were made at 77 K for all Fe-substituted series, and the corresponding Mössbauer spectra are shown in Figure 12b. At 77 K, below the T_N values of the Fe-substituted YCO compounds (see Figure 11), one would therefore expect a pure Zeeman nuclear interaction in all ^{57}Fe spectra. The spectra, in general, show all six expected absorption lines, but with different broadening and asymmetries depending on the Fe content. While the 77 K Mössbauer spectrum of the YFO (x = 1) compound can be perfectly described by a single magnetic component, those of RS5, RS3, and RS2 require at least two magnetic components. The refined values of the hyperfine parameters are given in Table 3. Thus, to fit these 77 K spectra, we have two magnetic sextets to account for, at least, two octahedral configurations of Fe^{3+} ions for samples with non-zero x. The results clearly show that B_{hf} values decrease with increasing content. Even at 77 K, the spectrum of the $YFe_{.25}Cr_{.75}O_3$ sample required an additional quadrupolar doublet, with a fraction of 5% of total spectra. Thus, considering that the magnetization data show a T_N value for this sample equal to 153 K, the quadrupolar doublet must be associated with Fe^{3+} ions with a Cr^{3+} ion-rich neighborhood.

According to the Néel temperature of the series, it is understandable why the samples with x = 1.0 and 0.75 show hysteresis cycles, although the $YFeO_3$ sample has a lower magnetic energy than the 50% sample. Similarly, for the other two samples with x = 0.25 and zero, the magnetic susceptibility is consistent with paramagnetic behavior, which is understandable due to their lower T_N values than 300 K. The series contains samples with weak ferromagnetism and paramagnetic behaviors.

In brief, 77 K Mössbauer spectra were fitted, at least, with two different octahedral environments for Fe ions, and the results suggest that the presence of Cr ions decreases the B_{hf} value, but the difference between the two sextets of the Fe-substituted YCO compounds increases, except for the pure YFO, where only the sextet was required to have a good fit of the spectrum. One explanation for this decrease may be due to competing mechanisms between the antiferromagnetic interactions between Fe–Fe, Fe–Cr, and Cr–Cr exchanges and the DM interaction. Indeed, the asymmetric DM interaction is known to be the main interaction responsible for the WFM observed in YFO, where the antiferromagnetic coupling mechanism is due to superexchange interactions between the t3–O–t3 and e2–O–e2 orbitals, whereas for the YCO compound, the mechanism is a coupling to the t–e orbitals [22–24]. Therefore, we have in the Fe-substituted YCO samples a mixed exchange mechanism that is enhanced by the atomic disorder naturally present in our samples. It can be expected that due to the Fermi contact and the transferred magnetic field contribution to the total hyperfine magnetic field depend on the s electrons and the superposition of 3d, s, and p electrons, respectively, there is increasing competition of Fe environments as the iron concentration of the sample increases. An appropriate calculation using the mean field theory gave the relationship $J_{Fe–Fe} > J_{Fe–Cr} > J_{Cr–Cr}$ [8].

4. Conclusions

In the present work, the structural and magnetic properties of the Fe-substituted perovskite series were studied in detail. Specifically, the average crystalline grain size of the $YFeO_3$ compound, calculated using the harmonic spherical approach in a Rietveld refinement, was 84 (8) nm, a size where weak ferromagnetism can occur in this compound. In the Fe-substituted $YFe_xCr_{1-x}O_3$ compounds, X-ray and neutron diffraction patterns collected at RT and 2 K gave a linear increase in the lattice parameters of the orthorhombic structure with increasing Fe concentration, where the c-parameter had the most pronounced increase. This increase obviously translates into an increase in the volume cell and consequently a change in the magnetocrystalline anisotropy of the samples, i.e., a magnetic anisotropy that depends on the Fe concentration. Considering that X-ray and neutron diffraction showed only one crystalline phase for all samples, and the above results of lattice parameters that showed a gradual increase with increasing iron content, we can highlight that autocombustion is a useful method for the synthesis of pure $YFeO_3$ with high stoichiometry. The 90 kOe $M(H)$ curves taken at RT and 5 K for all Fe-substituted samples suggest the presence of spin reorientation and magnetization reversal phenomena associated with homogenous (Fe–Fe, Cr–Cr) and non-homogeneous pairs of 3d-ions (Fe–Cr). The dependence of the H_C, M_r, and σ_{sat} with Fe concentration clearly showed the onset of WFM for x = 0.60–0.80 values. For x = 1.0, the high H_C value of 46.7 kOe was calculated after subtracting the AFM contribution (linear contribution of the paramagnetic phase). This latter result implies that an enhancement in the WFM is achieved due to chemical inhomogeneity of the $YFeO_3$ phase. Moreover, the values of M_r and σ_{sat} at 300 K are in agreement with the values commonly found in single crystals. The WFC and ZFC $M(T)$ curves recorded at low (50 Oe) and high (1000 Oe) probe field analysis allowed the magnetic properties and global magnetic response of the spin reorientation process to be tuned. The high field $M(T)$ curves allowed for accurate determination of T_N values and showed a nonlinear dependence of T_N on Fe concentration. The sample with x = 0.75 clearly exhibited a higher magnetic disorder, as corroborated by Mössbauer spectra recorded at 77 K and 300 K. The magnetization measurement performed in a low probe field of 50 Oe, for the sample with x = 0.50 showed a diamagnetic-like behavior near the compensation temperature of 245 K, where an inverse magnetization and the most intense remanence and saturation values at 300 K were found compared to the other samples. For the low Fe content (x = 0.25) and pure orthochromite samples, they showed paramagnetic-like behavior at RT, with a magnetic order only below 150 K. Mössbauer spectra allowed us to study the local Fe environment and visualize a weak enhanced ferromagnetism, due to a remarkable high canting angle (13°), estimated previously from neutron diffraction analysis, and its variation with Fe concentration. At least two octahedral Fe sites were identified in the non-pure Fe-substituted samples, whose evolution of the magnetic hyperfine field (B_{hf}) can be explained using the results of mean field theory reported in the literature. The sample with x = 0.50 showed good magnetic properties and is a suitable candidate for further study.

Author Contributions: Conceptualization, R.S.-R. and J.A.R.-G.; methodology, R.S.-R., N.-R.C.-H. and J.A.R.-G., validation, R.S.-R., N.-R.C.-H., E.C.P., J.-M.G. and J.A.R.-G.; formal analysis, R.S.-R., N.-R.C.-H. and J.A.R.-G.; investigation, R.S.-R., D.A.G., K.M.T., N.-R.C.-H., E.C.P., J.-M.G. and J.A.R.-G.; resources, R.S.-R., N.-R.C.-H. and J.A.R.-G.; data curation, R.S.-R., N.-R.C.-H. and J.A.R.-G.; writing—original draft preparation, R.S.-R. and J.A.R.-G.; writing—review and editing, R.S.-R., E.C.P. and J.A.R.-G.; visualization, R.S.-R. and J.A.R.-G.; supervision, R.S.-R. and J.A.R.-G.; project administration, R.S.-R.; funding acquisition, R.S.-R. All authors have read and agreed to the published version of the manuscript.

Funding: This research was funded by the "Universidad Nacional de Ingeniería-UNI" and the APC was funded by VRI-UNI. The part of this research conducted at ORNL's High Flux Isotope Reactor was sponsored by the Scientific User Facilities Division, Office of Basic Energy Sciences, US Department of Energy.

Data Availability Statement: The original data related to this research can be requested at any time from the corresponding author at the following email address: rsalazarr@uni.edu.pe.

Acknowledgments: We thank LABICER (UNI) for the XRD support and sample preparation analyzes. Juan A. Ramos-Guivar thanks the National Program of Scientific Investigation and Advanced Studies (PROCIENCIA), while Edson C. Passamani thanks FAPES and CNPq for their financial support.

Conflicts of Interest: The authors declare no conflict of interest.

References

1. Neusser, S.; Grundler, D. Magnonics: Spin Waves on the Nanoscale. *Adv. Mater.* **2009**, *21*, 2927–2932. [CrossRef]
2. Song, C.; You, Y.; Chen, X.; Zhou, X.; Wang, Y.; Pan, F. How to Manipulate Magnetic States of Antiferromagnets. *Nanotechnology* **2018**, *29*, 112001. [CrossRef]
3. Wu, A.; Shen, H.; Xu, J.; Wang, Z.; Jiang, L.; Luo, L.; Yuan, S.; Cao, S.; Zhang, H. Crystal Growth and Magnetic Property of $YFeO_3$ crystal. *Bull. Mater. Sci.* **2012**, *35*, 259–263. [CrossRef]

4. Zhang, W.; Fang, C.; Yin, W.; Zeng, Y. One-Step Synthesis of Yttrium Orthoferrite Nanocrystals Via Sol-Gel Auto-Combustion and Their Structural and Magnetic Characteristics. *Mater. Chem. Phys.* **2013**, *137*, 877–883. [CrossRef]

5. Popkov, V.I.; Almjasheva, O.V.; Semenova, A.S.; Kellerman, D.G.; Nevedomskiy, V.N.; Gusarov, V.V. Magnetic Properties of $YFeo_3$ Nanocrystals Obtained by Different Soft-Chemical Methods. *J. Mater. Sci. Mater. Electron.* **2017**, *28*, 7163–7170. [CrossRef]

6. Treves, D. Magnetic Studies of Some Orthoferrites. *Phys. Rev.* **1962**, *25*, 1843–1853. [CrossRef]

7. Mao, J.; Sui, Y.; Zhang, X.; Su, Y.; Wang, X.; Liu, Z.; Wang, Y.; Zhu, R.; Wang, Y.; Liu, W.; et al. Temperature- and Magnetic-Field-Induced Magnetization Reversal in Perovskite $YFe_{0.5}Cr_{0.5}O_3$. *Appl. Phys. Lett.* **2011**, *98*, 192510. [CrossRef]

8. Dasari, N.; Mandal, P.; Sundaresan, A.; Vidhyadhiraja, N.S. Weak Ferromagnetism and Magnetization Reversal in $YFe_{1−x}Cr_xO_3$. *EPL* **2012**, *99*, 17008. [CrossRef]

9. Calder, S.; An, K.; Boehler, R.; Dela Cruz, C.R.; Frontzek, M.D.; Guthrie, M.; Haberl, B.; Huq, A.; Kimber, S.A.J.; Liu, J.; et al. A Suite-Level Review of the Neutron Powder Diffraction Instruments at Oak Ridge National Laboratory. *Rev. Sci. Instrum.* **2018**, *89*, 092701. [CrossRef] [PubMed]

10. Canchanya-Huaman, Y.; Mayta-Armas, A.F.; Pomalaya-Velasco, J.; Bendezú-Roca, Y.; Guerra, J.A.; Ramos-Guivar, J.A. Strain and Grain Size Determination of CeO_2 and TiO_2 Nanoparticles: Comparing Integral Breadth Methods versus Rietveld, μ-Raman, and TEM. *Nanomaterials* **2021**, *11*, 2311. [CrossRef]

11. Salazar-Rodriguez, R.; Aliaga-Guerra, D.B.; Taddei, K.M. X-Ray Diffraction, Mössbauer Spectroscopy, Neutron Diffraction, Optical Absorption and Ab-Initio Calculation of Magnetic Process in Orthorhombic $YFe_xCr_{(1−x)}O_3$ ($0 \leq x \leq 1$) Compounds. *Hyperfine Interact.* **2019**, *240*, 82. [CrossRef]

12. Popa, N.C. The (hkl) Dependence of Diffraction-Line Broadening Caused by Strain and Size for all Laue Groups in Rietveld Refinement. *J. Appl. Cryst.* **1998**, *31*, 176–180. [CrossRef]

13. Pecharsky, V.J.; Zavalij, P.Y. *Fundamentals of Powder Diffraction and Structural Characterization of Materials*, 2nd ed.; Springer Science Business Media, LLC: Berlin/Heidelberg, Germany, 2009; pp. 269–292, ISBN 978-0-387-09578-3.

14. Shi, L.R.; Xia, Z.C.; Wei, M.; Jin, Z.; Shang, C.; Huang, J.W.; Chen, B.R.; Ouyang, Z.W.; Huang, S.; Xiao, G.L. Unusual Effects of Ho^{3+} Ion on Magnetic Properties of $YFe_{0.5}Cr_{0.5}O_3$. *Ceram. Int.* **2015**, *41*, 13455–13460. [CrossRef]

15. Tiwari, B.; Surendra, M.K.; Rao, M.S.R. $HoCrO_3$ and $YCrO_3$: A Comparative Study. *J. Phys. Condens. Matter* **2013**, *25*, 216004. [CrossRef]

16. Shang, M.; Zhang, C.; Zhang, T.; Yuan, L.; Ge, L.; Yuan, H.; Feng, S. The Multiferroic Perovskite $YFeO_3$. *Appl. Phys. Lett.* **2013**, *102*, 062903. [CrossRef]

17. Mathur, S.; Veith, M.; Rapalaviciute, R.; Shen, H.; Goya, G.F.; Filho, W.L.M.; Berquo, T.S. Molecule Derived Synthesis of Nanocrystalline $YFeO_3$ and Investigations on Its Weak Ferromagnetic Behavior. *Chem. Mater.* **2004**, *16*, 1906–1913. [CrossRef]

18. Wu, L.; Yu, J.C.; Zhang, L.; Wang, X.; Li, S. Selective Self-Propagating Combustion Synthesis of Hexagonal and Orthorhombic Nanocrystalline Yttrium Iron Oxide. *J. Solid State Chem.* **2004**, *177*, 3666–3674. [CrossRef]

19. Cristóbal, A.A.; Botta, P.M.; Bercoff, P.G.; Ramos, C.P. Hyperfine and Magnetic Properties of a $Y_xLa_{1−x}FeO_3$ series ($0 \leq x \leq 1$). *Mater. Res. Bull.* **2015**, *64*, 347–354. [CrossRef]

20. Shen, J.; Xu, J.; Wu, A.; Zhao, J.; Shi, M. Magnetic and thermal properties of perovskite $YFeO_3$ single crystals. *Mater. Sci. Eng. B* **2009**, *157*, 77–80. [CrossRef]

21. Jacobs, I.S.; Burne, H.F.; Levinson, L.M. Field-Induced Spin Reorientation in $YFeO_3$ and $YCrO_3$. *J. Appl. Phys.* **1971**, *42*, 1631. [CrossRef]

22. Dahmani, A.; Taibi, M.; Nogues, M.; Aride, J.; Loudghiri, E.; Belayachi, A. Magnetic Properties of the Perovskite Compounds $YFe_{1−x}Cr_xO_3$ ($0.5 \leq x \leq 1$). *Mater. Chem. Phys.* **2003**, *77*, 912–917. [CrossRef]

23. Nair, V.; Das, A.; Subramanian, V.; Santhosh, P.N. Magnetic Structure and Magnetodielectric Effect of $YFe_{0.5}Cr_{0.5}O_3$. *J. Appl. Phys.* **2013**, *113*, 213907. [CrossRef]

24. Zhou, J.S.; Alonso, J.A.; Pomjakushin, V.; Goodenough, J.B.; Ren, Y.; Yan, J.Q.; Cheng, J.G. Intrinsic Structural Distortion and Superexchange Interaction in the Orthorhombic Rare-Earth Perovskites $RCrO_3$. *Phys. Rev. B* **2010**, *81*, 214115. [CrossRef]

 *nanomaterials*

MDPI

Article

Synergistic Piezo-Photocatalysis of BiOCl/NaNbO$_3$ Heterojunction Piezoelectric Composite for High-Efficient Organic Pollutant Degradation

Li Li [1], Wenjun Cao [1], Jiahao Yao [1], Wei Liu [1], Feng Li [2,*] and Chunchang Wang [1,*]

1   Laboratory of Dielectric Functional Materials, School of Materials Science & Engineering, Anhui University, Hefei 230601, China; b20101003@stu.ahu.edu.cn (L.L.); caowenjun1206@126.com (W.C.); yjh1074457726@163.com (J.Y.); b20301114@stu.ahu.edu.cn (W.L.)
2   Information Materials and Intelligent Sensing Laboratory of Anhui Province, Institutes of Physical Science and Information Technology, Anhui University, Hefei 230601, China
*   Correspondence: fengli@ahu.edu.cn (F.L.); ccwang@ahu.edu.cn (C.W.)

Abstract: Piezo-photocatalytic technique is a new-emerging strategy to alleviate photoinduced charge recombination and thus enhance catalytic performance. The heterojunction construction engineering is a powerful approach to improve photocatalytic performance. Herein, the BiOCl/NaNbO$_3$ with different molar ratios piezoelectric composites were successfully synthesized by hydrothermal methods. The piezo/photodegradation rate (k value) of Rhodamine B (RhB) for BiOCl/NaNbO$_3$ (BN-3, 0.0192 min^{-1}) is 2.2 and 5.2 times higher than that of BiOCl (0.0089 min^{-1}) and NaNbO$_3$ (0.0037 min^{-1}), respectively. The enhanced performance of BN-3 composite can be attributed to the heterojunction construction between BiOCl and NaNbO$_3$. In addition, the piezo/photodecomposition ratio of RhB for BN-3 (87.4%) is 8.8 and 2.2 times higher than that of piezocatalysis (9.9%) and photocatalysis (40.4%), respectively. We further investigated the mechanism of piezocatalysis, photocatalysis, and their synergy effect of BN-3 composite. This study favors an in-depth understanding of piezo-photocatalysis, providing a new strategy to improve the environmental pollutant remediation efficiency of piezoelectric composites.

Keywords: BiOCl/NaNbO$_3$; heterojunction; piezocatalysis; photocatalysis; degradation

Citation: Li, L.; Cao, W.; Yao, J.; Liu, W.; Li, F.; Wang, C. Synergistic Piezo-Photocatalysis of BiOCl/ NaNbO$_3$ Heterojunction Piezoelectric Composite for High-Efficient Organic Pollutant Degradation. *Nanomaterials* **2022**, *12*, 353. https://doi.org/ 10.3390/nano12030353

Academic Editor: Dong-Joo Kim

Received: 30 December 2021
Accepted: 20 January 2022
Published: 22 January 2022

Publisher's Note: MDPI stays neutral with regard to jurisdictional claims in published maps and institutional affiliations.

1. Introduction

There is increasing social concern over the current energy and environmental issues, especially wastewater pollution (originating from rapid industrial development), which poses a direct threat to human health [1,2]. Researchers are seeking effective, cost-efficient, stable, and safe ways to degrade and remove hazardous compounds in water. Semiconductor photocatalysis is one of the prominent strategies, not only converting sustainable solar energy into hydrogen energy but also utilizing visible light to degrade organic pollutants. A large number of semiconductor photocatalysts and their composites have been reported to improve the water splitting or pollutant degradation performance, such as ZnO and WO$_3$ [3–5]. Among the various semiconductors, bismuth oxychloride (BiOCl) as a p-type bismuth oxyhalides semiconductor has an open layered structure composing of [Bi$_2$O$_2$]$^{2+}$ layers sandwiched between [Cl$_2$]$^{2-}$ plates, which can facilitate the separation of photo-produced electrons and holes due to the more space to polarize the atoms and orbitals involved [6]. However, BiOCl is limited by its wide band gap (~3.4 eV), representing favorable photocatalytic performance only under ultraviolet light irradiation. Therefore, heterojunction–construction engineering is one of the most powerful strategies to achieve a broader photoresponse and improved photocatalytic activity by constructing heterostructured BiOCl photocatalysts [7–9].

In addition to photocatalysis, piezoelectric catalysis is also an efficient and environmentally friendly dye degradation method [10–12]. Vibration is one of the most common sources of energy in our environment. In piezoelectric catalysis, mechanical vibration generates an electric charge on the surface of the piezoelectric catalyst through the piezoelectric effect, which in turn reacts with the dye molecules, leading to the decomposition of the dye [13,14]. Combining with photocatalysis, piezoelectric materials possess the ability to reduce the carrier recombination rate due to the existence of an internal piezoelectric field in piezoelectric materials. In past decades, perovskite niobates, including $NaNbO_3$, have attracted intensive attention because of their nonlinear optics, ionic conductive, piezoelectric, and photocatalytic properties [15–17]. In addition, the piezo-photocatalysis and pyroelectric catalysis of $NaNbO_3$ have been thoroughly studied in recent years. For example, Jia's group has found that $NaNbO_3$ nanofibers possess a highly efficient piezoelectrically and pyroelectrically bi-catalysis for decomposition of organic dye [18]. It is worth noting that BiOCl is a piezoelectric material with piezocatalytic activity in response to ultrasound. Jia's group have also reported that BiOCl shows highly efficient dye wastewater decomposition under the condition of light (300 W Xenon lamp) and ultrasound (120 W, 40 kHz) together, which is much greater than that of only with light or only with ultrasound, respectively [19]. However, there are not any reports for piezo-photocatalysis of $BiOCl/NaNbO_3$ heterojunction piezoelectric composite. Therefore, it is necessary to explore the mechanism of $BiOCl/NaNbO_3$ piezoelectric composite for enhancing degradation efficiency by using the synergistic effect of piezoelectric catalysis and photocatalysis.

In this work, the $BiOCl/NaNbO_3$ piezoelectric composites were prepared by the hydrothermal method. The polarization electric field hysteresis loop (P-E) and electric-field-induced strain (S-E) curves confirm the $BiOCl/NaNbO_3$ composite has good ferroelectric and piezoelectric properties. The catalytic performance of $BiOCl/NaNbO_3$ piezoelectric composite was remarkably enhanced by the heterojunction construction and the synergy effect of piezocatalysis and photocatalysis, which greatly promote the separation of electron-hole pairs under electric field.

2. Materials and Methods

2.1. Materials Fabrication

2.1.1. Synthesis of $NaNbO_3$

The $NaNbO_3$ powder was prepared by the hydrothermal method. Briefly, 1 g Nb_2O_5 was added into 30 mL sodium hydroxide (NaOH, 10 M) aqueous solution and stirred for 2 h. Then, transferred above solution into a 50 mL Teflon-lined autoclave and reacted at 180 °C for 48 h. The precipitate was washed thoroughly and dried at 80 °C.

2.1.2. Synthesis of BiOCl

In brief, 5 mmol bismuth nitrate ($Bi(NO_3)_3 \cdot 5H_2O$, 99.5%) and 5 mmol potassium chloride (KCl) were added into 30 mL deionized water (DI water) and stirred for 0.5 h. Then, the above solution was transferred into a 50 mL Teflon-lined autoclave and heated at 180 °C for 12 h. The precipitate was washed thoroughly and dried at 80 °C.

2.1.3. Synthesis of $BiOCl/NaNbO_3$ Composites

Different contents (1, 2, 3, 4 mmol) of $NaNbO_3$ (prepared in 2.1.1) were added into the 30 mL DI water and stirred for 0.5 h, and then, 5 mmol $Bi(NO_3)_3 \cdot 5H_2O$ and 5 mmol KCl were added into the solution and stirred for another 0.5 h. The above solution was transferred into a 50 mL Teflon-lined autoclave and reacted at 180 °C for 12 h. The corresponding composites were named as BN-1, BN-2, BN-3, BN-4.

2.2. Photocatalytic Performance Experiment

The piezo/photocatalytic activities were evaluated by the degradation of RhB under UV–vis light irradiation and ultrasound. The 0.1 g catalyst was added into the RhB aqueous solution (100 mL, 5 mg/L) in dark and stirred for 30 min to reach adsorption–desorption equilibrium. After that, the mixed solution was treated with UV–vis light irradiation (300 W Xe lamp) and ultrasonic (50 W, 40 kHz). Change the water every five minutes to avoid the effects of temperature. A 4 mL solution was taken out every 20 min and centrifuged to remove the catalyst. The residual amount of RhB was recorded by the UV–vis spectrophotometer (Yoke, N6000, Shanghai, China) within the range of 300–800 nm.

2.3. Characterization

X-ray diffractometer (XRD) patterns were obtained to validate the phase purity and crystallinity of the powders on the XRD equipment (Rigaku Smartlab Beijing Co, Beijing, China). Scanning electron microscope (SEM) images of the prepared catalysts, including energy dispersive X-ray spectroscopy (EDS) capabilities, were measured with an SEM Regulus 8230, Hitachi Co, Tokyo, Japan. The transmission electron microscope (TEM) was used by JEOL JEM-F200 (Tokyo, Japan). The absorption spectra of these powders were tested in a UV-vis spectrophotometer (PerkinElmer Lambda 35, Waltham, MA, USA). X-ray photoelectron spectroscopy (XPS) was carried out with a ESCALAB 250, Thermo-VG Scientific, Waltham, MA, USA to analyze the components and the valence states. The specific surface areas of the samples were tested by Micromeritics ASAP 2460 Brunauer-Emmet-Teller (BET, Shanghai, China) equipment with N_2 as the carrier gas. The polarization electric field (P-E) loops and electric-field-induced strain (S-E) were tested in silicone oil at room temperature with 1 Hz frequency using a MultiFerroic II, Radiant technologies Inc., Albuquerque, New Mexico. The sample powders were pressed in a pellet (1 cm diameter and 0.20 mm thick) with Polyvinyl Alcohol (PVA) solution as a binder and then annealed at 600 °C to burn out the PVA binder. The pellets were coated on both sides with Au electrodes.

3. Results and Discussion

X-ray diffraction patterns (XRD) of BiOCl, $NaNbO_3$, and a series of BiOCl/$NaNbO_3$ piezoelectric composites are shown in Figure 1a. The distinct diffraction peaks of pure BiOCl can be related to tetragonal BiOCl (PDF card no. 82–0485, space group: P4/nmm), and the diffraction peaks of pure $NaNbO_3$ can be indexed to orthorhombic $NaNbO_3$ (PDF card no. 77–0873, space group: $P2_1ma$). As for BiOCl/$NaNbO_3$ piezoelectric composites (BN-1, BN-2, BN-3, BN-4), there are both BiOCl and $NaNbO_3$ peaks can be observed. In addition, the crystallite sizes were calculated by Scherrer formula: $D = \frac{K\lambda}{\beta cos\theta}$, where D is crystallite size (nm), K is 0.9 (Scherrer constant), λ is 0.15406 nm (wavelength of the X-ray sources). The average crystallite sizes of BiOCl, $NaNbO_3$, and BN-3 are 57, 22, and 56 nm. With the increase of $NaNbO_3$ content, the diffraction peaks increased. The UV–vis diffuse reflectance spectra (DRS) of BiOCl, $NaNbO_3$, and BN-3 are exhibited in Figure 1b, which indicate the absorbance threshold of $NaNbO_3$, BiOCl, and BN-3 are the same. The estimated band gaps (E_g) of BiOCl and $NaNbO_3$ are computed in Figure 1c by $(Ahv)^{2/n} \sim hv - E_g$, where A is for absorbance, hv is for irradiation energy [20], and the obtained values are 3.44 and 3.52 eV, respectively. The valance band X-ray photoelectron spectroscopy (VB XPS) spectra in Figure 1d show that the valance band values of BiOCl and $NaNbO_3$ are 2.57 and 2.50 eV. Together with the band gaps, the conductive band (CB) position can be calculated by $E_{VB} = E_{CB} - E_g$, which are −0.87 (BiOCl) and −1.02 eV ($NaNbO_3$).

The morphology and microstructure of the BN-3 powder were investigated by scanning electron microscopy (SEM), element mapping and transmission electron microscope (TEM), and the results were shown in Figure 2. From Figure 2a, the irregular particles can be observed and the distribution of the corresponding main elements are shown in Figure 2b–f. The different colored areas suggest that Nb-, Na-, O-, Bi-, and Cl-enriched areas of the BN-3 composite, respectively. The TEM image of the BN-3 powder is displayed in Figure 2g. The lattice spacing of 0.343 and 0.273 nm in Figure 2h–i are corresponding to the (101) of BiOCl and (121) plane of $NaNbO_3$, respectively. The result agrees well with that in the XRD patterns as shown in Figure 1a.

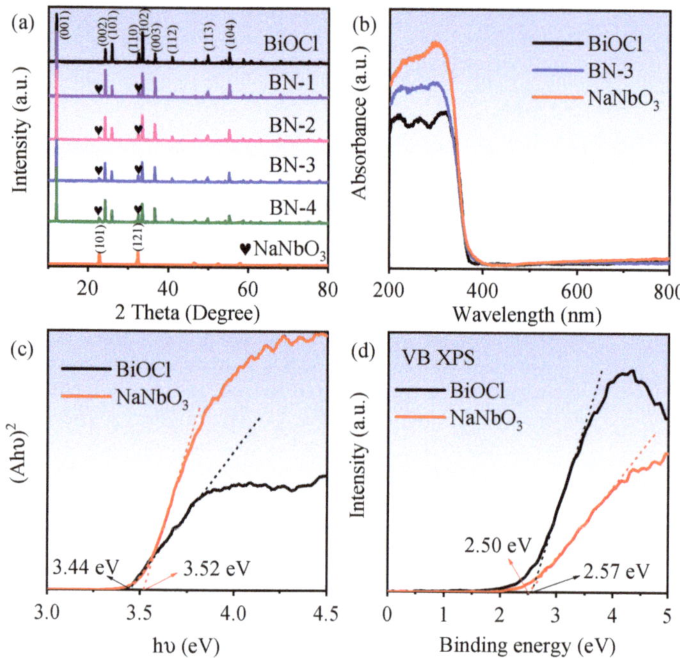

Figure 1. (**a**) XRD patterns of BiOCl, NaNbO$_3$, and BiOCl/NaNbO$_3$ piezoelectric composites; (**b**) UV–vis absorption spectra; (**c**) the estimated band gaps of BiOCl and NaNbO$_3$; (**d**) VB XPS spectra of BiOCl and NaNbO$_3$.

X-ray photoelectron spectroscopy (XPS) spectra of the BiOCl, NaNbO$_3$, and BN-3 piezoelectric composite are shown in Figure 3. From Figure 3a, the peaks of Bi 4f of BiOCl (BN-3) located at 159.60 (159.24 eV) and 164.90 eV (164.52 eV) can be assigned to Bi 4f$_{7/2}$ and Bi 4f$_{5/2}$, respectively, suggesting the Bi^{3+} exists in the BiOCl (BN-3). In Figure 3b, the Cl 2p peaks at 198.29 (197.90 eV) and 199.93 eV (199.54 eV) can be attributed to Cl 2p$_{3/2}$ and Cl 2p$_{1/2}$, respectively, which indicate the Cl$^-$ in BiOCl (BN-3) [21]. The peak at 1070.62 eV (1071.42 eV) in Figure 3c is ascribed to Na 1s in NaNbO$_3$ (BN-3). In Figure 3d, it is clearly seen that the binding energies located at 206.67 (207.05 eV) and 209.40 eV (209.78 eV) belong to Nb 3d$_{5/2}$ and Nb 3d$_{3/2}$, respectively, reflecting that Nb is in the Nb (+5) chemical state [22]. As shown in Figure 3e, the peaks located at 530.39 (BiOCl), 529.60 (NaNbO$_3$), and 529.98 eV (BN-3) correspond to O 1s. Compared with BiOCl, the blue shift of all peaks for BN-3 can be observed, while the red shift compared with NaNbO$_3$. The XPS survey spectra also indicate that BiOCl is composed of Bi, O, and Cl elements, and NaNbO$_3$ is mainly composed of Na, O, and Nb elements, while BN-3 contains all elements above, as shown in Figure 3f. In short, the XPS results demonstrate that the BN-3 piezoelectric composite is composed of BiOCl and NaNbO$_3$.

The polarization electric field hysteresis loop (P-E) and electric-field-induced strain (S-E) curves of BN-3 composite are displayed in Figure 4. From Figure 4a, a saturated and nearly squared P-E loop can be observed, and the remnant polarization (Pr) is 35.13 μC/cm^2 and the coercive field (Ec) is 8.72 kV/mm. The result shows that BN-3 composite has well ferroelectric properties, favoring the spatial separation and transportation of photo-induced carriers [23]. The S-E curve in Figure 4b exhibits an asymmetric butterfly shape, confirming the piezoelectricity of the BN-3 composite [24,25].

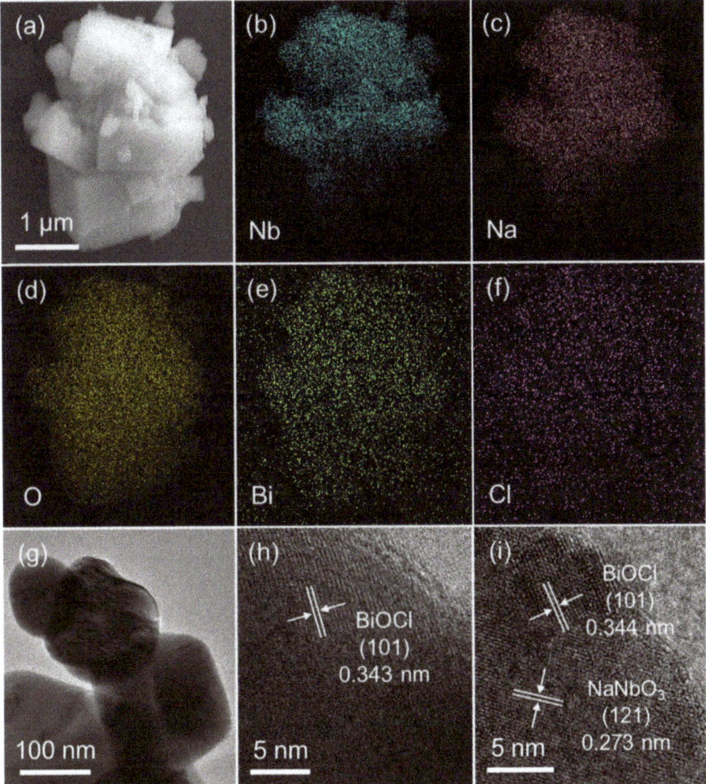

Figure 2. (**a**) SEM image; (**b–f**) EDS element mappings; (**g**) TEM image; (**h,i**) lattice fringes images of BN-3.

Consequently, the piezo/photocatalytic activities of BiOCl, NaNbO$_3$, and BiOCl/NaNbO$_3$ piezoelectric composites were evaluated by the degradation of Rhodamine B (RhB) under the condition of light irradiation and ultrasound. From Figure 5a, BN-3 exhibits better piezo/photocatalytic performance than that of BiOCl, NaNbO$_3$, and other content BiOCl/NaNbO$_3$ composites. The rate constant k values are obtained from Figure 5a via the pseudo-first-order equation [26]: $\ln(C_0/C_t) = -kt$, where C_0 is RhB concentration for initial and C_t is for after irradiation time t. And the decomposition ratio is calculated via the formula: $\eta = \left(1 - \frac{C_t}{C_0}\right) \times 100\%$. As shown in Figure 5b, the apparent reaction rate constant k for BiOCl, BN-1, BN-2, BN-3, BN-4, and NaNbO$_3$ is 0.0089, 0.0112, 0.0134, 0.0192, 0.0168, and 0.0037 min^{-1}, respectively. The piezo/photodegradation rate of RhB for BN-3 is 2.2 and 5.2 times higher than that of BiOCl and NaNbO$_3$, the histogram in Figure 5c reflects this directly. The degradation percentages of BiOCl, NaNbO$_3$ and BN-3 are 29.8%, 61.9% and 87.4%, respectively. In addition, the BET surface areas of BiOCl, NaNbO$_3$, and BN-3 are 0.24, 2.00, and 1.32 m^2/g, and the pore volumes are 0.0008, 0.0088, and 0.0037 cm^3/g, respectively. The BET surface areas of the samples are in the same order of magnitude, which means the BET surface areas not can decisive the catalytic activity. This result indicates the heterojunction in BN-3 exerts a tremendous advantage on the piezo/photocatalytic process. To investigate the most reactive species during the process of RhB decomposition, the radical trapping experiments were carried out in the presence of BN-3 as a catalyst. From Figure 5d, the piezo-photodegradation efficiency of RhB is remarkably inhibited while adding the triethanolamine (TEOA, 50 μL) scavenger for trapping hole (h$^+$) to the mixed solution, demonstrating an important role of h$^+$ in the piezo-photocatalytic process. While L-ascorbic acid (VC, 40 mg) for superoxide radical ($\cdot$O$_2^-$) was added, the degradation efficiency also decreased rapidly. The RhB degradation efficiency is decreased slightly by adding the isopropanol (IPA, 50 µL), reflecting the hydroxyl radical ($\cdot$OH) plays a secondary role in this process. These results indicate that the effect in this piezo-photocatalytic process is: h$^+$ > $\cdot$O$_2^-$ > $\cdot$OH.

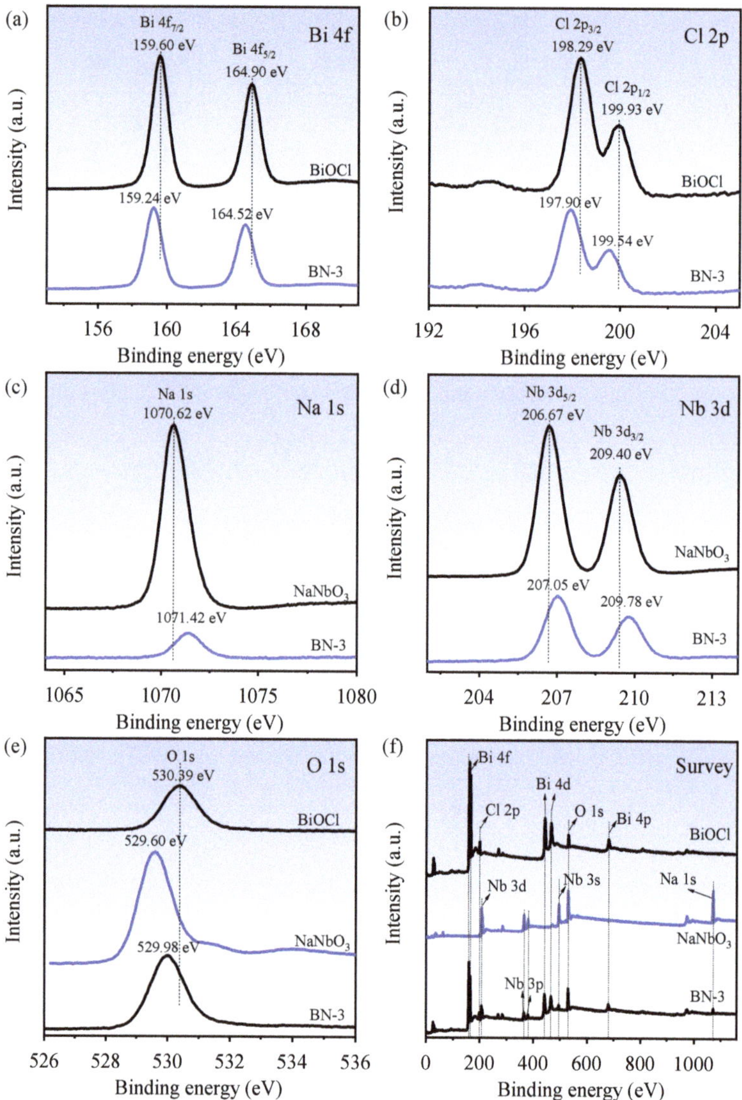

Figure 3. XPS survey spectra of BiOCl, NaNbO$_3$, and BN-3: (**a**) Bi 4f; (**b**) Cl 2p; (**c**) Na 1s; (**d**) Nb 3d; (**e**) O 1s; (**f**) Survey.

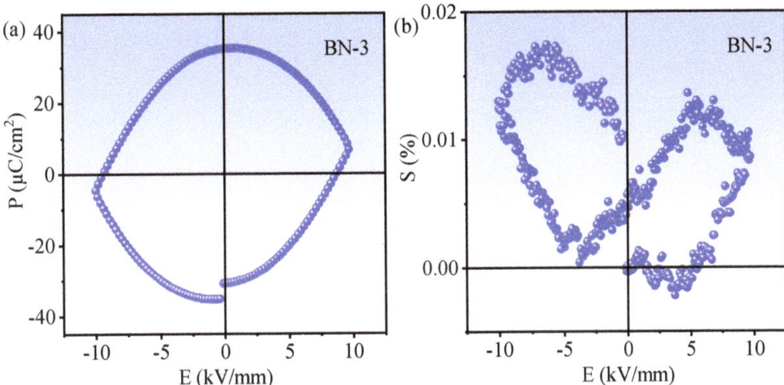

Figure 4. (a) The ferroelectric P-E loop and (b) electric field-induced S-E curve of BN-3 composite.

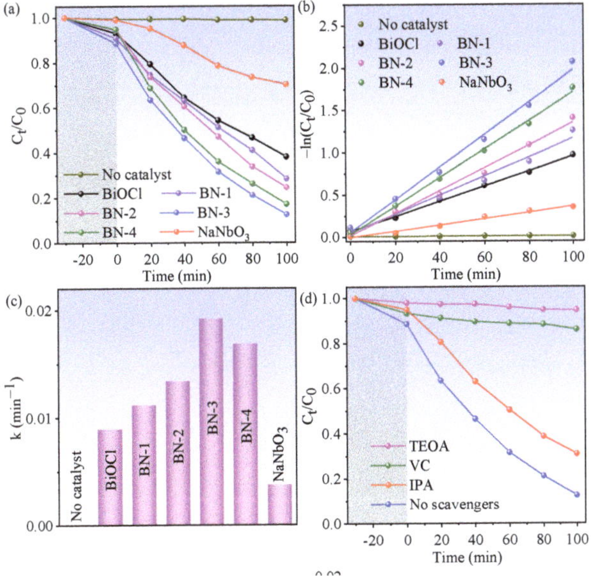

Figure 5. (**a**) The kinetic curves of piezo-photodegradation RhB performance for BiOCl, NaNbO$_3$, and BiOCl/NaNbO$_3$ piezoelectric composites; (**b**) the dynamics of degradation reaction [($-\ln(C_t/C_0)$)]; (**c**) the histogram of corresponding reaction rate constant; (**d**) piezo-photodegradation curves with disparate scavengers of BN-3 composite.

To demonstrate the piezocatalysis, photocatalysis, and the synergy effect of piezocatalysis and photocatalysis of BN-3 piezoelectric composite, the RhB degradation capability within 100 min was measured under the condition of ultrasound only, light only, and ultrasound + light together. UV–vis absorption spectra of RhB for BN-3 under different conditions are shown in Figure 6a,c,e, which correspond to light, ultrasound, and both light and ultrasound, respectively. From Figure 6b,d,f, the degradation rate is the lowest under the condition of only ultrasound, the rate constant k (0.0005 min^{-1}) and decomposition ratio (9.9%) are well below those of the condition of only light (0.0044 min^{-1} and 40.4%) and light + ultrasound together (0.0192 min^{-1} and 87.4%). The decomposition ratio of RhB under synergy of piezocatalysis and photocatalysis is 8.8 and 2.2 times higher than that of piezocatalysis and photocatalysis, respectively. The rate constant k under synergy of piezocatalysis and photocatalysis is 38.4 and 4.36 times higher than that of only ultrasound and only light. In addition, based on the same condition, compared to other piezoelectric materials past reported, the k value of BN-3 is higher than that of $NaNbO_3/CuBi_2O_4$ nanocomposites (0.0112 min^{-1}) [27], and closing to that of $BaTiO_3/KNbO_3$ heterostructure (0.01492 min^{-1}) [28]. The result confirms that the synergy effect of piezocatalysis and photocatalysis of $BiOCl/NaNbO_3$ piezoelectric composite plays an important role in the highly efficient degradation of RhB. One of the key parameters in the piezo-photocatalyst is reproducibility, and Figure 7 shows the cycling performance of the piezo-photocatalytic activity of BN-3 for degrading RhB. After three cycles, the degradation efficiency is just reduced a little. This result evidences that BN-3 possesses a high reproducibility.

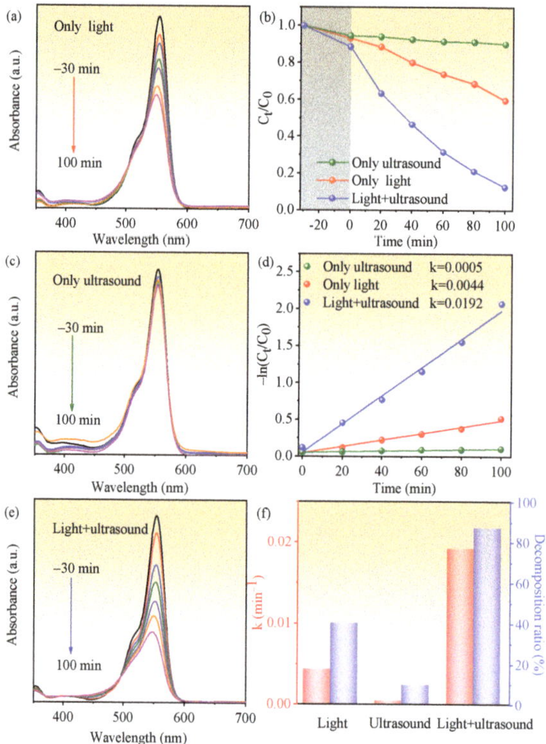

Figure 6. UV–vis spectral absorption of RhB for BN-3 under the condition of (**a**) only light, (**c**) only ultrasound and (**e**) light + ultrasound; (**b**) the kinetic curves of RhB degradation for BN-3 under these three control conditions; (**d**) the dynamics of degradation reaction [($-\ln(C_t/C_0)$)]; (**f**) the histogram of corresponding reaction rate constant and decomposition ratio.

On the basis of the above analysis, the possible mechanism for piezocatalytic, photocatalytic, and their synergetic catalytic process of $BiOCl/NaNbO_3$ piezoelectric composites are shown in Figure 8. According to our experiment, the valance band of $NaNbO_3$ is 2.50 eV, while BiOCl is 2.57 eV; the conductive band of $NaNbO_3$ is −1.02 eV, while BiOCl is −0.87 eV. In the condition of only light, the photoelectrons are excited from the valance band to the conductive band, and the electrons will transfer from the conductive band of $NaNbO_3$ to the conductive band of BiOCl, and thus build an inner electric field. The built-in electric field can promote the separation of electrons and holes. However, there is still a combination of electrons and holes in the inner of $BiOCl/NaNbO_3$ piezoelectric composites because the built-in electric field is easily prone to be screened by electrostatic compensated free space charges [29]. This reduces the degradation efficiency of RhB. In the condition of only ultrasound, the cavitation bubbles will form, expand, and burst, an amount of electric charge can be generated [30,31]. These positive and negative charges will transfer to the opposite directions under the influence of the alternating built-in electric field. In the condition of both light and ultrasound, the electrons and holes located at the conductive band and valance band will transfer to the opposite directions under the internal piezoelectric potential, causing electrons to accumulate in the conductive band of BiOCl and holes accumulate in the valance band of $NaNbO_3$ [32–34]. Subsequently, the electrons on the CB of BiOCl combined with the absorbed O_2 to produce $\cdot O_2^-$. Meanwhile, part holes on the VB of $NaNbO_3$ will oxidize hydroxyl to form $\cdot OH$. Finally, the reactive species $\cdot OH$, h^+, and $\cdot O_2^-$ will participate in the oxidative degradation of RhB. The combination rate of electrons and holes will be reduced significantly under the built-in electric field, thus the decomposition ratio of $BiOCl/NaNbO_3$ piezoelectric composite increased remarkably.

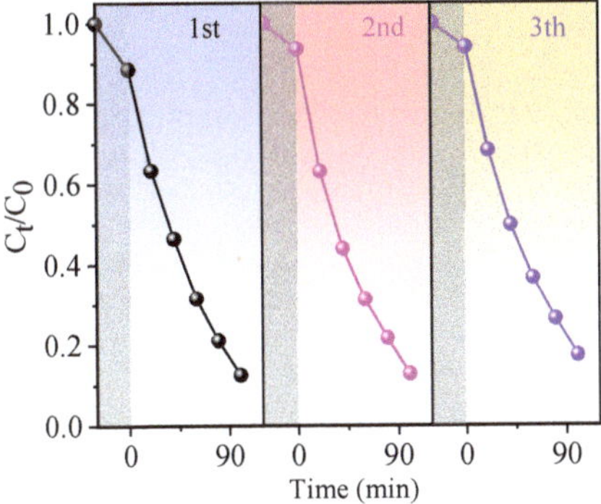

Figure 7. The cycling performance of the piezo-photocatalytic activity of BN-3 for degrading RhB solution.

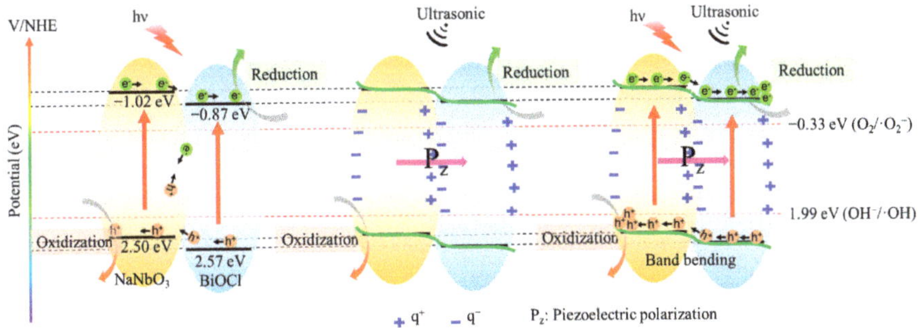

Figure 8. Possible piezocatalytic, photocatalytic, and piezo-photocatalytic mechanism of $BiOCl/NaNbO_3$ piezoelectric composites.

4. Conclusions

In conclusion, $BiOCl/NaNbO_3$ piezoelectric composites are synthesized via a two-step hydrothermal route. Under UV–vis light and ultrasonic exposure, the BN-3 for piezo-photocatalytic decomposition of RhB demonstrate remarkable piezo-photocatalytic performance than that of BiOCl and $NaNbO_3$ component due to the heterojunction construction. Furthermore, the piezo/photodegradation rate of RhB for BN-3 is higher than that of piezocatalysis and photocatalysis, indicating the synergistic effect of piezocatalysis and photocatalysis plays a significant role in the degradation process. Some issues, such as the specific promotion mechanism of the $NaNbO_3$ and BiOCl of the contribution of the piezo-photocatalytic performance, need to be further investigated for better understanding. However, this work provides a feasible approach for the development of efficient piezoelectric photocatalysts for heterojunction construction.

Author Contributions: Conceptualization, C.W. and F.L.; methodology, C.W.; investigation, L.L.; resources, C.W. and F.L.; data curation, L.L.; software, L.L.; W.C., J.Y. and W.L.; writing original draft preparation, L.L.; writing review and editing, L.L., C.W. and F.L.; supervision, C.W. and F.L. All authors have read and agreed to the published version of the manuscript.

Funding: This work was financially supported by the National Natural Science Foundation of China (Grant Nos. 12174001, 12104001, and 51872001).

Data Availability Statement: Data available in a publicly accessible repository. The data presented in this study are openly available in [repository name e.g., FigShare] at [doi], reference number [reference number].

Conflicts of Interest: The authors declare no conflict of interest.

References

1. Hu, L.; Yang, S.; Zhao, Y.; He, J.; Jiang, S.; Sun, C.; Song, S. Spontaneous polarization electric field briskly boosting charge separation and transfer for sustainable photocatalytic H_2 bubble evolution. *Appl. Catal. B* **2021**, *283*, 119631. [CrossRef]
2. Wang, Y.C.; Wu, J.M. Effect of controlled oxygen vacancy on H_2-production through the piezocatalysis and piezophototronics of ferroelectric R3C $ZnSnO_3$ nanowires. *Adv. Funct. Mater.* **2019**, *30*, 1907619. [CrossRef]
3. El-Bindary, A.A.; El-Marsafy, S.M.; El-Maddah, A.A. Enhancement of the photocatalytic activity of ZnO nanoparticles by silver doping for the degradation of AY99 contaminants. *J. Mol. Struct.* **2019**, *1191*, 76–84. [CrossRef]
4. Kiwaan, H.A.; Atwee, T.M.; Azab, E.A.; El-Bindary, A.A. Efficient photocatalytic degradation of Acid Red 57 using synthesized ZnO nanowires. *J. Chin. Chem. Soc.* **2019**, *66*, 89–98. [CrossRef]
5. Wang, Y.; Cai, J.; Wu, M.; Chen, J.; Zhao, W.; Tian, Y.; Ding, T.; Zhang, J.; Jiang, Z.; Li, X. Rational construction of oxygen vacancies onto tungsten trioxide to improve visible light photocatalytic water oxidation reaction. *Appl. Catal. B* **2018**, *239*, 398–407. [CrossRef]
6. Ye, L.; Zan, L.; Tian, L.; Peng, T.; Zhang, J. The {001} facets-dependent high photoactivity of BiOCl nanosheets. *Chem. Commun.* **2011**, *47*, 6951–6953. [CrossRef]
7. Li, Q.; Guan, Z.; Wu, D.; Zhao, X.; Bao, S.; Tian, B.; Zhang, J. Z-scheme BiOCl-Au-CdS heterostructure with enhanced sunlight-driven photocatalytic activity in degrading water dyes and antibiotics. *ACS Sustain. Chem. Eng.* **2017**, *5*, 6958–6968. [CrossRef]
8. Li, T.B.; Chen, G.; Zhou, C.; Shen, Z.Y.; Jin, R.C.; Sun, J.X. New photocatalyst BiOCl/BiOI composites with highly enhanced visible light photocatalytic performances. *Dalton Trans.* **2011**, *40*, 6751–6758. [CrossRef]
9. Yang, Y.; Zhang, C.; Lai, C.; Zeng, G.; Huang, D.; Cheng, M.; Wang, J.; Chen, F.; Zhou, C.; Xiong, W. BiOX (X=Cl, Br, I) photocatalytic nanomaterials: Applications for fuels and environmental management. *Adv. Colloid Interface Sci.* **2018**, *254*, 76–93. [CrossRef]
10. Yuan, B.; Wu, J.; Qin, N.; Lin, E.; Bao, D. Enhanced piezocatalytic performance of (Ba,Sr)TiO_3 nanowires to degrade organic pollutants. *ACS Appl. Nano Mater.* **2018**, *1*, 5119–5127. [CrossRef]
11. Liu, X.; Xiao, L.; Zhang, Y.; Sun, H. Significantly enhanced piezo-photocatalytic capability in $BaTiO_3$ nanowires for degrading organic dye. *J. Materiomics* **2020**, *6*, 256–262. [CrossRef]
12. Li, L.; Ma, Y.; Chen, G.; Wang, J.; Wang, C. Oxygen-vacancy-enhanced piezo-photocatalytic performance of AgNbO. *Scr. Mater.* **2022**, *206*, 114234. [CrossRef]
13. Lei, H.; Zhang, H.; Zou, Y.; Dong, X.; Jia, Y.; Wang, F. Synergetic photocatalysis/piezocatalysis of bismuth oxybromide for degradation of organic pollutants. *J. Alloys Compd.* **2019**, *809*, 151840. [CrossRef]
14. Zhou, X.; Wu, S.; Li, C.; Yan, F.; Bai, H.; Shen, B.; Zeng, H.; Zhai, J. Piezophototronic effect in enhancing charge carrier separation and transfer in ZnO/$BaTiO_3$ heterostructures for high-efficiency catalytic oxidation. *Nano Energy* **2019**, *66*, 104127. [CrossRef]
15. Singh, S.; Khare, N. Coupling of piezoelectric, semiconducting and photoexcitation properties in $NaNbO_3$ nanostructures for controlling electrical transport: Realizing an efficient piezo-photoanode and piezo-photocatalyst. *Nano Energy* **2017**, *38*, 335–341. [CrossRef]
16. Ma, J.; Jia, Y.; Chen, L.; Zheng, Y.; Wu, Z.; Luo, W.; Jiang, M.; Cui, X.; Li, Y. Dye wastewater treatment driven by cyclically heating/cooling the poled $(K_{0.5}Na_{0.5})NbO_3$ pyroelectric crystal catalyst. *J. Clean. Prod.* **2020**, *276*, 124218. [CrossRef]
17. Zhang, A.; Liu, Z.; Geng, X.; Song, W.; Lu, J.; Xie, B.; Ke, S.; Shu, L. Ultrasonic vibration driven piezocatalytic activity of lead-free $K_{0.5}Na_{0.5}NbO_3$ materials. *Ceram. Int.* **2019**, *45*, 22486–22492. [CrossRef]
18. You, H.; Ma, X.; Wu, Z.; Fei, L.; Chen, X.; Yang, J.; Liu, Y.; Jia, Y.; Li, H.; Wang, F.; et al. Piezoelectrically/pyroelectrically-driven vibration/cold-hot energy harvesting for mechano-/pyro- bi-catalytic dye decomposition of $NaNbO_3$ nanofibers. *Nano Energy* **2018**, *52*, 351–359. [CrossRef]
19. Ismail, M.; Wu, Z.; Zhang, L.; Ma, J.; Jia, Y.; Hu, Y.; Wang, Y. High-efficient synergy of piezocatalysis and photocatalysis in bismuth oxychloride nanomaterial for dye decomposition. *Chemosphere* **2019**, *228*, 212–218. [CrossRef] [PubMed]

20. Kamat, P.V. Photochemistry on Nonreactive and Reactive (Semiconductor) Surfaces. *Chem. Rev.* **1993**, *93*, 267–300. [CrossRef]
21. Pan, J.; Liu, J.; Zuo, S.; Khan, U.A.; Yu, Y.; Li, B. Structure of Z-scheme CdS/CQDs/BiOCl heterojunction with enhanced photocatalytic activity for environmental pollutant elimination. *Appl. Surf. Sci.* **2018**, *444*, 177–186. [CrossRef]
22. Liu, Q.; Chai, Y.; Zhang, L.; Ren, J.; Dai, W.L. Highly efficient Pt/NaNbO$_3$ nanowire photocatalyst: Its morphology effect and application in water purification and H$_2$ production. *Appl. Catal. B* **2017**, *205*, 505–513. [CrossRef]
23. Yu, Z.; Zhan, B.; Ge, B.; Zhu, Y.; Dai, Y.; Zhou, G.; Yu, F.; Wang, P.; Huang, B.; Zhan, J. Synthesis of high efficient and stable plasmonic photocatalyst Ag/AgNbO$_3$ with specific exposed crystal–facets and intimate heterogeneous interface via combustion route. *Appl. Surf. Sci.* **2019**, *488*, 485–493. [CrossRef]
24. Liu, L.; Shi, D.; Knapp, M.; Ehrenberg, H.; Fang, L.; Chen, J. Large strain response based on relaxor–antiferroelectric coherence in Bi$_{0.5}$Na$_{0.5}$TiO$_3$–SrTiO$_3$–(K$_{0.5}$Na$_{0.5}$)NbO$_3$ solid solutions. *J. Appl. Phys.* **2014**, *116*, 184104. [CrossRef]
25. Zhu, M.; Li, S.; Zhang, H.; Gao, J.; Kwok, K.W.; Jia, Y.; Kong, L.B.; Zhou, W.; Peng, B. Diffused phase transition boosted dye degradation with Ba (Zr$_x$Ti$_{1-x}$)O$_3$ solid solutions through piezoelectric effect. *Nano Energy* **2021**, *89*, 106474. [CrossRef]
26. Turchi, C.S.; Ollis, D.F. Photocatalytic degradation of organic water contaminants: Mechanisms involving hydroxyl radical attack. *J. Catal.* **1990**, *122*, 178–192. [CrossRef]
27. Rajan, K.D.; Gotipamul, P.P.; Khanna, S.; Chidambaram, S.; Rathinam, M. Piezo-photocatalytic effect of NaNbO$_3$ interconnected nanoparticles decorated CuBi$_2$O$_4$ nanocuboids. *Mater. Lett.* **2021**, *296*, 129902. [CrossRef]
28. Zhang, Y.; Shen, G.; Sheng, C.; Zhang, F.; Fan, W. The effect of piezo-photocatalysis on enhancing the charge carrier separation in BaTiO$_3$/KNbO$_3$ heterostructure photocatalyst. *Appl. Surf. Sci.* **2021**, *562*, 150164. [CrossRef]
29. Li, H.; Sang, Y.; Chang, S.; Huang, X.; Zhang, Y.; Yang, R.; Jiang, H.; Liu, H.; Wang, Z.L. Enhanced ferroelectric–nanocrystal–based hybrid photocatalysis by ultrasonic–wave–generated piezophototronic effect. *Nano Lett.* **2015**, *15*, 2372–2379. [CrossRef]
30. Hu, C.; Huang, H.; Chen, F.; Zhang, Y.; Yu, H.; Ma, T. Coupling piezocatalysis and photocatalysis in Bi$_4$NbO$_8$X (X = Cl, Br) polar single crystals. *Adv. Funct. Mater.* **2019**, *30*, 1908168. [CrossRef]
31. Sharma, M.; Halder, A.; Vaish, R. Effect of Ce on piezo/photocatalytic effects of Ba$_{0.9}$Ca$_{0.1}$Ce$_x$Ti$_{1-x}$O$_3$ ceramics for dye/ pharmaceutical waste water treatment. *Mater. Res. Bull.* **2020**, *122*, 110647. [CrossRef]
32. Zhao, T.; Wang, Q.; Du, A. High piezo–photocatalytic efficiency of H$_2$ production by CuS/ZnO nanostructure under solar and ultrasonic exposure. *Mater. Lett.* **2021**, *294*, 129752. [CrossRef]
33. Yang, B.; Chen, H.; Yang, Y.; Wang, L.; Bian, J.; Liu, Q.; Lou, X. Insights into the tribo–/pyro–catalysis using Sr–doped BaTiO$_3$ ferroelectric nanocrystals for efficient water remediation. *Chem. Eng. J.* **2021**, *416*, 128986. [CrossRef]
34. Zhou, X.; Yan, F.; Wu, S.; Shen, B.; Zeng, H.; Zhai, J. Remarkable piezophoto coupling catalysis behavior of BiOX/BaTiO$_3$ (X = Cl, Br, Cl$_{0.166}$Br$_{0.834}$) piezoelectric composites. *Small* **2020**, *16*, 2001573. [CrossRef] [PubMed]

*nanomaterials*

MDPI

Article

Humidity Sensitivity Behavior of CH₃NH₃PbI₃ Perovskite

Xuefeng Zhao, Yuting Sun, Shuyu Liu, Gaifang Chen, Pengfei Chen, Jinsong Wang, Wenjun Cao and Chunchang Wang *

Laboratory of Dielectric Functional Materials, School of Materials Science & Engineering, Anhui University, Hefei 230601, China; b51814003@stu.ahu.edu.cn (X.Z.); b51814004@stu.ahu.edu.cn (Y.S.); b51814001@stu.ahu.edu.cn (S.L.); b19201039@stu.ahu.edu.cn (G.C.); b19201035@stu.ahu.edu.cn (P.C.); b20301123@stu.ahu.edu.cn (J.W.); b20201064@stu.ahu.edu.cn (W.C.)
* Correspondence: ccwang@ahu.edu.cn

Abstract: The $CH_3NH_3PbI_3$ (MAPbI₃) powders were ground by PbI_2 and CH_3NH_3I prepared by ice bath method. The humidity sensitive properties of an impedance-type sensor based on MAPbI₃ materials were systematically studied. Our results indicate that the MAPbI₃-based sensor has superior sensing behaviors, including high sensitivity of 5808, low hysteresis, approximately 6.76%, as well as good stability. Water-molecule-induced enhancement of the conductive carrier concentration was argued to be responsible for the excellent humidity sensitive properties. Interestingly, the humidity properties can be affected by red light sources. The photogenerated carriers broke the original balance and decreased the impedance of the sensor. This work promotes the development of perovskite materials in the field of humidity sensing.

Keywords: humidity sensing; impedance-type sensors; organometallic halide perovskite

1. Introduction

Perovskites with the formula of AMX_3 (where A stands for an organic group or inorganic cation with twelve neighboring, X is a halide anion, and M is a metal cation) have been evidenced to be exciting solar absorber materials [1]. Besides the light harvesting performance, these compounds have recently demonstrated several intriguing properties, such as high dielectric constant [2], ferroelecticity [3], photorestriction [4], resistive switching [5], and optical cooling [6]. However, the AMX_3 perovskites suffer from the major obstacle of chemical and structural instability because they are sensitive to environmental factors. Previous work indicated that the perovskite lattice can interact with the polar water molecules due to the formation of strong hydrogen bonds between water molecules and the halide lattice [7]. This characteristic makes the properties of AMX_3 heavily dependent on the environment moisture. For example, Alberto García-Fernández et al. [8] investigated the electric properties of $CH_3NH_3PbI_3$ (MAPbI₃) in wet and dry environments. The results showed that both capacitance and conductivity in wet condition were several orders of magnitude larger than those in dry condition. This humidity sensitive feature gives the materials tremendous promise as probes for sensing of humidity. Truly, outstanding humidity performances of large sensitivity, remarkable fast response/recovery time, small hysteresis loop, and good linearity were reported in humidity sensors based on Cs_2PdBr_6 [9], $CH_3NH_3PbI_{3-x}Cl_x$ [10,11], $Cs_2BiAgBr_6$ [12], and $CsPbBr_3$ [7].

MAPbI₃, being an important member of the AMX_3 family, has properties that strongly depend on environment humidity [8], indicating that it can be used as a humidity sensing material. Although humidity detection by MAPbI₃ was attempt by Ilin and co-authors [13], details about its humidity performances have not been studied. Additionally, because of the low-cost solution-based method and cheaper CH_3NH_2, perovskite MAPbI₃ is less expensive than most humidity sensing materials such as $CsPbBr_3$ [14], Cs_2PdBr_6 [9], and $Cs_2BiAgBr_6$ [12]. Moreover, it was proved that illumination could lead to a giant dielectric

Citation: Zhao, X.; Sun, Y.; Liu, S.; Chen, G.; Chen, P.; Wang, J.; Cao, W.; Wang, C. Humidity Sensitivity Behavior of CH₃NH₃PbI₃ Perovskite. *Nanomaterials* **2022**, *12*, 523. https://doi.org/10.3390/nano12030523

Academic Editor: Sergey Ya. Istomin

Received: 14 December 2021
Accepted: 31 January 2022
Published: 2 February 2022

Publisher's Note: MDPI stays neutral with regard to jurisdictional claims in published maps and institutional affiliations.

constant of lead halide perovskite [2]. Hence, we herein present a systematical investigation on the humidity performances of an impedance-type humidity sensor based on MAPbI$_3$.

In this work, MAPbI$_3$ was fabricated by grinding CH$_3$NH$_3$I made from an ice bath with PbI$_2$. The humidity sensing performance of the humidity sensor based on perovskite MAPbI$_3$ were tested at room temperature in the range of 11–94% relative humidity. Our results show that the MAPbI$_3$-based humidity sensor exhibits high humidity sensitivity performance. The possible mechanism of the humidity sensitivity and photoinduced changes were also discussed.

2. Materials and Methods

2.1. Powder Preparation

Methylamine iodide (MAI) was synthesized from 24 mL methylamine (MA) solution and 10 mL hydroiodic acid (HI) solution [15]. MA solution and HI solution were firstly mixed in a smaller beaker that was placed in a larger beaker at 0 °C. The mixed solution was continuously stirred in an ice bath for 120 min until evenly mixed to form a slightly yellowish solution. Then, the resulting mixture was evaporated in a vacuum oven at 90 °C for 7 h until a white precipitate was obtained. After naturally cooling down to room temperature, the obtained precipitate was washed 3 to 4 times with ether in a centrifuge. Eventually, the precipitate was dried at 60 °C for 24 h to obtain white crystalline powder MAI.

Preparation of lead methyl iodide triiodide (MAPbI$_3$). An equal molar amount of MAI powder and PbI$_2$ powder were ground in an agate mortar for 20 min until a visually black powder was obtained.

2.2. Characterization of Materials

The crystalline phase composition and structure of all powders were examined by X-ray diffraction (XRD, Rigaku Smartlab Beijing Co. Beijing, China) using Cu K α radiation. The morphology of MAPbI$_3$ powders was analyzed by scanning electron microscope (SEM, Model S 4800, Hitachi Co, Tokyo, Japan). The possible ferroelectricity of MAPbI$_3$ was investigated by measuring polarization-electric field (*P-E*) hysteresis loop using a TF analyzer (MultiFerroicII, Radiant technologies Inc, Albuquerque, New Mexico). In doing so, Au top electrodes were sputtered onto the surface of MAPbI3 film to form a sandwich structure of Au/MAPbI$_3$/Au. The ultraviolet photoelectron spectroscopy (UPS) measurement was carried out using He-I radiation (21.2 eV) in vacuum (10^{-8} mbar) to measure the highest filled energy level of valence band (VBM) of MAPbI$_3$ film. The sample was biased at -10 V for measurement in the secondary electron cut off region.

2.3. Humidity Sensor Fabrication and Performances Measurements

The MAPbI$_3$-based humidity sensor was fabricated by aerosol deposition method on Al$_2$O$_3$ substrate covered with Au interdigitated electrodes. Figure 1 presents the photo images of the Al$_2$O$_3$ substrate and fabricated sensor. The sensor fabrication processes are as follows: firstly, a little amount of MAPbI$_3$ powder was mixed with anhydrous ethanol, and ultrasound for 15 min to form a homogeneous paste. Then, the paste was uniformly sprayed onto a Al$_2$O$_3$ substrate using a 0.2 mm caliber spray pen (Sao Tome V130). Two copper wires were fixed on the electrodes with silver glue. Finally, the sensor was dried at 100 °C for 10 min and naturally cooled down to indoor temperature. The whole process of sample preparation was shown in Figure 2.

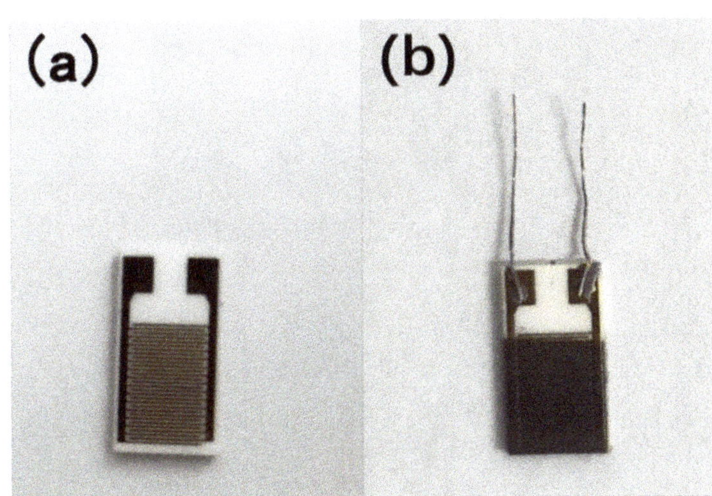

Figure 1. (**a**) Photo images of the Al$_2$O$_3$ substrate with Au interdigitated electrodes; and (**b**) the MAPbI$_3$-based humidity sensor.

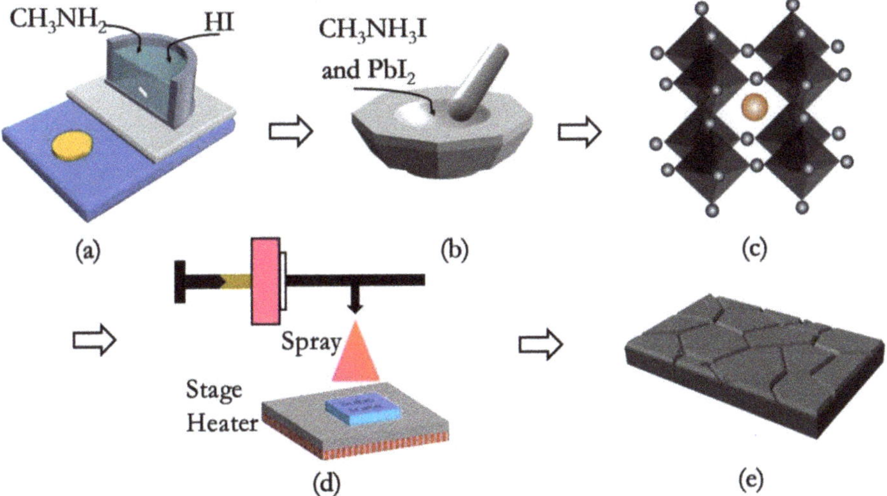

Figure 2. (**a**) Ice bath method; (**b**) grinding; (**c**) the structure of MAPbI$_3$ powder; (**d**) aerosol deposition method; and (**e**) the polycrystalline layer structure of MAPbI$_3$ thin film.

The different relative humidity (RH) environments were obtained by saturated salt solutions of LiCl, MgCl$_2$, Mg (NO$_3$)$_2$, NaCl, KCl, and KNO$_3$ in closed containers. The environments above these solutions can provided RH levels of 11, 33, 54, 75, 85, and 94%, respectively. The impedance of the sensor was measured by an impedance analyzer (Hioki 3532-50 LCR). The experimental setup was illustrated in Figure 3. The influence of illumination on the MAPbI$_3$-based sensor was reflected by impedance variation with and without light provided by 10 small red LED lights.

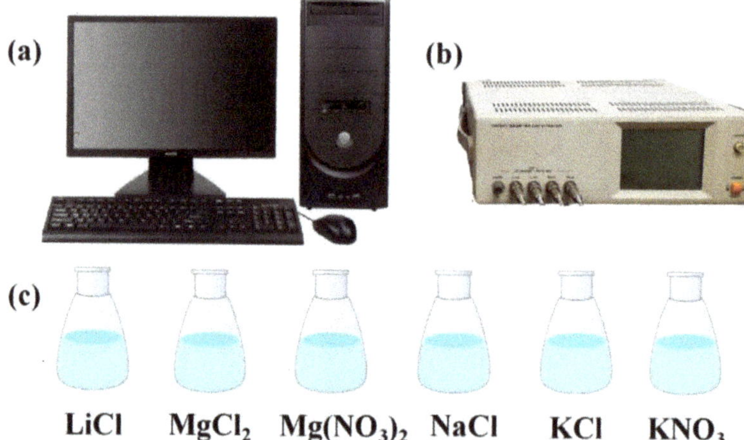

Figure 3. Schematic of the humidity sensing experimental setup: (**a**) PC; (**b**) Hioki 3532-50 LCR; and (**c**) different humidity environments.

3. Results

3.1. Structure and Morphology Characterizations

The XRD pattern of the MAPbI$_3$ powder was shown in the Figure 4a. The peaks at 14.1°, 28.4°, 31.7°, 40.5°, and 43.0° are assigned to (110), (220), (310), (224) and (314) planes, respectively. These peaks agree perfectly with MAPbI$_3$ perovskite tetragonal structure (The COD ID of MAPbI$_3$ is 2107954) [16]. The average grain size of the MAPbI$_3$ powder was clarified to be ~1.39 μm from the SEM image shown in the inset of Figure 4a. Figure 4b shows the *P-E* loops of the MAPbI$_3$ measured at room temperature with a frequency of 10 Hz under various voltages of 15, 30, and 45 V. The thickness of the film is 450 mm. The ellipse-shaped *P-E* loops indicate that the MAPbI$_3$ exhibits no ferroelectric behavior macroscopically at room temperature. This result is consistent with that reported in Refs. [3,17].

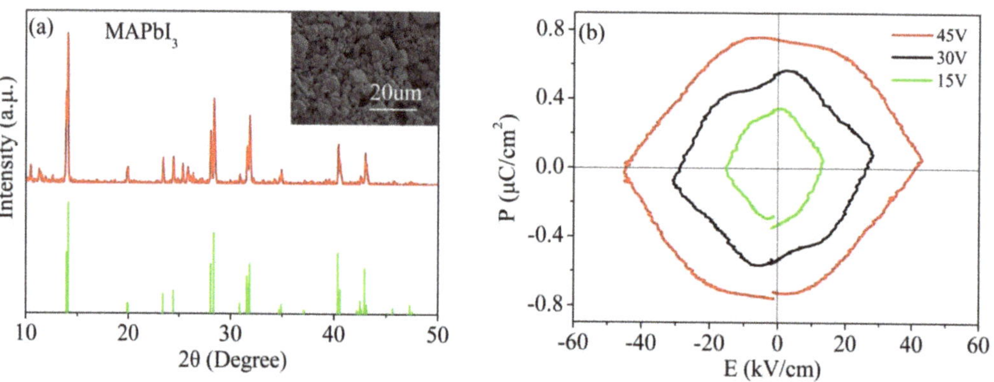

Figure 4. (**a**) XRD pattern and SEM image of MAPbI$_3$ powder; and (**b**) *P-E* loops of MAPbI$_3$ film measured at the frequency of 10 Hz under the voltage of 15, 30, and 45 V.

3.2. Humidity Sensitive Properties

The humidity sensitive performance of the MAPbI$_3$-based resistive sensor was studied by testing the impedance upon exposing the sensor to various RH levels with the a duration time of 5 min in each level. Figure 5a presents the impedance as a function of RH level of the MAPbI$_3$-based sensor at different frequencies. The impedance decreases with increasing measurement frequency. The curve measured with 100 Hz shows the largest impedance variation. Hence, 100 Hz is chosen as the optimum measuring frequency and will be used in the following part.

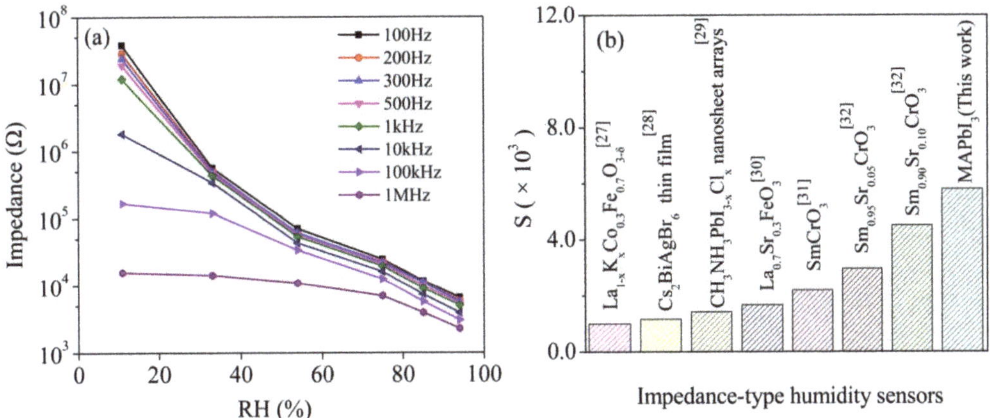

Figure 5. (**a**) The impedance as a function of RH level of the MAPbI$_3$-based sensor at different frequencies; (**b**) the contrast between the MAPbI$_3$-based sensor and other impedance-type humidity sensors in published literature.

Based on the data in Figure 5a, the humidity sensitive response (S) of the sensor can be calculated according to the relation [18,19]:

$$S = Z_d/Z_h \tag{1}$$

where Z_d and Z_h are the impedance values measured at 11%RH and at a specific RH level, respectively. The MAPbI$_3$-based sensor showed a superior sensitivity of S = 5808. Figure 5b displays a contrast of sensitive response between the MAPbI$_3$-based sensor and other impedance-type sensors reported in the literature [12,20–24]. The comparison highlights that the MAPbI$_3$ shows the largest sensitivity.

For the purpose of fully characterizing the performance of the MAPbI$_3$-based sensor, the hysteresis and recovery/response curves of the MAPbI$_3$-based sensor were tested at 100 Hz. The results were given in Figure 6a,b, respectively. The hysteresis curve was acquired by switching the sensor between the containers with the different RH levels of 11, 33, 54, 75, 85 and 94% in turn, and then shifting back. After an exposure duration of 5 min in each of the RH levels, the impedance was recorded under the optimum frequency of 100 Hz. Based on the measured impedance values, the humidity hysteresis values can be reckoned by the following formula [25]:

$$\left[\frac{\log(Z_{ads}) - \log(Z_{des})}{\log(Z_{ads})}\right] \times 100\% \tag{2}$$

where Z_{des} and Z_{ads} represent the impedance value of the desorption and the adsorption processes, respectively. A hysteresis value of 6.76% is obtained at 11%RH for the MAPbI$_3$-based sensor. The physical adsorption, which is toilless to be desorbed in MAPbI$_3$ material, is much larger than the chemical adsorption that can be weakly influenced by environment humidity at low humidity level, which results in good hysteresis behavior of the sensor at low RH levels. Studies have shown that there is a steep increase in the water uptake in the RH level beyond 80%RH [26]. Therefore, the sensor still exhibits notable hysteresis under high humidity levels because the physical adsorption is greatly increased. Figure 6b displays the response and recovery time for the MABbI$_3$-based sensor recorded between 11% and 94%RH. The result shows that the response and recovery times are 31 s and 148 s, respectively. Response/recovery time under other relative humidity levels are listed in Table 1.

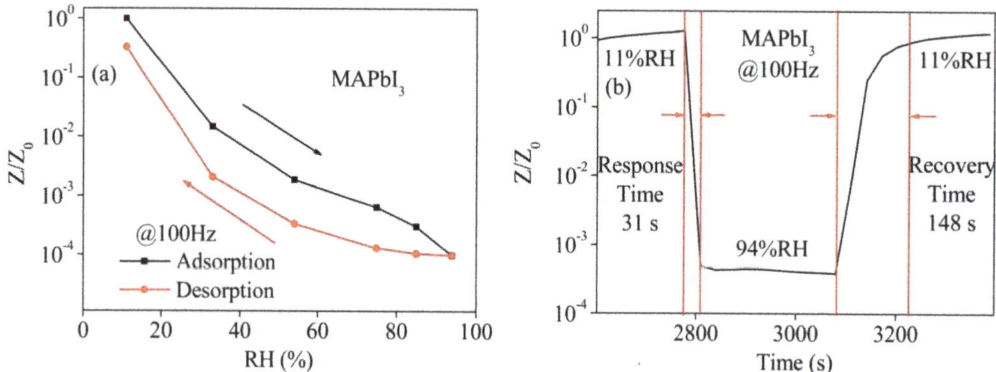

Figure 6. (a) Hysteresis behavior and (b) recovery/response time of the MAPbI$_3$-based sensor (Z/Z_0 is the normalized impedance to the initial impedance Z_0).

Table 1. Hysteresis and response/recovery time under different RH levels of the MAPbI$_3$-based sensor.

	Hysteresis	Response Time (s)	Recovery Time (s)
MAPbI$_3$-based sensor	Max = 18.61% (75%RH) Min = 6.76% (11%RH)	64 (11%→33%) 56 (11%→54%) 77 (11%→75%) 56 (11%→85%) 31 (11%→94%)	90 (33%→11%) 89 (54%→11%) 131 (75%→11%) 134 (85%→11%) 148 (94%→11%)

The repeatability of the sensor was measured by switching the sensor between 11 and 94%RH levels for 5 cycles. The stability test was conducted over a period of 135 days. The results of repeatability and stability were shown in Figure 7a,b, respectively. After 5 cycles, the impedance under 11%RH remains 88.13% of the initial value, while under 11%RH, the impedance hardly changes. This result indicates that the sensor shows satisfactory repeatability. The impedance curves shown in Figure 7b remain almost constant, revealing that the sensor exhibits good long-term stability.

3.3. Influence of Light on the Humidity Sensing Properties

Previous work has indicated that red light affects the electric properties of MAPbI$_3$ [26]. To investigate the influence of illumination on the humidity sensing properties of the MAPbI$_3$-based sensor, 10 red LED lights were used as light source. The humidity sensor was placed in the environments of 11%, 54%, and 94%RH successively, and the impedance values were recorded during the three stages of darkness–lightness–darkness with the duration time of 300 s in each stage. The results were plotted in Figure 8. It is clearly shown that under a low humidity level of 11%RH, the impedance can be notably reduced by the light because the photogenerated carrier breaks the original carrier balance and increases the electron concentration, leading to the enhancement of electron conductivity. In the medium and high humidity levels, a similar situation of illumination-induced impedance decrease to that of the low humidity level was observed when the lights were on. However, this impedance variation becomes much weaker than that in 11%RH. However, in the later stage of lighting, the electron concentration decreases due to electron–hole combination, so the impedance increases again. After the light source was switched off, considering that the non-equilibrium carriers does not disappear immediately, the impedance value increases at low and medium humidity levels due to the electron–hole interaction. However, impedance remarkably decreases at high humidity level, probably because the interaction between the light source and the continuous layer of water that forms on the surface of the sample.

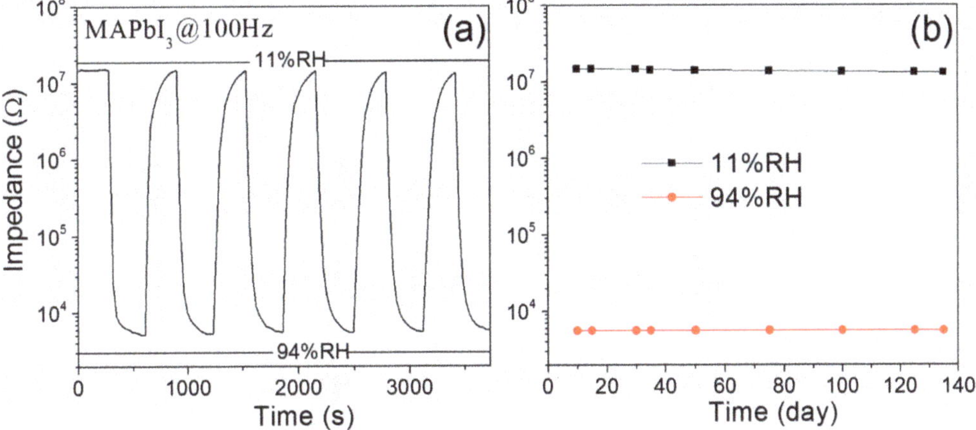

Figure 7. Repeatability (**a**) and long-term stability (**b**) of the MAPbI$_3$-based sensor.

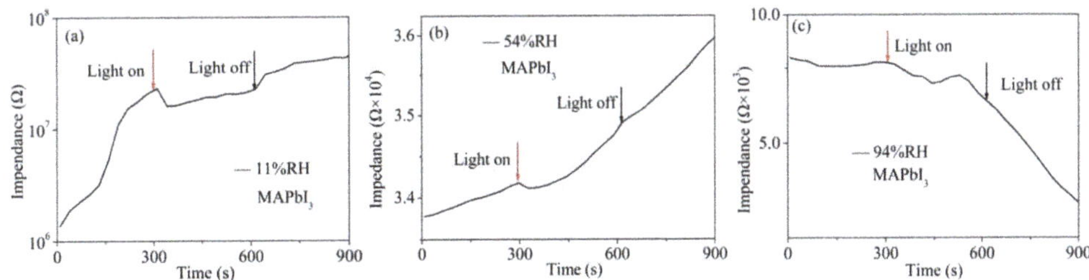

Figure 8. Illumination-induced impedance variations of the MAPbI$_3$-based sensor in the environment of (**a**) 11%, (**b**) 54%, and (**c**) 94%RH.

4. Discussion

To investigate the sensing mechanism, complex impedance diagrams of MAPbI$_3$-based sensor were measured with different RH levels. Figure 9 displays the complex impedance plots (Nyquist plots) obtained in different RH environments. At the low RH levels of 11 and 33%, the Nyquist plots are arc-like curves with large radius. The large radius indicates the high resistance at low humidity levels. The arc shrinks with the increasing of RH from 33% to 54%. A whole semicircle followed by a liner line in the low-frequency range is visible within the measuring frequency range when the RH beyond 75%RH. Further increasing the RH level, the semicircle becomes smaller, but the linear line becomes longer. This phenomenon indicates that both the resistance and reactance underwent a large decrease as the RH increases.

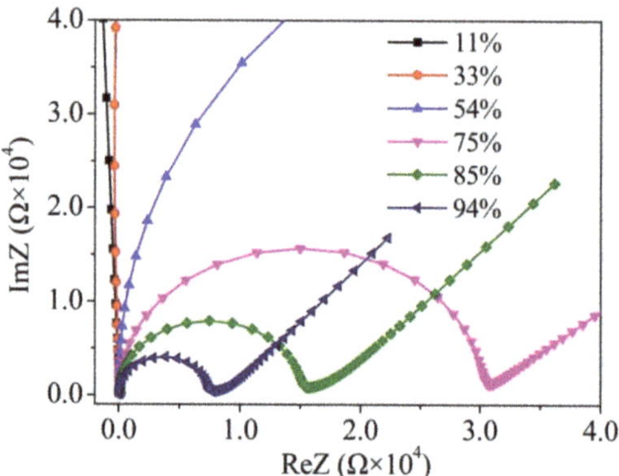

Figure 9. Complex impedance diagrams of the MAPbI$_3$-based sensor with different RH levels.

Previous work has manifested that the absorbed water molecules can deform the crystal lattice. As the content of water molecules increases to one molecule in one unit cell, this deformation becomes prominent, but the crystal structure is still not broken down [11]. This implies that perovskite materials have good capability accommodating water molecules. Figure 10a presents the UPS spectrum of the material, where the VBM is 1.35 eV below Ef, and its secondary electron truncation is located at 19.5 eV. In addition to the band gap value calculated from the absorption spectrum, the band structure is acquired and depicted in Figure 10b (the Eg of MAPbI$_3$ is 1.55 eV), where a standard n type feature could be concluded. With more water molecules absorbed by the structure, more electrons would be injected. This would lift the E_f of the material and enhance the conductivity of the material [11]. However, perovskite MAPbI$_3$ would form the hydrate-CH$_3$NH$_3$PbI$_3$·H$_2$O under high humidity [27]. It is clearly shown in Figure 5a that the impedance of MAPbI$_3$-based sensor decreases rapidly at low and moderate humidity levels, while the impedance value decreases slowly at high humidity level. Besides, the color of MAPbI$_3$-based sensor changes from black to yellow and then back to black when the sensor was successively exposed to 11, 94%RH and then back to 11%RH for 5 min at each RH level, which indicates the degradation is reversible. Therefore, under high humidity, most of the water molecules entering the crystal lattice were used to form hydrate, which leads to slow increase of electrons in the conduction band and slow increase of electrical conductivity. Moreover, due to the limitation of perovskite crystal structure, the hydrate formed is limited, so the electrical conductivity of this material increases slowly.

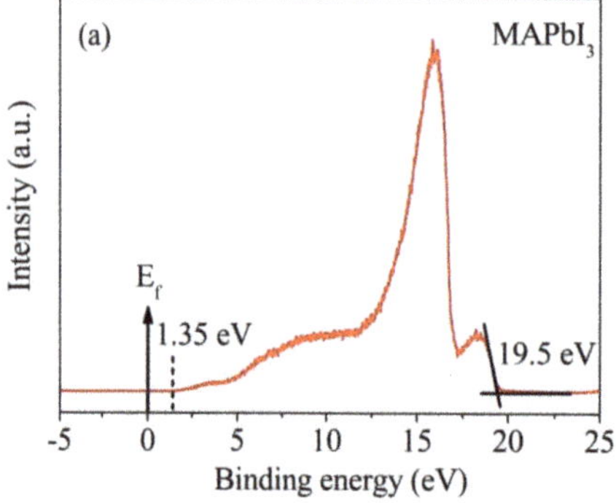
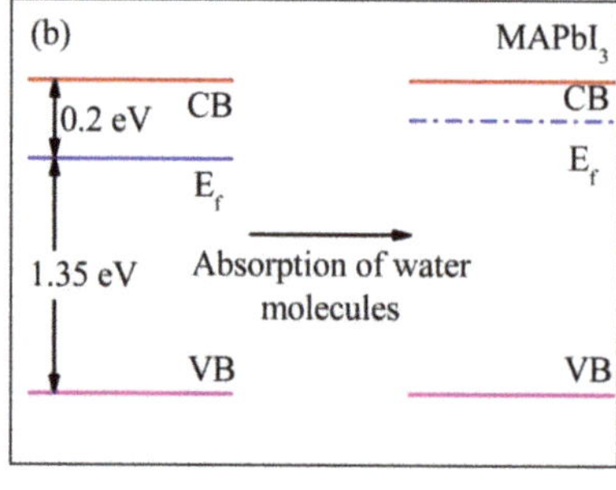

Figure 10. (a) The direct UPS spectrum of the as-grown bare MAPbI$_3$; and (b) the energy level diagrams of as-grown MAPbI$_3$ in the cases before and after absorbing water molecules.

5. Conclusions

In this work, we present investigation on the humidity properties on the MAPbI$_3$-based sensor. The results show that the sensor exhibits high sensitivity, superior to that of many perovskite materials. The humidity performances were argued to be caused by the fact that water molecules entered the perovskite lattice, serving as a strong *n*-type dopant that greatly enhances the concentration of conductive electrons. Our results emphasize that the MAPbI$_3$ perovskite has a great promise as humidity sensing material and could have widespread applications in humidity sensor.

Author Contributions: Conceptualization, X.Z. and Y.S.; methodology, S.L.; software, P.C.; validation, X.Z., Y.S. and S.L.; formal analysis, G.C.; investigation, X.Z.; resources, J.W.; data curation, X.Z.; writing—original draft preparation, X.Z.; writing—review and editing, C.W.; visualization, W.C.; supervision, C.W.; project administration, X.Z.; funding acquisition, C.W. All authors have read and agreed to the published version of the manuscript.

Funding: This research was funded by National Natural Science Foundation of China, grant number 51872001 and 12174001.

Institutional Review Board Statement: Not applicable.

Informed Consent Statement: Not applicable.

Data Availability Statement: The data is available on reasonable request from the corresponding author.

Conflicts of Interest: The authors declare no conflict of interest.

References

1. Ni, Z.Y.; Bao, C.X.; Liu, Y.; Jiang, Q.; Wu, W.Q.; Chen, S.S.; Dai, X.Z.; Chen, B.; Hartweg, B.; Yu, Z.S.; et al. Resolving spatial and energetic distributions of trap states in metal halide perovskite solar cells. *Science* **2020**, *367*, 1352–1358. [CrossRef] [PubMed]
2. Juarez-Perez, E.J.; Sanchez, R.S.; Badia, L.; Garcia-Belmonte, G.; Kang, Y.S.; Mora-Sero, I.; Bisquert, J. Photoinduced giant dielectric constant in lead halide perovskite solar cells. *J. Phys. Chem. Lett.* **2014**, *5*, 2390–2394. [CrossRef] [PubMed]
3. Fan, Z.; Xiao, J.X.; Sun, K.; Chen, L.; Hu, Y.T.; Quyuan, J.Y.; Ong, K.P.; Zeng, K.Y.; Wang, J. Ferroelectricity of CH$_3$NH$_3$PbI$_3$ perovskite. *J. Phys. Chem. Lett.* **2015**, *6*, 1155–1161. [CrossRef] [PubMed]
4. Zhou, Y.; You, L.; Wang, S.W.; Ku, Z.L.; Fan, H.J.; Schmidt, D.; Rusydi, A.; Chang, L.; Wang, L.; Ren, P.; et al. Giant photostriction in organic-inorganic lead halide perovskites. *Nat. Commun.* **2016**, *7*, 11193. [CrossRef] [PubMed]
5. Zhang, X.H.; Zhao, X.N.; Shan, X.Y.; Tian, Q.L.; Wang, Z.Q.; Lin, Y.; Xu, H.Y.; Liu, Y.C. Humidity effect on resistive switching characteristics of the CH$_3$NH$_3$PbI$_3$ memristor. *ACS Appl. Mater. Inter.* **2021**, *13*, 28555–28563. [CrossRef] [PubMed]
6. Ha, S.T.; Shen, C.; Zhang, J.; Xiong, Q.H. Laser cooling of organic-inorganic lead halide perovskites. *Nat. Photonics* **2016**, *10*, 115–121. [CrossRef]
7. Cho, M.Y.; Kim, S.; Kim, I.S.; Kim, E.S.; Wang, Z.J.; Kim, N.Y.; Kim, S.W.; Oh, J.M. Perovskite-induced ultrasensitive and highly stable humidity sensor systems prepared by aerosol deposition at room temperature. *Adv. Funct. Mater.* **2019**, *30*, 1907449. [CrossRef]
8. García-Fernández, A.; Moradi, Z.; Bermúdez-García, J.M.; Sánchez-Andújar, M.; Gimeno, V.A.; Castro-García, S.; Señaris-Rodriguez, M.A.; Mas-Marzá, E.; Garcia-Belmonte, G.; Fabregat-Santiago, F. Effect of environmental humidity on the electrical properties of lead halide perovskites. *J. Phys. Chem. C* **2018**, *123*, 2011–2018. [CrossRef]
9. Ye, W.; Cao, Q.; Cheng, X.F.; Yu, C.; He, J.H.; Lu, J.M. Lead-free Cs$_2$PdBr$_6$ perovskite-based humidity sensor for artificial fruit waxing detection. *J. Mater. Chem. A* **2020**, *8*, 17675–17682. [CrossRef]
10. Hu, L.; Shao, G.; Jiang, T.; Li, D.B.; Lv, X.L.; Wang, H.Y.; Liu, X.S.; Song, H.S.; Tang, J.; Liu, H. Investigation of the interaction between perovskite films with moisture via in situ electrical resistance measurement. *ACS Appl. Mater. Inter.* **2015**, *7*, 25113–25120. [CrossRef]
11. Ren, K.K.; Huang, L.; Yue, S.Z.; Lu, S.D.; Liu, K.; Azam, M.; Wang, Z.J.; Wei, Z.M.; Qu, S.C.; Wang, Z.G. Turning a disadvantage into an advantage: Synthesizing high-quality organometallic halide perovskite nanosheet arrays for humidity sensors. *J. Mater. Chem. C* **2017**, *5*, 2504–2508. [CrossRef]
12. Weng, Z.H.; Qin, J.J.; Umar, A.A.; Wang, J.; Zhang, X.; Wang, H.L.; Cui, X.L.; Li, X.G.; Zheng, L.R.; Zhan, Y.Q. Lead-free Cs$_2$BiAgBr$_6$ double perovskite-based humidity sensor with superfast recovery time. *Adv. Funct. Mater.* **2019**, *29*, 1902234. [CrossRef]
13. Ilin, A.S.; Forsh, P.A.; Martyshov, M.N.; Kazanskii, A.G.; Forsh, E.A.; Kashkarov, P.K. Humidity sensing properties of organometallic perovskite CH$_3$NH$_3$PbI$_3$. *Chemistryselect* **2020**, *5*, 6705–6708. [CrossRef]

14. Wu, Z.L.; Yang, J.; Sun, X.; Wu, Y.J.; Wang, L.; Meng, G.; Kuang, D.L.; Guo, X.Z.; Qu, W.J.; Du, B.S.; et al. An excellent impedance-type humidity sensor based on halide perovskite $CsPbBr_3$ nanoparticles for human respiration monitoring. *Sens. Actuators B Chem.* **2021**, *337*, 129772. [CrossRef]

15. Jassim, S.M.; Bakr, N.A.; Mustafa, F.L. Synthesis and characterization of $MAPbI_3$ thin film and its application in C-Si/perovskite tandem solar cell. *J. Mater. Sci. Mater. Electron.* **2020**, *31*, 16199–16207, Correction in **2020**, *31*, 16208. [CrossRef]

16. Yang, J.L.; Yu, Y.; Zeng, L.Y.; Li, Y.T.; Pang, Y.Q.; Huang, F.Q.; Lin, B.L. Effects of aromatic ammoniums on methyl ammonium lead iodide hybrid perovskite materials. *J. Nanomater.* **2017**, 1640965. [CrossRef]

17. Li, W.W.; Zeng, J.G.; Zheng, L.Y.; Zeng, H.R.; Li, C.B.; Kassiba, A.; Park, C.; Li, G.R. Apparent ferroelectric-like and dielectric properties of $CH_3NH_3PbI_3$ synthesized in ambient air. *Ferroelectrics* **2019**, *553*, 95–102. [CrossRef]

18. Patil, D.; Seo, Y.K.; Hwang, Y.K.; Chang, J.S.; Patil, P. Humidity sensitive poly (2,5-dimethoxyaniline)/WO_3 composites. *Sens. Actuators B Chem.* **2008**, *132*, 116–124. [CrossRef]

19. Suri, K.; Annapoorni, S.; Sarkar, A.K.; Tandon, R.P. Gas and humidity sensors based on iron oxide-polypyrrole nanocomposites. *Sens. Actuators B Chem.* **2002**, *81*, 277–282. [CrossRef]

20. Wang, Z.Y.; Shi, L.Y.; Wu, F.Q.; Yuan, S.; Zhao, Y.; Zhang, M.H. Structure and humidity sensing properties of $La_{1-x}K_xCo_{0.3}Fe_{0.7}O_{3-\delta}$ perovskite. *Sens. Actuators B Chem.* **2011**, *158*, 89–96. [CrossRef]

21. Yao, P.J.; Wang, J.; Ji, M. Study on dielectric and humidity sensing properties of $La_{1-x}Sr_xFeO_3$ materials. *Ferroelectrics* **2010**, *402*, 79–88. [CrossRef]

22. Rao, M.C.; Srikumar, T.; Satish, D.V.; Krishna, J.S.R.; Rao, C.S.; Rao, C.R. Effect of zinc stannate nanofiller on PVA-CH_3COONa polymer electrolyte for humidity sensor application. *AIP Conf. Proc.* **2018**, *1992*, 040022.

23. Farahani, H.; Wagiran, R.; Urban, G.A. Investigation of room temperature protonic conduction of perovskite humidity sensors. *IEEE Sens. J.* **2020**, *21*, 9657–9666. [CrossRef]

24. Sundaram, R.; Raj, E.S.; Nagaraja, K.S. Microwave assisted synthesis, characterization and humidity dependent electrical conductivity studies of perovskite oxides, $Sm_{1-x}Sr_xCrO_3$ ($0 \leq x \leq 0.1$). *Sens. Actuators B Chem.* **2004**, *99*, 350–354. [CrossRef]

25. Lin, W.D.; Liao, C.T.; Chang, T.C.; Chen, S.H.; Wu, R.J. Humidity sensing properties of novel graphene/TiO_2 composites by sol-gel process. *Sens. Actuators B Chem.* **2015**, *209*, 555–561. [CrossRef]

26. Haque, M.A.; Syed, A.; Akhtar, F.H.; Shevate, R.; Singh, S.; Peinemann, K.V.; Baran, D.; Wu, T. Giant humidity effect on hybrid halide perovskite microstripes: Reversibility and sensing mechanism. *ACS Appl. Mater. Inter.* **2019**, *11*, 29821–29829. [CrossRef]

27. Prochowicz, D.; Franckevičius, M.; Cieślak, A.M.; Zakeeruddin, S.M.; Grätzel, M.; Lewiński, J. Mechanosynthesis of the hybrid perovskite $CH_3NH_3PbI_3$: Characterization and the corresponding solar cell efficiency. *J. Mater. Chem. A* **2015**, *3*, 20772–20777. [CrossRef]

nanomaterials

Article

Interface Optimization and Transport Modulation of Sm$_2$O$_3$/InP Metal Oxide Semiconductor Capacitors with Atomic Layer Deposition-Derived Laminated Interlayer

Jinyu Lu [1], Gang He [1,*], Jin Yan [1], Zhenxiang Dai [2], Ganhong Zheng [1,*], Shanshan Jiang [3], Lesheng Qiao [1], Qian Gao [1] and Zebo Fang [4,*]

[1] School of Materials Science and Engineering, Anhui University, Hefei 230601, China; lu19971210@outlook.com (J.L.); B01914070@ahu.edu.cn (J.Y.); B19201052@ahu.edu.cn (L.Q.); gaoq@ahu.edu.cn (Q.G.)
[2] School of Physics and Optoelectronics Engineering, Anhui University, Hefei 230601, China; physdai@ahu.edu.cn
[3] School of Integration Circuits, Anhui University, Hefei 230601, China; jiangshanshan@ahu.edu.cn
[4] School of Mathematical Information, Shaoxing University, Shaoxing 312000, China
* Correspondence: ganghe01@issp.ac.cn (G.H.); ghzheng@ahu.edu.cn (G.Z.); csfzb@usx.edu.cn (Z.F.)

Abstract: In this paper, the effect of atomic layer deposition-derived laminated interlayer on the interface chemistry and transport characteristics of sputtering-deposited Sm$_2$O$_3$/InP gate stacks have been investigated systematically. Based on X-ray photoelectron spectroscopy (XPS) measurements, it can be noted that ALD-derived Al$_2$O$_3$ interface passivation layer significantly prevents the appearance of substrate diffusion oxides and substantially optimizes gate dielectric performance. The leakage current experimental results confirm that the Sm$_2$O$_3$/Al$_2$O$_3$/InP stacked gate dielectric structure exhibits a lower leakage current density than the other samples, reaching a value of 2.87×10^{-6} A/cm^2. In addition, conductivity analysis shows that high-quality metal oxide semiconductor capacitors based on Sm$_2$O$_3$/Al$_2$O$_3$/InP gate stacks have the lowest interfacial density of states (D_{it}) value of 1.05×10^{13} cm^{-2} eV^{-1}. The conduction mechanisms of the InP-based MOS capacitors at low temperatures are not yet known, and to further explore the electron transport in InP-based MOS capacitors with different stacked gate dielectric structures, we placed samples for leakage current measurements at low varying temperatures (77–227 K). Based on the measurement results, Sm$_2$O$_3$/Al$_2$O$_3$/InP stacked gate dielectric is a promising candidate for InP-based metal oxide semiconductor field-effect-transistor devices (MOSFET) in the future.

Keywords: MOS capacitors; Sm$_2$O$_3$ high-k gate dielectric; atomic layer deposition; conduction mechanisms; interface state density

Citation: Lu, J.; He, G.; Yan, J.; Dai, Z.; Zheng, G.; Jiang, S.; Qiao, L.; Gao, Q.; Fang, Z. Interface Optimization and Transport Modulation of Sm$_2$O$_3$/InP Metal Oxide Semiconductor Capacitors with Atomic Layer Deposition-Derived Laminated Interlayer. *Nanomaterials* **2021**, *11*, 3443. https://doi.org/10.3390/nano11123443

Academic Editor: Sergio Brutti

Received: 14 November 2021
Accepted: 16 December 2021
Published: 19 December 2021

Publisher's Note: MDPI stays neutral with regard to jurisdictional claims in published maps and institutional affiliations.

1. Introduction

As the integration of IC continues to increase, CMOS feature sizes will also continue to decrease in order to reduce the cost of individual transistors, increase the switching speed of transistors, and reduce the power consumption of the circuit. As the feature size of CMOS devices continues to decrease, it will also cause the SiO$_2$ gate dielectric and Si substrate used in conventional processes to decrease in size as well. When the size is smaller than a certain limit, the gate leakage current will grow exponentially, while the device will not operate properly due to the laws of quantum physics [1,2]. The use of high-k materials for the replacement of SiO$_2$ gate dielectrics is an option that has been shown to be feasible [3]. Among these high k materials, samarium oxide (Sm$_2$O$_3$) is considered as the next potential gate dielectric due to its high dielectric constant (~15) [4], sufficiently large band gap (5.1 eV) [5], low hygroscopicity, and high chemical and thermal stability [6]. Coulomb scattering and phonon scattering at the interface between the high-k gate dielectric and the channel material lead to a significant reduction in channel mobility, which severely affects the further increase in the speed of CMOS logic devices. Selecting channel materials with high mobility is an effective way to solve this problem [7]. Compared with conventional Si-based material CMOS devices, III-V group semiconductors have advantages due to their large switching speed and small dynamic power consumption [8]. Among the group III-V semiconductors, InP has received more attention due to its higher carrier mobility and smaller band gap [9].

However, InP is prone to the formation of interfacial defects, which can limit the operating performance of the device [10]. Also, a surface with many chemical impurities can have a considerable impact on the performance of InP MOS capacitors [11]. High D_{it} leads to the frequency dispersion of the Fermi energy level pegging and capacitance, which also prevents the formation of inverse or accumulation layers in CMOS devices [12].

Different InP surface passivation methods have been investigated for a long time, including low-temperature processes [13], ozone treatment [14], chemical etching [15], and sulfide solution passivation [16,17]. A great deal of work has also been devoted to atomic layer deposition (ALD) passivation layers to modulate the InP interface [18]. It has been demonstrated that ALD-derived Al_2O_3 films can effectively suppress the interfacial diffusion from the substrate to the high-k films. More importantly, the operating temperature can be kept low (~200 °C) when the Al_2O_3 film is on the passivated substrate surface. There are previous reports confirming that the insertion of Al_2O_3 between the high-k gate dielectric and GaAs can improve the thermal stability. However, even in the presence of an Al_2O_3 passivation layer to improve the interface, the diffusion of In and P elements into the gate dielectric still has an impact on the electrical characteristics of the device when fabricating InP MOS capacitors. R. V. Galatage et al. reported that the In-O and P-O states at the interface lead to a degradation of the electrical characteristics [19]. Their results also demonstrated the effectiveness of ALD-derived Al_2O_3 passivation layers between the gate dielectric and the InP substrate. Chee-Hong An et al. systematically analyzed that Al_2O_3 can inhibit dissociation and reactant diffusion in InP substrates [20]. However, the effect of the position of the Al_2O_3 passivation layer on the electrical properties and interfacial bonding state of InP MOS devices has not been reported systematically.

In this work, we deposited Sm_2O_3 films by magnetron sputtering and obtained Al_2O_3 passivation layers by ALD equipment to fabricate three different gate stacks on InP substrates, corresponding to $Al_2O_3/Sm_2O_3/InP$, $Al_2O_3/Sm_2O_3/Al_2O_3/InP$, and $Sm_2O_3/Al_2O_3/InP$, respectively. X-ray photoelectron spectroscopy (XPS) and electrical measurements were used to investigate the effect of Al_2O_3 passivation position on the chemical composition and electrical parameters of the interface. In addition, the leakage current conduction mechanisms (CCMs) of InP-based MOS capacitors with three different laminated gate electrical stacks measured at room temperature and low temperature (77–227 K) were systematically investigated.

2. Materials and Methods

In this work, we chose sulfur-doped n-type InP wafers as the substrate for fabricating MOS capacitors. Before depositing the Sm_2O_3 gate dielectric, the wafers were subjected to a standard degreasing process by sequential immersion in ethanol and acetone for 5 min each. After that, the wafers were immersed in 20% ammonium sulfide solution for 15 min to remove the native oxides. Then, the wafers are rinsed with deionized water and then blown dry with high purity nitrogen gas. The cleaned wafers are transferred to an ALD system (MNT-PD100Oz-L6S1G2, MNT Micro and Nanotech). On the ALD process, plasma O_2 and trimethylaluminum (TMA) were selected as the oxidant and aluminum metal precursor, and a 2 nm Al_2O_3 passivation layer was deposited on the InP substrate. The Al_2O_3 passivation layers were deposited by using 30 pulse cycles of plasma O_2 precursors [O_2(2s)/Ar purge (25s)] and 15 pulse cycles of trimethylaluminum (TMA) and plasma O_2 [TMA (0.03s)/O_2 2s/Ar purge (25s)], respectively. During this process, the chamber pressure and the deposition temperature were maintained at 35 Pa and 200 °C. After ALD Al_2O_3 passivation, the wafers were transferred to a sputtering chamber to deposit Sm_2O_3 gate dielectrics by sputtering samarium target with purity of 99.9%. When the chamber pressure was 0.8 Pa, Sm_2O_3 thin film was deposited under an Ar/O_2 (50/10 sccm) atmosphere. To explore the electrical characteristics of Sm_2O_3/InP MOS capacitors with different stacking positions of Al_2O_3 passivation layers, a 200-μm-diameter Al electrode was deposited by thermal evaporation, while an aluminum electrode was grown on the back side to form an ohmic contact. Figure 1 demonstrates the schematics of InP-MOS capacitors based on different stacked gate dielectric structures. Sample S1 corresponds to Al_2O_3 (2 nm)/Sm_2O_3 (8 nm)/InP, sample S2 corresponds to Al_2O_3 (2 nm)/Sm_2O_3 (6 nm)/Al_2O_3 (2 nm)/InP, and sample S3 corresponds to Sm_2O_3 (8 nm)/Al_2O_3 (2 nm)/InP, respectively. By using the ESCALAB 250Xi system, XPS (X-ray photoelectron spectroscopy) measurements were performed at Al Ka (1486.7 eV) to investigate the interfacial chemical properties of the Sm_2O_3/InP gate stack and the chemical function of the Al_2O_3 passivation layer. Furthermore, the escape angle used in obtaining the XPS profiles is 50° and the corresponding probing depth is about 1–10 nm. Ultraviolet-visible spectroscopy (Shimadzu, UV-2550) was performed to obtain the samples' optical band gap. The physical thickness of the above samples was extracted by using spectroscopic ellipsometry measurements (SANCO Inc., Shanghai, China, SC630) with the help of the Cauchy-Urbach model. The Cascade Probe Station was connected to the semiconductor analysis equipment (Agilent B1500A) for capacitance-voltage (C-V), transconductance-voltage (G-V), and leakage current-voltage (I-V) measurements at room temperature. For low temperature (77–227 K) leakage current testing, the Lake Shore Cryotronics Vacuum Probing Station was used.

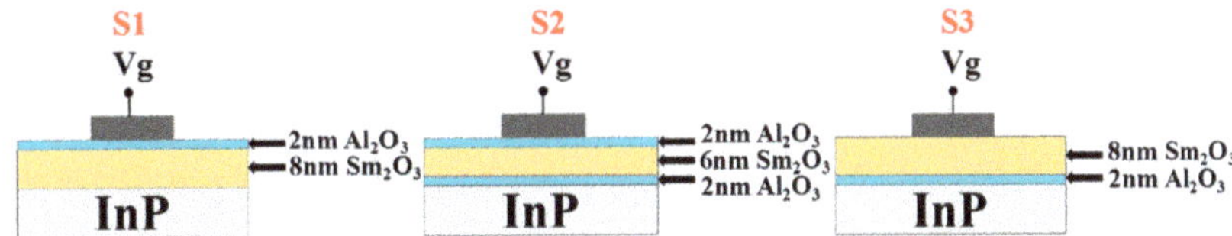

Figure 1. Schematics of InP-based MOS capacitors based on different stacked gate dielectrics.

3. Results and Discussion

3.1. XPS Analyses

To evaluate the chemical bonding states of various stacked gate dielectrics, XPS measurements were carried out. Figure 2 displays the In 3d, P 2p, and O 1s XPS spectra of three samples with various stacked gate dielectrics. It can be noted that In 3d spectra can be deconvoluted into four components that represent the InP, $InPO_4$, $In(PO_3)_3$, and In_2O_3, respectively. The relative intensity values of the different components have been extracted and are shown in Figure 3a. For S2 and S3, the contents of $In(PO_3)_3$ and In_2O_3 shows a decreasing trend, indicating that the ALD-derived Al_2O_3 passivation layers prior to Sm_2O_3 deposition can significantly prohibit the formation of In and P suboxides, which can be attributed to the interface cleaning function of plasma O_2 [21]. Compared to S2, the peak areas of $InPO_4$ corresponding to S1 and S3 remain approximate at about 7.89% and 5.20%, which is much lower than that of S2 (19.41%), indicating that double deposition of ALD-derived Al_2O_3 may accelerate the diffusion of oxygen in the substrate and the formation of indium phosphate. During the secondary deposition of Al_2O_3, more oxygen vacancies may generate, which can be ascribed to plasma O_2 acting as an oxygen source, and promote the oxygen interdiffusion between Al_2O_3 passivation layers and the InP substrate. $In(PO_3)_3$ can react with In to produce InP and $InPO_4$ using the following reaction Equation [22].

$$8In + 4In(PO_3)_3 \rightarrow 3InP + 9InPO_4 \tag{1}$$

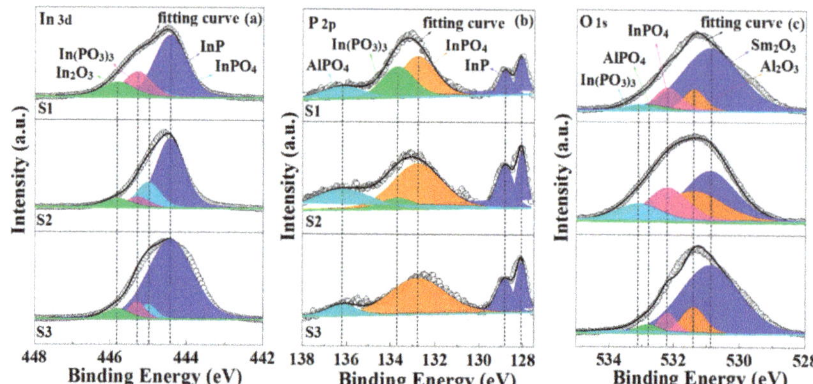

Figure 2. (**a**) In 3d, (**b**) P 2p, and (**c**) O 1s XPS spectra for S1, S2, and S3 sample.

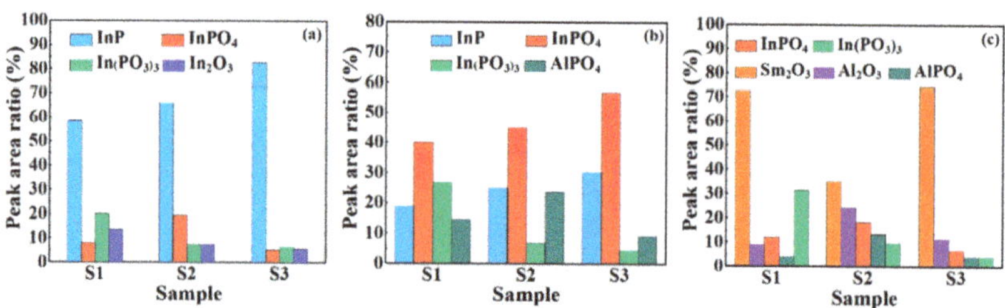

Figure 3. Peak area ratio histograms of (**a**) In 3d, (**b**) P 2p, and (**c**) O 1s spectra.

More importantly, sample S3 shows a tendency to decrease the content of $In(PO_3)_3$ and $AlPO_4$ compared to sample S2, which can give a detailed illustration from the phenomenon of P 2p spectral changes. As shown in Figure 2b, it can be noted that P 2p spectra can be deconvoluted into four components, which represent InP, $InPO_4$, $In(PO_3)_3$, and $AlPO_4$, respectively. No P_2O_5 was detected in all samples, which can be attributed to the fact that gaseous P_2O_5 generated during the deposition can easily diffuse through the defects in the gate dielectric [22]. The peak area ratio of $In(PO_3)_3$ for S2 and S3 showed a significant decreasing trend compared to S1, indicating that Al_2O_3 prior to the deposition of Sm_2O_3 gate dielectric can inhibit the formation of P-O bound states and improve the interfacial quality. The detection of $AlPO_4$ in P 2p spectra can be attributed to the reaction equation described below [23].

$$4Al + 7O_2 + 2In(PO_3)_3 + 2InP \rightarrow 4AlPO_4 + 4InPO_4 \tag{2}$$

Based on the mentioned reaction above, it can be inferred that two depositions of Al_2O_3 passivation layers increase the formation of $AlPO_4$, which is confirmed by the change in peak area ratio shown in Figure 3c. In order to systematically explore the interfacial chemistry of various stacked gate dielectrics, O 1s spectra were investigated and are shown in Figure 2c. O 1s spectra can be deconvoluted into Sm_2O_3, Al_2O_3, $InPO_4$, $In(PO_3)_3$, and $AlPO_4$. According to the reaction Equation (2), in the plasma O_2 atmosphere, $In(PO_3)_3$ can react with O_2 to produce $AlPO_4$ and $InPO_4$ and leads to the disappearance of $In(PO_3)_3$. In agreement with the previous In 3d and P 2p spectra, S1 has the largest $In(PO_3)_3$ content, leading to a decrease in interfacial quality and deterioration of electrical properties. Meanwhile, $AlPO_4$ of S2 is the highest, originating from the second deposition of Al_2O_3. For S3 sample, the contents of $InPO_4$, $AlPO_4$, and In_2O_3 were significantly controlled, indicating that the addition of an ALD-derived Al_2O_3 layer prior to the deposition of Sm_2O_3 gate dielectric could reduce the generation of suboxides and improve the interfacial quality.

3.2. Band Alignment Characteristics

To assess the optical characteristics of the three various stacked gate dielectrics, UV-Vis spectroscopy was used to obtain the absorption spectra and the optical bandgap values (Figure 4) of samples S1, S2, and S3 were determined to be 5.49, 5.51, and 5.63 eV, respectively, based on the Tauc relationship [24]. Compared with pure Sm_2O_3 and pure Al_2O_3, the band gaps of three various stacked gate dielectrics showed a value balance [25]. Also, this section investigates the valence band maximum (VBM) of various stacked gate dielectrics, as the valence band alignment is crucial for assessing the interface quality. As shown in Figure 5a, the band gap values of InP substrates were derived from XPS measurements, while the valence bands of samples S1, S2, and S3 were deduced from the absorption spectra by linear extrapolation. Based on Kraut's method [26], we also calculated the valence band shift (ΔE_V) to evaluate the valence band electronic structure of the samples. By using Sm $3d_{5/2}$ and In 3d core-level spectra, the ΔE_V of high-k/InP gate stacks was determined based on the following formula:

$$\Delta E_V = (E_{In\ 3d} - E_V)_{InP} - (E_{Sm\ 3d} - E_V)_{high-k} - (E_{In\ 3d} - E_{Sm\ 3d})_{high-k/InP} \qquad (3)$$

where $E_{In\ 3d}$ (InP) and $E_{Sm\ 3d}$ (high-k) corresponding to the core-level positions are extracted to be 445.8 and 1084.3 eV. In addition, the E_v (InP) and E_v (high-k) represent the VBM (Valence-Band Maximum) of the bulk materials. The values of ΔE_v are calculated as 1.87, 1.82, and 1.76 eV, respectively, based on the binding energy difference in the high-k/InP structure. Meanwhile, the value of the conduction band offset (ΔE_c) is obtained by subtracting the extracted ΔE_v and the band gap of the InP (1.34 eV) from the band gap of dielectric layers [27].

$$\Delta E_C\ (high-k/InP) = E_g\ (high-k) - \Delta E_V(High-k/InP) - E_g(InP) \qquad (4)$$

As shown in Figure 5b, the ΔE_c values for the three samples were calculated as 2.28, 2.35, and 2.53 eV. According to previous reports in the literature, ΔE_c is related with the tunneling leakage current. The higher ΔE_c indicates that the leakage current of the $Sm_2O_3/Al_2O_3/InP$ samples is smaller.

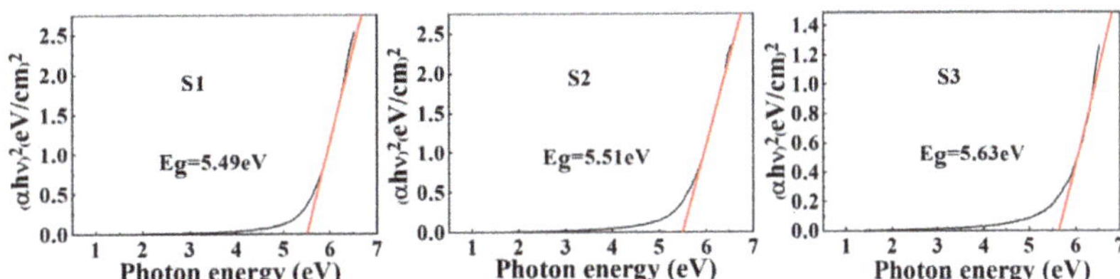

Figure 4. The determination of band gaps for sample S1, S2, and S3.

3.3. Electrical Properties of InP-MOS Capacitors

3.3.1. Capacitance-Voltage Measurements

The frequency dependent capacitance-voltage curves of sample S1, S2, and S3 with double sweep mode are shown in Figure 6a–c. When the frequency increases, all samples show a decreased accumulation capacitance.

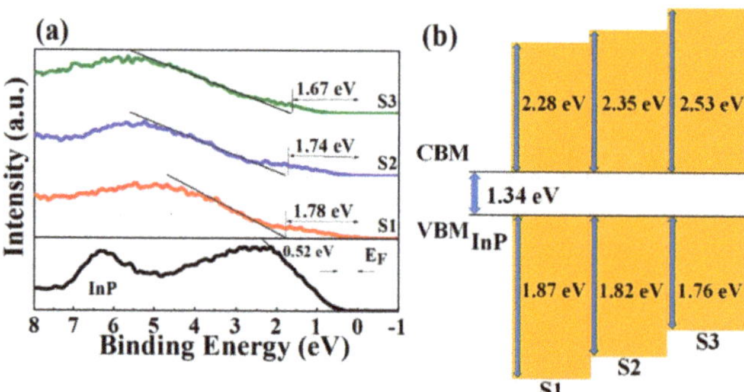

Figure 5. (**a**) Valence band spectra; (**b**) Schematic band diagram of S1, S2, and S3 sample.

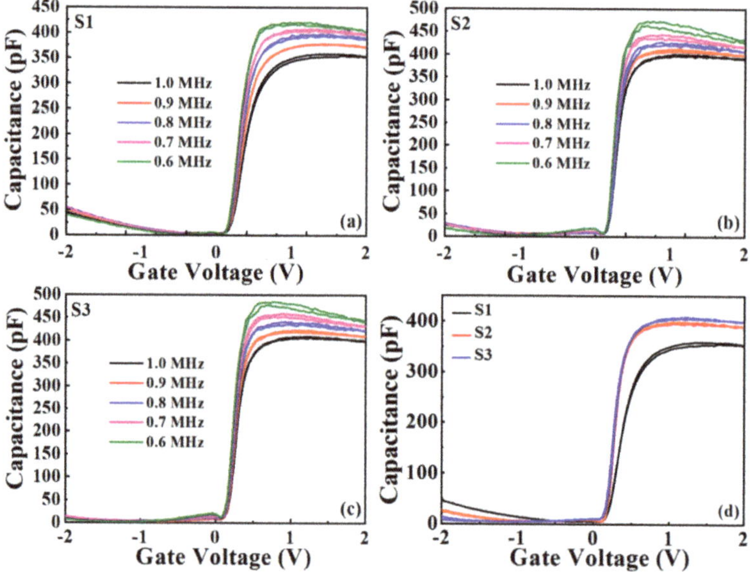

Figure 6. (**a**–**c**) Capacitance–voltage (C–V) curves for S1–S3 measured at different frequency (0.6–1 MHz). (**d**) Capacitance–voltage (C–V) curves for all samples measured at 1 MHz.

Meantime, at high frequency conditions, the series resistance will deviate from the predetermined theoretical value due to the disappearance of the interface trap charge. On the contrary, at low frequencies, when the oxide capacitance (C_{ox}) connects with the space charge capacitance (C_{sc}), the value of the accumulation region increases with the series resistance due to the interface state showing frequency-dependent properties [28–30].

The decrease in the accumulation capacitance can be attributed to the fact that the interfacial traps do not have enough time to respond to the voltage frequency [30]. The maximum accumulation capacitance and minimum hysteresis voltage were observed in the S3 sample, indicating that the Al_2O_3 passivation layer suppressed the appearance of In and P oxides and the formation of low-K interfacial layers. To evaluate the interface quality, important electrical parameters such as equivalent oxide thickness (EOT), dielectric constant (k), flat band voltage (V_{fb}), hysteresis voltage (ΔV_{fb}), oxidation charge density (Q_{ox}), and boundary trapped oxide charge density (N_{bt}) were extracted from the test curves, and these data are presented in Table 1. The variation of V_{fb} depends on the values of oxide capacitance and bulk oxide charge [31]. The k values corresponding to samples S1, S2, and S3 are calculated to be 12.96, 14.39, and 14.75, which are consistent with the previous investigation [4].

Table 1. MOS capacitors electrical parameters obtained from C–V and J–V Curves.

Sample	EOT (nm)	k	V_{fb} (V)	$\triangle V_{fb}$ (mV)	Q_{ox} (cm^{-2})	N_{bt} (cm^{-2})	J (A/cm^{-2})
S1	3.01	12.96	0.25	3.44	-1.62×10^{12}	-2.46×10^{10}	1.07×10^{-5}
S2	2.71	14.39	0.21	5.16	-1.43×10^{12}	-4.11×10^{10}	8.42×10^{-6}
S3	2.65	14.75	0.19	1.55	-1.30×10^{12}	-1.26×10^{10}	2.87×10^{-6}

A small V_{fb} of 0.19 V was observed for sample S3. This phenomenon can be explained by the following statement: electrons are easily captured by oxygen vacancies to form negatively charged interstitial oxygen atoms [32] and as fewer oxygen vacancies exist at the interface, the smaller the positive flat voltage required to maintain the band unbent [33]. Also, the hysteresis voltage depends on the boundary trap caused by the intermixing of the high K layer and the interfacial layer [34]. The value of the hysteresis voltage reaches a minimum (1.55 mV) for S3, indicating that the boundary trapping charge becomes weaker after the insertion of Al_2O_3 between the gate dielectric and the substrate. The Q_{ox} and N_{bt} values were calculated from the obtained V_{fb} and ΔV_{fb} values by the following equations [35].

$$Q_{ox} = -\frac{C_{max}\left(v_{fb} - \varphi_{ms}\right)}{qA} \tag{5}$$

$$N_{bt} = -\frac{C_{max} \triangle V_{fb}}{qA} \tag{6}$$

where φ_{ms} is the contact potential difference between Al electrode and InP substrate, q is the electronic charge, and A is the Al electrode areas. According to Table 1, it can be noticed that S3 has the lowest Q_{ox} and N_{bt}, which implies the reduction of interfacial trap defects and the optimization of interfacial properties.

3.3.2. Conductivity-Voltage Measurements

Moreover, to quantify the interface defect distribution for all samples, the interface state density (D_{it}) has been extracted by the conductivity-voltage measurements with frequencies varying from 100 kHz to 1 MHz. D_{it} is related to the parallel interfacial trap capacitance (C_{it}) and parallel conductivity (G). At the same time, C_{it} can be related by the following equation. $C_{it} = qD_{it}$, while the condition is that the position of the energy level does not change D_{it}. The basic principle of conductivity measurements is to analyze the losses due to the diversity of charge states at the trap level. Near the Fermi level, the synchronous conductivity occupancy is mobilized by the interfacial traps to produce a regular variation. The maximum loss occurs when the interface trap is resonantly shifted with the applied AC signal ($\omega\tau = 1$). The response time of the characteristic trap changes the frequency, $\tau = 2\pi/\omega$. The capture and emission rates from Shockley-Redhall theory modulate the response time [36]:

$$\tau = \frac{exp[\Delta E/k_B T]}{\sigma v_{th} D_{dos}} \tag{7}$$

where there is an energy difference ΔE between the trap level E_T and the edge of the majority carrier band, v_{th} is the majority carrier being thermally activated to obtain the average velocity, D_{dos} is the effective density of states of the majority carrier band, k_B is the Boltzmann constant, and T is the temperature [37]. The curves between conductivity (G/ω) and gate voltage for all samples are shown in Figure 7a–c. The apparent shift of the conductivity peak proves the validity of the Fermi-level shift and confirms the existence of the Fermi-level deconvolution effect [38]. Assuming that the underlying surface oscillations can be neglected, the value of D_{it} is inferred using the normalized parallel conductivity peak $(G_P/\omega)_{max}$ [39].

$$D_{it} \approx \frac{2.5}{Aq} \left(\frac{G_p}{\omega} \right)_{max} \tag{8}$$

where A is the device area. It is necessary to confirm the transformation law between the band bending potential of the energy location E_T and the trap energy level distribution. Furthermore, the values of E_T can be determined by the frequency of $(G_P/\omega)_{max}$, where Equation (8) is used to calculate D_{it} and to correspond its value to ΔE [40].

$$\Delta E = (E_C - E_T) = \frac{k_B T}{q} In \left(\frac{\sigma v D_{dos}}{2\pi f_{max}} \right) \tag{9}$$

Figure 7d shows the variation of D_{it} for the three samples. With the increase of ΔE, the value of D_{it} shows an increasing trend. However, S3 possesses a lower density of interfacial states compared to S1 and S2, which indicates that the insertion of an Al_2O_3 passivation layer between the Sm_2O_3 gate dielectric and the InP substrate can suppress the formation of In and P suboxides and improve the quality of MOS capacitors.

We compared some of the data obtained from this work with some previously published work. As can be seen in Table 2, the Sm_2O_3 dielectric has a smaller leakage current density than TiO_2 and HfO_2, indicating that the Sm_2O_3 stacked gate dielectric has a larger conduction band shift, resulting in an increased barrier height and thus a reduced leakage current density. The Sm_2O_3 stacked gate dielectric has the smallest hysteresis value, indicating that the trapped charge in the gate dielectric is not very sensitive to the frequency response of the voltage, and will trap fewer electrons to keep the energy band from being bent, while the device maintains a consistent response to different test voltages in the antipattern region. In the interface state density, it is smaller than HfO_2 as the gate dielectric directly deposited in InP, but it seems to be higher than TiO_2 gate dielectric, considering the different testing methods, the interface state density of the current work is obtained directly by conductivity method with accuracy, the previous work is by C–V curve, there may be some differences.

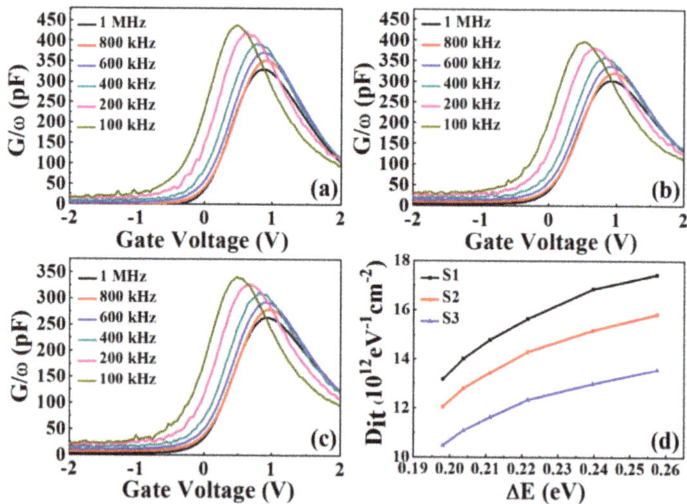

Figure 7. Multi-frequency G–V characteristics of InP-based MOS capacitors of (**a**) S1, (**b**) S2, and (**c**) S3. (**d**) Energy distributions of Dit for S1, S2, and S3.

Table 2. Comparison of different InP MOS capacitor parameters.

	$Sm_2O_3/Al_2O_3/InP$ (This Work)	PMA-TiO$_2$/S-InP [41]	TiO$_2$/S-InP [41]	10 Å Si IPL/51 Å HfO$_2$/InP [42]	70 Å HfO$_2$/InP [42]	HfO$_2$ (10 nm)/Al$_2$O$_3$ (0.2 nm)/InGaAs/InP [43]
Leakage current density (A/cm^2)	2.87×10^{-6} at 1 V	1.9×10^{-7} at 2 V 2.7×10^{-5} at -2 V	5.01×10^{-6} at 2 V 1.5×10^{-2} at -2 V	1.32×10^{-3} at 1 V	3.94×10^{-2} at 1 V	2.4×10^{-2}
k	14.75	39	34	/	/	/
ΔVfb (mV)	1.55	40	250	240	280	/
D$_{it}$ (cm^{-2}eV^{-1})	(G-V) 1.05×10^{13}	(C-V) 3.1×10^{11}	(C-V) 5×10^{11}	(C-V) 3-8×10^{12}	(C-V) 2-9×10^{13}	(C-V) 2×10^{12}

3.3.3. $J-V$ Analyses and Conduction Mechanisms at Room Temperature

Figure 8a shows the leakage current characteristics of all samples measured at room temperature. The leakage current density (*J*) values for S1, S2, and S3 at 1 V are 1.07×10^{-5}, 8.42×10^{-6}, and 2.87×10^{-6} A/cm^2, respectively. It can be seen that S1 has a higher leakage current density, which can be attributed to larger interface traps and the border traps that deteriorate the interface quality and degrade the device performance [44]. For the S3 sample, the minimum leakage current density has been observed, which is due to the higher ΔE_c and the suppressed tunneling in the $Sm_2O_3/Al_2O_3/InP$ gate stack [45].

To investigate the leakage current characteristics of various stacked gate dielectrics, we systematically studied three different current conduction mechanisms (CCMs) under substrate injection, as shown in Figure 8b–d. The extracted important electrical parameters are listed in Table 3.

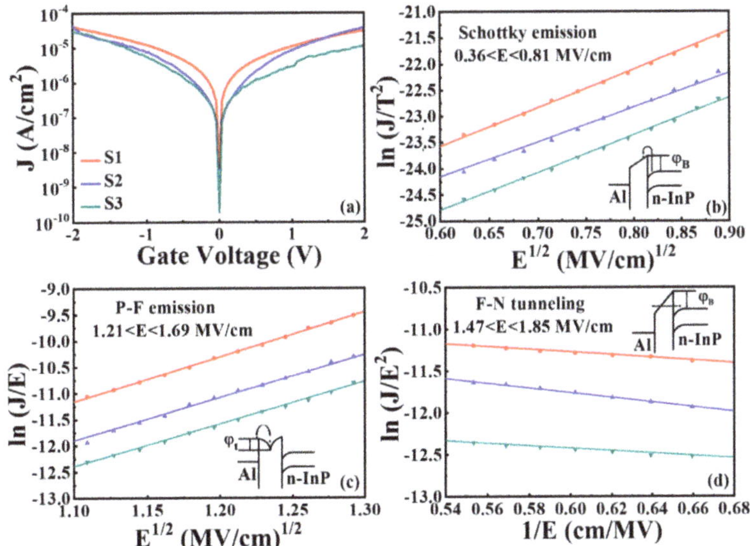

Figure 8. (**a**) J–V characteristics measured at room temperature. (**b**) SE emission, (**c**) PF emission, and (**d**) FN tunneling plots for all the samples under substrate injection.

Table 3. Extracted MOS capacitors electrical parameters measured at room temperature.

Sample	J (A/cm^2)	ε_r	n	ε_{ox}	φ_t (eV)
S1	1.07×10^{-5}	4.00	2.00	11.90	0.53
S2	8.42×10^{-6}	4.96	2.23	13.01	0.54
S3	2.87×10^{-6}	4.23	2.06	13.41	0.55

Schottky emission (*SE*) is a typical type of thermal ionization emission in which charges gain energy to overcome barriers to migration into the dielectric. The standard *SE* can be described as [46]:

$$J_{SE} = A^* T^2 exp \left[\frac{-q\left(\varphi_B - \sqrt{qE/4\pi\varepsilon_0\varepsilon_r}\right)}{k_B T} \right] \tag{10}$$

$$A^* = \frac{4\pi q k_B^2 m_{ox}^*}{h^3} = 120\frac{m_{ox}^*}{m_0} \tag{11}$$

where A^* is the effective Richardson constant, the free electron mass and the effective mass of electrons in the gate dielectric correspond to m_0 and m_{ox^*}, E is the electric field, $q\varphi_B$ is the Schottky barrier height, and ε_0 and ε_r represent the vacuum dielectric constant and the optical dielectric constant, respectively [47]. It is observed in Figure 8b that at lower electric fields (0.36–0.81 MV/cm), there is a good linear relationship between $\ln(J/T^2)$ and $E^{1/2}$ for S1, S2, and S3. The slope of the SE diagram is denoted as $\sqrt{q^3/4\pi\varepsilon_0\varepsilon_r}/k_B T$. The fitted ε_r and the refractive index n ($n = \varepsilon_r^{1/2}$) for S1, S2, and S3 are (4, 2), (4.96, 2.23), and (4.23, 2.06), respectively. All the fits are consistent with the previously reported values [48], revealing that CCM (current conduction mechanism) at room temperature is dominated by SE emission in the low electric field region.

The Poole–Frenkel (*PF*) emission can be ascribed to the thermally excited electrons obtaining sufficient energy to escape from traps into the conduction band of the dielectric at a higher electric field, which can be expressed by the following formula [49]:

$$J_{PF} = AE\,exp \left[\frac{-q\left(\varphi_t - \sqrt{qE/\pi\varepsilon_0\varepsilon_{ox}}\right)}{k_B T} \right] \tag{12}$$

where A represents a constant, the trap energy level of the conduction band corresponds to φ_t, and ε_{ox} represents the dielectric constant. According to the previous theory, $\ln(J/E)$ should have a good proportionality with $E^{1/2}$, as shown in Figure 8c. The ε_{ox} extracted from the slope of the fitted line for all samples was calculated as 11.90, 13.01, and 13.41, which is in agreement with the reported reference [4]. It can be concluded that at higher electric fields (1.21–1.69 MV/cm), the PF emission dominates the CCM of all samples. Also, the value of the trap energy level (φ_t) can be extracted based on the intercept point of the fitted curve described as $lnB - \frac{q\varphi_t}{k_B T}$. As shown in Figure 8c, the calculated values of φ_t are 0.53, 0.54, and 0.55 eV, corresponding to S1, S2, and S3. S3 has the largest φ_t value in the three samples, indicating that the electrons obtain more energy to cross the trap, leading to present the smallest leakage current density in the S3 sample.

The high-field dependent conduction mechanism is represented by Fowler-Nordheim tunneling, which is manifested by the fact that the insulating layer can be penetrated by electrons, which enter the conduction band of the gate dielectric in a high electric field. The leakage current density is linked to other parameters of Fowler-Nordheim (*FN*) tunneling and is described by the following Equation [46]:

$$J_{FN} = \frac{q^3 E^2}{16\pi^2 \hbar \varphi_{ox}}\,exp\left[-\frac{4\sqrt{2m_T^* \varphi_B^{3/2}}}{3\hbar qE} \right] \tag{13}$$

where φ_{ox} is oxide barrier height; m_T^* is the tunneling effective electron mass in the gate oxide film, and the other notations remain unchanged from the previous definitions. Figure 8d shows the curve of $\ln(J/E2)$ versus $1/E$. The slope of the linear fit for the above samples shows an increase in current with increasing electric field, indicating that at high electric fields (1.47–1.85 MV/cm), all three samples are consistent with the *FN* tunneling conduction mechanism. Based on the previous analysis, it can be concluded that all samples are dominated by three main conduction mechanisms. In the lower electric fields,

SE emission dominates, however, in the higher electric fields, PF emission dominates together with FN tunneling.

3.3.4. Low Temperature *J–V* Analyses and Conduction Mechanisms

To investigate the variation of CCMs in $Sm_2O_3/Al_2O_3/InP$ MOS capacitor, low temperature (77–227 K) measurements were performed. The leakage current densities of $Sm_2O_3/Al_2O_3/InP$ MOS capacitors measured at 1 V were extracted as 4.64×10^{-9}, 1.48×10^{-8}, 1.13×10^{-7}, and 1.02×10^{-6} A/cm^2, corresponding to the temperature range of 77–227 K, respectively. By observing the leakage current densities at different temperatures, the $Sm_2O_3/Al_2O_3/InP$ gate stack exhibits nearly three orders of magnitude lower leakage current density at 77 K than that measured at room temperature, indicating that the low temperature is favorable for the MOS capacitor to exhibit optimized *J–V* characteristics. Figure 9b–d show the variation of the CCM under substrate injection along with the temperature trend. The extracted important electrical parameters are listed in Table 4. Figure 9b shows the fitted lines for the vertical temperature range suitable for SE emission at lower electric fields (0.49–0.90 MV/cm). The extracted important electrical parameters are listed in Table 4. With increasing temperature, the values of ε_r and n calculated from the slope and intercept are (20.39, 4.52), (18.53, 4.31), (10.40, 3.22), and (6.10, 2.47). It can be noted that at extremely low temperature of 77–177 K, these values are completely different from the theoretical values, indicating that SE emission is not the dominant conduction mechanism at lower temperatures. Figure 9c shows the curves in the temperature range 77–227 K compatible with PF emission at higher electric fields (0.64–1.44 MV/cm). Again, it can be noted that φ_t and ε_{ox} are not in the expected range of values, indicating that the PF emission is not compatible for all samples at intermediate electric fields of 0.64–1.44 MV/cm. FN tunneling is a potential conduction mechanism because of its dependence on the electric field at low temperatures. Figure 9d shows the fitted line of FN tunneling with a temperature range of 77–227 K at higher electric fields (1.11–1.67 MV/cm), and the established slope indicates that FN tunneling is dominant at low temperatures. In conclusion, the effects of SE emission and PF emission are attenuated due to low temperature, and FN tunneling is used to explain the $Sm_2O_3/Al_2O_3/InP$ stacked gate dielectric structure showing low drain current density.

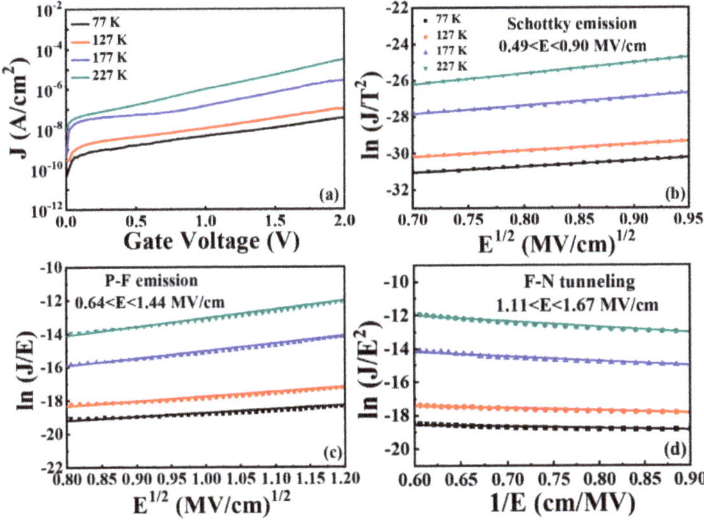

Figure 9. (**a**) J–V characteristics measured at low temperature. (**b**) SE emission, (**c**) PF emission, and (**d**) FN tunneling plots for all the samples under substrate injection.

Table 4. S3's MOS capacitors electrical parameters measured at low temperature.

T	J (A/cm^2)	ε_r	n	ε_{ox}	φ_t (eV)
77 K	4.64×10^{-9}	20.39	4.52	164.88	0.54
127 K	1.13×10^{-8}	18.53	4.31	108.14	0.53
177 K	1.48×10^{-7}	10.40	3.22	42.66	0.50
227 K	1.02×10^{-6}	6.10	2.47	31.05	0.47

Additionally, the integrated dielectric properties in MOS capacitors can be estimated from two important values, including the electron effective mass m_{ox}^* and the barrier height $q\varphi_B$ [50]. The intercept of the SE emission fitting curve described as $\ln\left(120\frac{m_{0x}^*}{m_0}\right) - \frac{q\varphi_B}{k_BT}$ and the slope of the FN tunneling fitting curve expressed as $-6.83 \times 10^7 \sqrt{\left(\frac{m_T^*}{m_0}\right)\varphi_B^3}$ can be calculated together with the above two values. By setting the equation $m_{0x}^* = m_T^*$, the two key physical quantities m_{0x}^* and $q\varphi_B$ of Sm$_2$O$_3$/Al$_2$O$_3$/InP MOS capacitor are obtained by applying mathematical analysis, which are calculated as 0.23 m$_0$ and 0.95 eV, respectively. Figure 10 shows the determination of the electron effective mass and barrier height for S3 sample. The smaller m_{0x}^* and the higher $q\varphi_B$ are beneficial to obtain better electrical properties and optimized interface quality.

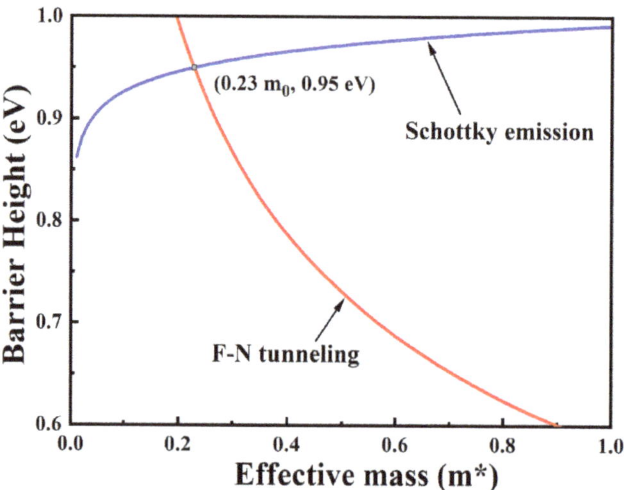

Figure 10. The determination of the electron effective mass and barrier height for S3 sample under substrate injection.

4. Conclusions

In this work, we explore in detail the effect of ALD-derived laminated interlayers on the interfacial chemistry and transport properties of sputter-deposited Sm$_2$O$_3$/InP gate stacks. It has been found that Sm$_2$O$_3$/Al$_2$O$_3$/InP gate stack can obviously prevent the diffusion of the substrate diffusion oxide and substantially optimize the electrical properties of MOS capacitors, including a larger dielectric constant of 14.75, a larger accumulation capacitance, and a lower leakage current density of 2.87×10^{-6} A/cm^2. Three different stacked gate dielectric structures are also evaluated by means of conductivity of the interfacial density of states. The results show that the Sm$_2$O$_3$/Al$_2$O$_3$/InP stacked gate dielectric achieves the lowest interfacial density of states of 1.05×10^{13} cm^{-2}eV^{-1}. According to the analysis of CCMs, SE emission is dominant in lower electric fields and higher temperature environments, and PF emission as well as F-N tunneling is dominant in higher electric fields. Meanwhile, FN tunneling is the only dominant mechanism at lower temperatures. Also, to evaluate the properties of the whole MOS capacitor in low temperature environment, m_{0x}^* and $q\varphi_B$ have been determined by a self-consistent method. These findings are of crucial importance for the future fabrication of high mobility InP-based MOSFET (Metal Oxide Semiconductor Field Effect Transistor) devices.

Author Contributions: G.H. and Z.F. conceived and designed the experiments; J.L., J.Y. and L.Q. performed the experiments; S.J., Z.D., Q.G. and G.Z. analyzed the data; J.L. and G.H. wrote the paper. All authors have read and agreed to the published version of the manuscript.

Funding: National Natural Science Foundation of China (11774001) and Anhui Project (Z010118169).

Institutional Review Board Statement: Not applicable.

Informed Consent Statement: Not applicable.

Data Availability Statement: The study did not report any data.

Acknowledgments: The authors acknowledge the support from National Natural Science Foundation of China (11774001) and Anhui Project (Z010118169).

Conflicts of Interest: The authors declare no conflict of interest.

References

1. Pelella, M.M.; Fossum, J.G.; Suh, D.; Krishnan, S.; Jenkins, K.A.; Hargrove, M.J. Low-voltage transient bipolar effect induced by dynamic floating-body charging in scaled PD/SOI MOSFET's. *IEEE Electron Dev. Lett.* **1996**, *17*, 196–198. [CrossRef]
2. Lauer, I.; Antoniadis, D.A. Enhancement of electron mobility in ultrathin-body silicon-on-insulator MOSFETs with uniaxial strain. *IEEE Electron Dev. Lett.* **2005**, *26*, 314–316. [CrossRef]
3. Robertson, J.; Wallace, R.M. High-K materials and metal gates for CMOS applications. *Mater. Sci. Eng. R* **2015**, *88*, 1–41. [CrossRef]
4. Chin, W.C.; Cheong, K.Y. Effects of post-deposition annealing temperature and ambient on RF magnetron sputtered Sm_2O_3 gate on n-type silicon substrate. *J. Mater. Sci. Mater. Electron.* **2011**, *22*, 1816–1826. [CrossRef]
5. Stewart, A.D.; Gerger, A.; Gila, B.P.; Abernathy, C.R.; Pearton, S.J. Determination of Sm_2O_3 GaAs heterojunction band offsets by X-ray photoelectron spectroscopy. *Appl. Phys. Lett.* **2008**, *92*, 153511. [CrossRef]
6. Dakhel, A.A. Dielectric and optical properties of samarium oxide thin films. *J. Alloy. Compd.* **2004**, *365*, 233–239. [CrossRef]
7. Mahata, C.; Oh, I.K.; Yoon, C.M.; Lee, C.W.; Seo, J.; Algadi, H. The impact of atomic layer deposited SiO_2 passivation for high-k Ta1-xZrxO on the InP substrate. *J. Mater. Chem. C* **2015**, *3*, 10293–10301. [CrossRef]
8. Sonnet, A.M.; Hinkle, C.L.; Jivani, M.N.; Chapman, R.A.; Pollack, G.P.; Wallace, R.M.; Vogel, E.M. Performance enhancement of n-channel inversion type $In_xGa_{1-x}As$ metal-oxide-semiconductor field effect transistor using ex situ deposited thin amorphous silicon layer. *Appl. Phys. Lett.* **2008**, *93*, 122109. [CrossRef]
9. Yuan, Y.; Yu, B.; Ahn, J.; McIntyre, P.C.; Asbeck, P.M.; Rodwell, M.J.W.; Taur, Y. A Distributed Bulk-Oxide Trap Model for Al_2O_3 InGaAs MOS Devices. *IEEE Trans. Electron Devices* **2012**, *59*, 2100–2106. [CrossRef]
10. Yen, C.F.; Yeh, M.Y.; Chong, K.K.; Hsu, C.F.; Lee, M.K. InP MOS capacitor and E-mode n-channel FET with ALD Al_2O_3-based high-k dielectric. *Appl. Phys. A* **2016**, *122*, 1–9. [CrossRef]
11. Sun, Y.; Liu, Z.; MacHuca, F.; Pianetta, P.; Spicer, W.E. Optimized cleaning method for producing device quality InP(100) surfaces. *J. Appl. Phys.* **2005**, *97*, 124902. [CrossRef]
12. He, G.; Gao, J.; Chen, H.S.; Cui, J.B.; Sun, Z.Q.; Chen, X.S. Modulating the Interface Quality and Electrical Properties of HfTiO/InGaAs Gate Stack by Atomic- Layer-Deposition-Derived Al_2O_3 Passivation Layer. *ACS Appl. Mater. Interfaces* **2014**, *6*, 22013–22025. [CrossRef] [PubMed]
13. Wang, S.K.; Cao, M.; Sun, B.; Li, H.; Liu, H. Reducing the interface trap density in Al_2O_3/InP stacks by low-temperature thermal process. *Appl. Phys. Exp.* **2015**, *8*, 091201. [CrossRef]
14. Ingrey, S.; Lau, W.M.; McIntyre, N.S.; Sodhi, R. An X-ray photoelectron spectroscopy study on ozone treated InP surfaces. *J. Vac. Sci. Technol. A* **1987**, *5*, 1621–1624. [CrossRef]
15. Çetin, H.; Ayyildiz, E. The electrical properties of metal-oxide-semiconductor devices fabricated on the chemically etched n-InP substrate. *Appl. Surf. Sci.* **2007**, *253*, 5961–5966. [CrossRef]
16. Lebedev, M.V.; Serov, Y.M.; Lvova, T.V.; Endo, R.; Masuda, T.; Sedova, I.V. InP(1 0 0) surface passivation with aqueous sodium sulfide solution. *Appl. Surf. Sci.* **2020**, *533*, 147484. [CrossRef]
17. Carpenter, M.S.; Melloch, M.R.; Lundstrom, M.S.; Tobin, S.P. Effects of Na_2S and $(NH_4)_2S$ edge passivation treatments on the dark current-voltage characteristics of GaAs pn diodes. *Appl. Phys. Lett.* **1988**, *52*, 2157–2159. [CrossRef]
18. Driad, R.; Sah, R.E.; Schmidt, R.; Kirste, L. Passivation of InP heterojunction bipolar transistors by strain controlled plasma assisted electron beam evaporated hafnium oxide. *Appl. Phys. Lett.* **2012**, *100*, 014102. [CrossRef]
19. Galatage, R.V.; Dong, H.; Zhernokletov, D.M.; Brennan, B.; Hinkle, C.L.; Wallace, R.M.; Vogel, E.M. Effect of post deposition anneal on the characteristics of HfO_2/InP metal-oxide-semiconductor capacitors. *Appl. Phys. Lett.* **2011**, *99*, 172901. [CrossRef]

20. An, C.H.; Byun, Y.C.; Cho, M.H.; Kim, H. Thermal instability of HfO$_2$ on InP structure with ultrathin Al$_2$O$_3$ interface passivation layer. *Phys. Status Solidi Rapid Res. Lett.* **2012**, *6*, 247–249. [CrossRef]

21. Kakiuchi, H.; Ohmi, H.; Harada, M.; Watanabe, H.; Yasutake, K. Significant enhancement of Si oxidation rate at low temperatures by atmospheric pressure Ar/O$_2$ plasma. *Appl. Phys. Lett.* **2007**, *90*, 151904. [CrossRef]

22. Hollinger, G.; Bergignat, E.; Joseph, J.; Robach, Y. On the nature of oxides on InP surfaces. *J. Vac. Sci. Technol. A* **1985**, *3*, 2082–2088. [CrossRef]

23. Aguirre-Tostado, F.S.; Milojevic, M.; Hinkle, C.L.; Vogel, E.M.; Wallace, R.M.; McDonnell, S.; Hughes, G.J. Indium stability on InGaAs during atomic H surface cleaning. *Appl. Phys. Lett.* **2008**, *92*, 171906. [CrossRef]

24. Murphy, A.B. Band-gap determination from diffuse reflectance measurements of semiconductor films, and application to photoelectrochemical water-splitting. *Sol. Energy Mater. Sol. Cells* **2007**, *91*, 1326–1337. [CrossRef]

25. Jaggernauth, A.; Mendes, J.C.; Silva, R.F. Atomic layer deposition of high-: κ layers on polycrystalline diamond for MOS devices: A review. *J. Mater. Chem. C* **2020**, *8*, 13127–13153. [CrossRef]

26. Kraut, E.A.; Grant, R.W.; Waldrop, J.R.; Kowalczyk, S.P. Precise Determination of the Valence-Band Edge in X Ray Photoemission Spectra. *Phys. Rev. Lett.* **1980**, *44*, 1620–1623. [CrossRef]

27. Mahata, C.; Byun, Y.-C.; An, C.-H.; Choi, S.; An, Y.; Kim, H. Comparative study of atomic-layer-deposited stacked (HfO$_2$/Al$_2$O$_3$) and nanolaminated (HfAlO$_x$) dielectric on In0.53Ga0.47As. *ACS Appl. Mater. Interfaces* **2013**, *5*, 4195–4201. [CrossRef]

28. Çiçek, O.; Durmuş, H.; Altındal, Ş. Identifying of series resistance and interface states on rhenium/n-GaAs structures using C-V-T and G/ω-V-T characteristics in frequency ranged 50 kHz to 5 MHz. *J. Mater. Sci. Mater. Electron.* **2020**, *31*, 704–713. [CrossRef]

29. Yoshioka, H.; Nakamura, T.; Kimoto, T. Accurate evaluation of interface state density in SiC metal-oxide-semiconductor structures using surface potential based on depletion capacitance. *J. Appl. Phys.* **2012**, *111*, 04C100. [CrossRef]

30. Mutale, A.; Deevi, S.C.; Yilmaz, E. Effect of annealing temperature on the electrical characteristics of Al/Er$_2$O$_3$/n-Si/Al MOS capacitors. *J. Alloy. Compd.* **2021**, *863*, 158718. [CrossRef]

31. Varzgar, J.B.; Kanoun, M.; Uppal, S.; Chattopadhyay, S.; Tsang, Y.L.; Escobedo-Cousins, E.; Olsen, S.H.; O'Neill, A.; Hellström, P.E.; Edholm, J.; et al. Reliability study of ultra-thin gate oxides on strained-Si/SiGe MOS structures. *Mater. Sci. Eng. B* **2006**, *135*, 203–206. [CrossRef]

32. Foster, A.S.; Lopez Gejo, F.; Shluger, A.L.; Nieminen, R.M. Vacancy and interstitial defects in hafnia. *Phys. Rev. B* **2002**, *65*, 1741171–17411713. [CrossRef]

33. Liu, J.W.; Oosato, H.; Da, B.; Koide, Y. Fixed charges investigation in Al$_2$O$_3$/hydrogenated-diamond metal-oxide-semiconductor capacitors. *Appl. Phys. Lett.* **2020**, *117*, 163502. [CrossRef]

34. Shahinur Rahman, M.; Evangelou, E.K.; Konofaos, N.; Dimoulas, A. Gate stack dielectric degradation of rare-earth oxides grown on high mobility Ge substrates. *J. Appl. Phys.* **2012**, *112*, 094501. [CrossRef]

35. Wang, D.; He, G.; Hao, L.; Qiao, L.; Fang, Z.; Liu, J. Interface Chemistry and Dielectric Optimization of TMA-Passivated high-k/Ge Gate Stacks by ALD-Driven Laminated Interlayers. *ACS Appl. Mater. Interfaces* **2020**, *12*, 25390–25399. [CrossRef] [PubMed]

36. Engel-Herbert, R.; Hwang, Y.; Stemmer, S. Comparison of methods to quantify interface trap densities at dielectric/III-V semiconductor interfaces. *J. Appl. Phys.* **2010**, *108*, 124101. [CrossRef]

37. Lin, H.C.; Wang, W.E.; Brammertz, G.; Meuris, M.; Heyns, M. Electrical study of sulfur passivated In$_{0.53}$Ga$_{0.47}$As MOS capacitor and transistor with ALD Al$_2$O$_3$ as gate insulator. *Microelectron. Eng.* **2009**, *86*, 1554–1557. [CrossRef]

38. Martens, K.; Wang, W.; De Keersmaecker, K.; Borghs, G.; Groeseneken, G.; Maes, H. Impact of weak Fermi-level pinning on the correct interpretation of III-V MOS C-V and G-V characteristics. *Microelectron. Eng.* **2007**, *84*, 2146–2149. [CrossRef]

39. Qiao, L.; He, G.; Hao, L.; Lu, J.; Gao, Q.; Zhang, M.; Fang, Z. Interface Optimization of Passivated Er$_2$O$_3$/Al$_2$O$_3$/InP MOS Capacitors and Modulation of Leakage Current Conduction Mechanism. *IEEE Trans. Electron Devices* **2021**, *68*, 2899–2905. [CrossRef]

40. Carter, A.D.; Mitchell, W.J.; Thibeault, B.J.; Law, J.J.M.; Rodwell, M.J.W. Al$_2$O$_3$ growth on (100) In$_{0.53}$Ga$_{0.47}$ as initiated by cyclic trimethylaluminum and hydrogen plasma exposures. *Appl. Phys. Exp.* **2011**, *4*, 091102. [CrossRef]

41. Yen, C.-F.; Lee, M.-K. Very Low Leakage Current of High Band-Gap Al$_2$O$_3$ Stacked on TiO$_2$/InP Metal–Oxide–Semiconductor Capacitor with Sulfur and Hydrogen Treatments. *Jpn. J. Appl. Phys.* **2012**, *51*, 081201.

42. Chen, Y.-T.; Zhao, H.; Yum, J.H.; Wang, Y.; Lee, J.C. Metal-oxide-semiconductor field-effect-transistors on indium phosphide using HfO$_2$ and silicon passivation layer with equivalent oxide thickness of 18 Å. *Appl. Phys. Lett.* **2009**, *94*, 213505. [CrossRef]

43. Suzuki, R.; Taoka, N.; Yokoyama, M.; Lee, S.; Kim, S.H. 1-nm-capacitance-equivalent-thickness HfO$_2$/Al$_2$O$_3$/InGaAs metal-oxide-semiconductor structure with low interface trap density and low gate leakage current density. *Appl. Phys. Lett.* **2012**, *100*, 132906. [CrossRef]

44. Shahrjerdi, D.; Rotter, T.; Balakrishnan, G. Fabrication of Self-Aligned Enhancement-Mode MOSFETs With Gate Stack. *IEEE Electron Dev. Lett.* **2008**, *29*, 557–560. [CrossRef]

45. He, G.; Zhang, L.D.; Liu, M.; Sun, Z.Q. HfO$_2$-GaAs metal-oxide-semiconductor capacitor using dimethylaluminumhydride-derived aluminum oxynitride interfacial passivation layer. *Appl. Phys. Lett.* **2010**, *97*, 223501. [CrossRef]

46. Yang, M.; Wang, H.; Ma, X.; Gao, H.; Wang, B. Effect of nitrogen-accommodation ability of electrodes in SiN$_x$-based resistive switching devices. *Appl. Phys. Lett.* **2017**, *111*, 223510. [CrossRef]

47. Kim, J.; Krishnan, S.A.; Narayanan, S.; Chudzik, M.P.; Fischetti, M.V. Thickness and temperature dependence of the leakage current in hafnium-based Si SOI MOSFETs. *Microelectron. Reliab.* **2012**, *52*, 2907–2913. [CrossRef]

48. Sadeq, M.S.; Morshidy, H.Y. Effect of samarium oxide on structural, optical and electrical properties of some alumino-borate glasses with constant copper chloride. *J. Rare Earths* **2020**, *38*, 770–775. [CrossRef]
49. Paskaleva, A.; Bauer, A.J.; Lemberger, M.; Zürcher, S. Different current conduction mechanisms through thin hlgh-k Hf $_x$Ti$_y$Si$_z$O films due to the varying Hf to Ti ratio. *J. Appl. Phys.* **2004**, *95*, 5583–5590. [CrossRef]
50. Chiu, F.C. Interface characterization and carrier transportation in metal/ HfO$_2$/silicon structure. *J. Appl. Phys.* **2006**, *100*, 114102. [CrossRef]

MDPI

St. Alban-Anlage 66

4052 Basel

Switzerland

Tel. +41 61 683 77 34

Fax +41 61 302 89 18

www.mdpi.com

Nanomaterials Editorial Office

E-mail: nanomaterials@mdpi.com

www.mdpi.com/journal/nanomaterials

9 783036 559438